The Future Is Abundant

Edited by Larry Korn, Barbara Snyder, and Mark Musick
Designed and Produced by Barbara Snyder and Mark Musick
Principal authors:
Shery Litwin, Michael Pilarski, Mark Musick, Mike Maki, and Sego Jackson
Cover Drawing: Maia Massion
Cover Lettering: Jean Streinz

The Future Is Abundant

A Guide to Sustainable Agriculture

Tilth

Grateful acknowledgement is made to the following for permission to reprint previously published material:

Tim McNulty, for "Myths and a Sense of Place," as reprinted in *The Northwest Experience, Volume 1*, © 1980 Madrona Publishers.
Gary Snyder, from *The Old Ways*, © 1977 Gary Snyder.
Planet Drum Foundation, for logo appearing on page 17 and quote from *Reinhabiting a Separate Country* by Peter Berg, © 1978 Planet Drum Foundation.
International Federation of Organic Agriculture Movements (IFOAM), for "Towards A Sustainable Agriculture," by Colin Fisher. From *Towards A Sustainable Agriculture*, © 1978 Verlag Wirz AG, Aarau.
Earthcare Information Centre, for *Earthcare, Ecological Agriculture in Saskatchewan*, Paul Hanley, ed. © 1980 Earthcare Information Centre.
Oxford University Press, Inc, for "Cheat Takes Over," from *A Sand County Almanac: And Sketches from Here and There*, by Aldo Leopold. © 1949 & 1977 Oxford University Press, Inc.
University of California Press, for *Plants, Man and Life*, by Edgar Anderson. © 1952 Edgar Anderson.
Island Press, for *Tree Talk, The People and Politics of Timber*, by Ray Raphael. © 1981 Ray Raphael.
The Food and Agriculture Organization of the United Nations (FAO), for *Forestry for Local Community Development*, FAO Forestry Paper #7.
Ananda Publications, for *Sharing Nature with Children*, by Joseph Bharat Cornell. © 1979 Joseph Bharat Cornell.
Frank Porter Graham Center, for "Reaping What We Sow," from *The Graham Center Seed Directory*, by Cary Fowler. © 1979 Cary Fowler and Rural Advancement Fund.

Grateful acknowledgement is made to the following for permission to reprint previously published graphic art:

A.T. Times, January-February 1980, for graphic appearing on page 39.
Permaculture, Spring 1978, for graphic appearing on page 40.
Charles Scribner's Sons, for graphic appearing on page 41. From *The Owner-Built Homestead: A How-To-Do-It Book*. © 1974 Ken Kern; © 1977 Ken and Barbara Kern (New York: Charles Scribner's Sons, 1977).
British Columbia Provincial Museum, for graphics appearing on pages 21 and 83, from *The Lily Family of British Columbia*, by T.M.C. Taylor.
Environmental Design Press and Susan Van Gieson, for graphic appearing on page 113, from *Environmental Form Studies*.
Rodale Press, for graphic appearing on page 34. From *Rodale's Color Handbook of Garden Insects*, by Anna Carr. © 1979 Rodale Press.

This book was edited on an Apple II and typeset on an IBM 50 Electronic Typewriter. Text is in Boldface, references in Title. Headlines were set on a Compugraphic in Garamond and Garamond Bold.
Printed at Ad Pro Printing, Lynnwood, Washington, USA

1 2 3 4 5 6 7 8 9

Library of Congress Cataloging in Publication Data

The future is abundant.

Includes bibliographies and index.
1. Agriculture— Northwest, Pacific. 2. Organic farming— Northwest, Pacific. I. Tilth (Association) II. Title: Sustainable Agriculture.
S506.N6F87 1982 620'.9795 82-10351
ISBN 0-931380-01-4

Contents

Acknowledgements 7
Foreword 9
Introduction 11
Physiographic Map 14

Chapter 1—Knowing Home 15
- Native People 20
- Native Plants 21
- Public Gardens 22
- Garden in the Desert 23

Chapter 2—Sustainable Agriculture 25
- Ecology and the Farm 28
- Insectary Plants 31
- Insectary Plant Index 34
- Nitrogen-Fixing Plants 36
- Book Sources 39
- Farming with Computers 39
- Design 40
- Research Centers 42
- Appropriate Technology 45

Chapter 3—The Farm 49
- Perennial Grain 50
- Soils 53
- Laboratories 53
- No-Till Farming 54
- Weeds 56
- Wind Shelter 58
- Grassland Agriculture 60
- Rangelands 62
- Arid Lands 63
- New Crops 65
- Marketing 66
- Sources of Farm Credit 68

Chapter 4—The Garden 69

Dump Heaps 70
Community Agriculture 72
Horticulture Therapy 74
Eden in the City 75
Edible Flowers 76
Let Your Worms do the Digging 78
The Herb Garden 79
Extending the Season 80
The Wild Garden 82
Seeds of the Future 84
Bamboo 86
Uncommon Mushrooms 88

Chapter 5—Fruits of Sustainable Agriculture 89

Native Berries 90
Kiwi and Grapes 94
Northwest Tree Fruits 95
Northwest Nut Crops 99
Wild Fruit & Nut Forests 102
A Wealth of Trees 104
Chopaka Nut Orchard 106

Chapter 6—The Forest 107

Urban Forestry 108
Trees of the Northwest 110
Truffles 115
Forest Management 116
Community Forestry 120
Site-Specific Forestry 122
Coppicing 122
Alder & Hardwoods 124

Chapter 7—Animals 125

Wildlife 126
Bees 129
Animals at Work 131
Rare Breeds 133
Water Culture 135

Chapter 8—Community 139

Sharing Nature with Children 140
Land in Trust 142
Politics 145
The Green Front 147
Tilth People 149

Plant Species Index 151
Seed and Nursery List 171
Glossary 181
Index 184

Acknowledgements

Two streams converged to bring about this book. In the summer of 1980 Larry Korn and Mark Musick discussed the idea of compiling into a pamphlet the series of articles on natural farming which had been published in *Tilth* over the previous few years. A few months later Michael Pilarski proposed organizing two conferences on tree crops for the Pacific Northwest and publishing the proceedings as a reference for people seeking to create a permanent agriculture. The two ideas soon merged, and *The Future Is Abundant* is the result.

Publication of this book would not have been possible without the support and assistance of a large and sustaining community of friends throughout the region. Principal writers were Shery Litwin, Michael Pilarski, Mark Musick, Mike Maki, and Sego Jackson. Other writers included Kate Burroughs, Jim Bauermeister, Karen Hayden, Jim Orjala, Forest Shomer, Regina Hugo, Michael Soule, Phil Andrus, and Louise Dix. Larry Korn and Barbara Snyder coordinated the editing of hundreds of pages of manuscript, and Barbara Snyder, Mark Musick and Phil Andrus transformed a mountain of text and research material into the book you now hold.

In addition to the primary writers and editors, dozens of people assisted with research. Included were Lane Morgan, Sharon Hart, Burt Webb, Lynn Coody, Lee Kirschner, Maryanne Caruthers-Akin, Anne Maggs, James Bergdahl, Bob Madar, Chris and Bruce Fortin, Tom Ward, Clay Olson, Terry Brady, Andrea Patten, Charles Holzhauer, Carl Nielson, John Quinney, Larry and Barbara Geno, Jesse Schwartz, Arachim Israel, Netsah Bruvold, Robert Kourik, Cheryl Clayton, Tom Cleveland, Robert Papendick, and our wonderful friends at *RAIN Magazine*. Several dozen of the over 400 people who attended the two conferences which kicked off the book project contributed with advice and information. We also received help from several librarians who generously responded to our many questions.

People who reviewed the manuscript and provided much needed criticism included Greg Williams, Willow Rain, Les Abbenhouse, Miles Merwin, Laura Stuchinsky, Larry and Barbara Geno, Wayne Olson, Fred Euphrat, Alfred Reynolds, Bill Murphy, Vivian and Irving Korn, Binda Colebrook, Hide, Nash Huber, Prabuddha, Bob Cooperrider, and Ron Engeland.

A number of photographers and artists contributed their talents to this book. Included were Karen Hayden, Maryanne Caruthers-Akin, Jim Hallas, Kay Hoffman, Maia

Massion, Mike Karpinski, Kari Berger, John Stamets, Robert Williamson, Jean Streinz, Imagination Israel, Mary Morgan, Robert C. Wells, Beverly Reed, George Ballis, Jan Kepley, and Nancy Walz. Rodale Press, the *Permaculture* journal, and the U.S. Bureau of Land Management also assisted with permission to reproduce graphics and photographs.

Michele Heiderer, who typeset and produced camera-ready copy for most of this book, goes at the top of our "it wouldn't have been possible without" list. She was assisted by Marilyn O'Rear, who will be celebrating the birth of her third child about the time this book comes off the press. Keetje McConnell rounded out the office staff, balancing the checkbook and keeping up with correspondence as more and more of our time was devoted to the book. Marcia Buser-Moletore spent many volunteer hours compiling the index, and Patti Simon assisted with final indexing. We are also deeply indebted to the hard-working crew at Ad Pro Printing, the shop where our graphics were produced and the book was printed.

Financing and managing a project of this scale required extraordinary support and patience on the part of the Tilth Executive Board—Margaret McCrea, Tilth President, Bob Cooperrider and Laura Stuchinsky, Vice-Presidents, Regina Hugo, Secretary, and Randy Lee, Treasurer. Working capital loans for a project that went eight months and several thousand dollars over budget came from the Evergreen Land Trust and three other friends.

As always in a project of this scope there are probably people we have forgotten to mention. We hope that they will accept our apologies, and that everyone who has helped will accept our heartfelt thanks. As *The Future Is Abundant* goes to press we realize more than ever that this is just a beginning, not an end. It's great to have so many friends with whom to share the path ahead.

Foreword

The Future Is Abundant is a guide to creating a sustainable agriculture in the Pacific Northwest. Although set in one small corner of North America, we see this work as part of a planetary movement, and thus as a model for other regions. This book is part of the global effort to enhance and sustain the world's resources and make them available to all people.

The Future Is Abundant is a sourcebook, not a blueprint. In it we sketch the elements and principles of a sustainable agriculture. We provide references to books, periodicals, organizations, and resource people to guide you in your own research. The Plant Species Index describes the characteristics of over 300 useful species, and the Seed and Nursery List tells where these plants, and many others, can be obtained.

THE PROBLEM OF ABUNDANCE

When we chose *The Future Is Abundant* as our title we knew it would be controversial. Abundance as a goal is paradoxical. In *The Unsettling of America*, Wendell Berry notes that "any abundance, in any amount, is illusory if it does not safeguard its producers, and in American agriculture it is now virtually the accepted rule that abundance will destroy its producers." Indeed, abundance is driving many North American farmers to bankruptcy. In the farm press the word "abundance" is most often used in the context of overproduction. Today, farmers in the Northwest produce surpluses of wheat, apples, peas, milk, beef, potatoes, mint, beans, and turkeys.

While farmers are struggling to produce more crops with dwindling resources, the emphasis on exporting crops to balance foreign trade deficits is creating serious *im*balances throughout the world. The cruel irony is that the producers of abundance suffer from declining prices while millions of people around the world starve.

Although nation after hungry nation has become dependent on the industrialized countries for food, they retain their inherent capacity to feed themselves. As Joseph Collins and Frances Moore Lappé stress in *Food First: Beyond the Myth of Scarcity*, the world *is* abundant, with sufficient resources to provide for everyone's needs. They argue that satisfying those needs will require the redistribution of power and equal access to the resources essential for survival. The goal of this book is to provide access to some of those resources.

Worldwatch

Since 1975 the Worldwatch Institute has published over 50 incisive and influential reports on global resource trends. Worldwatch Papers consistently break new ground, documenting social and environmental issues that affect the future of our planet. The Worldwatch series includes papers on the firewood crisis, population trends, family planning, desertification, energy alternatives, soil erosion, land reform, employment, and community forestry.

In Building a Sustainable Society, Lester Brown, Worldwatch director, has summarized the institute's latest findings on prospects for the future. He concludes that human demand is now outstripping the sustainable yield of the world's oceans, grasslands, forests, and crop lands, and that output per person for every major commodity is now declining. While prospects appear bleak, Brown goes on to describe nations that are cutting population growth, developing energy alternatives, replanting their forests, and providing full employment. Brown maintains that we know what is required to build an enduring world society. What is needed now is the vision and the will to create a sustainable future.

For a listing of publications and more information on their programs, contact the Worldwatch **Institute, 1776 Massachusetts Avenue NW, Washington, DC 20036.**

In *The Future Is Abundant* , we seek to break down barriers—between farming and forestry, between producer and consumer, between urban and rural, and between rich and poor. Our goal is to create agricultural landscapes with the stability and resilience of natural systems, landscapes which provide food, fuel, clothing, and shelter sufficient for everyone's needs. The knowledge and techniques are available. What is required now is their application.

Cornucopia Project

North American agriculture is facing crises: soil erosion, soaring fossil fuel costs, the loss of agricultural land, chemical pollution, and the increasing concentration of land in fewer and fewer hands. Rodale Press has initiated The Cornucopia Project to analyze the current food system and seek alternatives which will lead to regional self-reliance. The goal is to create a food system that is both abundant and enduring. Everyone is encouraged to become involved.

A free Cornucopia Project Newsletter and listing of publications is available by writing **The Cornucopia Project, 33 E Minor Street, Emmaus, PA 18049.**

HOW TO USE THIS BOOK

The Future Is Abundant is a tool. In essence, it is an annotated bibliography and sourcebook. While we have provided addresses for lesser-known publishers and organizations, the reader will need to turn to local libraries and bookstores for access to many of the publications listed. We have emphasized books currently in print, but have included several classics that may be hard to locate. These can be obtained either through inter-library loan or through one or more of the Book Sources listed on page 39. Because prices are always subject to change, we have included the cost of publications only in exceptional cases.

Many of the groups listed in this book are quite small, with limited budgets. When inquiring as to the availabilty of publications, or when requesting additional information on programs or services, please include a stamped, self-addressed return envelope.

KEEP US INFORMED

The Future Is Abundant is the product of a participatory effort involving people throughout the Pacific Northwest. Wherever possible we have personally reviewed the literature referenced in this volume. Because of the explosion of research on biological agriculture, however, there must be sources of information of which we are not aware. Also, given the amount of information with which we are dealing, errors have undoubtedly crept into this work. Please recognize that this book is just the beginning of an ongoing process of discovery, and that we welcome your additions and corrections. Keep us informed of ideas or information that you would like to share.

BARBARA SNYDER & MARK MUSICK

Introduction

The Future Is Abundant is a statement of hope and an invitation to join us in creating a new agriculture. The agriculture we envision is based on sustainable, ecologically sound food production and forestry methods, which are responsive to local conditions and human needs.

We are inspired in this task largely by the work of Japanese farmer Masanobu Fukuoka, author of *The One-Straw Revolution*, and Australian environmental scientist Bill Mollison, author of *Permaculture One* and *Permaculture Two*. Although these men live and work far from our region, they share a common perspective in their approaches to agriculture. Both emphasize the practical value of minimum tillage, tree crops, perennial plants, and soil-building combinations of grasses, legumes and nitrogen-fixing trees and shrubs. They also stress the importance of maintaining diversity and complexity, and of integrating plants, animals and human society into whole, self-sustaining agricultural landscapes. We believe that the principles put forth by Fukuoka and Mollison offer a solid starting point from which we can establish a sustainable regional agriculture.

Masanobu Fukuoka grew up in a traditional Japanese village where his family had lived and farmed for more than 1,400 years. He was trained as a plant pathologist, and at the age of 25 he experienced an inspiration which caused him to question the fundamental principles of modern society and led to his life's work in natural agriculture. He believed that many of our environmental and social problems arose from people living out of touch with nature. He returned home to his father's farm where, over the past 40 years, he has been demonstrating the soundness of his ideas by applying them to agriculture.

Fukuoka's approach is to interfere as little as possible with natural processes. He broadcasts his seeds in the season they would naturally fall, or lets plants reseed themselves directly. He does not plow the soil, but instead keeps it soft and enriched with a continuous groundcover of white clover. Instead of using tillage or herbicides to control weeds, Fukuoka uses the clover, a mulch of rice and barley straw, and a rotation of crops carefully timed to discourage weeds. As he explains in *The One-Straw Revolution*, "I ultimately reached the conclusion that there was no need to plow, no need to make prepared compost, no need to apply chemical fertilizer or pesticide. And so my farming has become this very simple way. It's really no more than tossing out seeds and spreading straw, but it has taken me more

BIBLIOGRAPHY

Fukuoka, Masanobu. The One-Straw Revolution, An Introduction to Natural Farming. Edited by Larry Korn. Emmaus, PA: Rodale Press, 1978.
On fields that haven't been tilled or received artificial fertilizers in over 30 years, Masanobu Fukuoka achieves grain yields equal to the highest in Japan. In this eloquent translation, Fukuoka describes the philosophy and techniques which make possible his unique and inspiring approach to farming.

Entropy

In a recent interview (in The Mother Earth News, No. 66), Bill Mollison stated that only two texts should be issued to students of agricultural design, Masanobu Fukuoka's The One-Straw Revolution and Jeremy Rifkin's Entropy, A New World View (New York: Viking, 1980). What is "entropy" and what does it have to do with agriculture?

Entropy is the consequence of the second law of thermodynamics, a physical law which states that matter and energy can only be changed in one direction, from an ordered and usable form to a disordered and unusable form. While this may sound purely theoretical, the Entropy Law does have a profound effect on everyday life. According to Rifkin, the law defines the ultimate physical limits of the natural world, and finite stores of natural resources are now being rapidly depleted.

As fossil fuels, in particular, are depleted, Rifkin contends that the Industrial Era will come to an end. "Our survival," he states, "and the survival of all other forms of life now depend on our willingness to make peace with nature and begin to live cooperatively with the rest of our ecosystem."

Rifkin explains that the Entropy Law is important because it provides a framework for understanding the period of transition which the world has entered. It provides, too, solid guidelines for the design of a truly sustainable agriculture.

Mollison, Bill and Holmgren, David. Permaculture One: A Perennial Agriculture for Human Settlements. 1981. International Tree Crops Institute, Box 666, Winters, CA 95694
-and-
Mollison, Bill. Permaculture Two: Practical Design for Town and Country in Permanent Agriculture. Tagari Books, 1979. (Available from ITCI above.)
Sourcebooks for the conscious design of self-sustaining agricultural landscapes. The first volume introduces the permaculture concept, while the second one expands on design applications.

Permaculture Journal
c/o Terry White, 37 Goldsmith Street, Maryborough, Victoria 3465, Australia
Official journal of the Australian Permaculture Association. This is the best way to keep in touch with the world permaculture movement.

Maia Massion

ORGANIZATIONS

Maritime Permaculture Institute
Box 7611
Olympia, WA 98507
The Maritime Permaculture Institute is a regional link with the international permaculture movement. They offer workshops and provide design consultation. The Institute has also established a bird hatchery to provide selected breeds for homesteaders and small farmers.

North American Permaculture Institute
Box 1100
Winters, CA 95694
North American network for permaculture designers. They publish a newsletter and are initiating The Permaculture Observer, a design journal.

than thirty years to achieve this simplicity."

The greater meaning of natural farming, according to Fukuoka, is to learn to observe and accept the natural cycles of plants, animals, soil, water and climate so closely that they become part of you, and you a part of them. It is difficult, however, to come to know nature simply by looking at it. Some form of interaction is necessary. By sowing seeds and closely watching the results, we get subtle clues from nature. From these clues we can learn to garden and farm within natural cycles. Therefore, it is important to walk through the garden, fields or forest regularly, watching developments and applying this knowledge to future plantings.

In the year and a half that I worked on Fukuoka's farm in southern Japan, I came to realize that his simple, yet sophisticated, growing methods were molded by local crops and conditions. Although many of his techniques would be difficult to apply directly to the Pacific Northwest, his work demonstrates how much can be accomplished in a lifetime of persistent effort and patient observation of nature.

Bill Mollison has devised a concept of consciously designed, self-sustaining ecosystems which he refers to as "permaculture." His system, involving a great diversity of plant and animal species, emphasizes the importance of tree crops and other perennial plants which have multiple beneficial uses in the landscape. The elements are designed into an integrated system which takes advantage of the unique conditions and attributes of each site. They are arranged in such a way that the species which require the greatest attention and care are located closest to the dwelling site, while the plants and animals that require less attention are placed on the periphery. By careful design, energies which enter the area from the outside, such as wind, sunlight, water, fire and wildlife, are encouraged or screened so they work to the benefit of the whole system. The idea is to design a perennial, highly productive ecosystem which, once established, will operate with a minimum of maintenance. Neither Mollison's nor Fukuoka's systems are "no-work" methods, but both are guided by the principle of reducing *unnecessary* work as much as possible.

Mollison's ideas have generated a great deal of excitement, and some controversy. Some people question how such a diversity of native and exotic species will interact when placed together, perhaps for the first time. Others doubt that the planting of aggressive nitrogen-fixing species such as white clover, acacia and alder is compatible with the principle of minimum maintenance. It has also been pointed out that a greater number of interrelationships among species does not necessarily mean that the system will be more stable, as Mollison claims. This basic assumption of permaculture is still being actively debated among ecologists.

The reason for the controversy, I think, is that there are few models of a working permaculture. Although based on years of careful research and observation,

Mollison's ideas are still largely theoretical, awaiting the authority of practical demonstration. The situation is similar to Fukuoka's, following his inspiration for natural farming, but prior to his return to the farm. The proof of the usefulness of the permaculture concept will come with its implementation in farms and gardens in the Pacific Northwest and other parts of the world.

The various models of sustainable agriculture which are being developed in the Pacific Northwest and other parts of North America cannot be exactly like Fukuoka's rice farming or Mollison's examples of Australian permaculture, since specific techniques need to be adapted to local conditions. They are, instead, expressions of people's diverse attempts to heal the damage caused by past exploitation and to establish a way of life and a way of farming which helps to fully realize human potential. More than simply gardening or farming techniques, natural farming and permaculture encompass an attitude and a way of life.

Developing a sustainable agriculture in the Pacific Northwest will help restore the health of the land and build stable human communities, and will provide a model to people in other regions who share our goals. It will take time, patience, and a great deal of effort, but by working together, sharing our experiences, and continuously clarifying our vision, we can share a future of sustained abundance.

LARRY KORN

The Clover Project

The Clover Project was begun in January, 1979 to gather information on natural farming techniques for the Northwest. The fruit of the project was a series of six reports in the Tilth journal.

The project was introduced in the Spring, 1979 issue (Vol 5, No 1) with an article by Larry Korn. The Summer, 1979 issue (Vol 5, No 2) included an essay on the role of grass and clover in agriculture. The Fall/Winter, 1979 issue (Vol 5, No 3 & 4) featured a report on Masanobu Fukuoka's visit to North America, seed pelletizing, and intercropping grains and clover. The Spring, 1980 issue (Vol 6, No 1) continued the report on Masanobu Fukuoka's visit, while the Fall/Winter, 1980 issue (Vol 6, No 3) reported on natural farming tools and techniques. The Clover Project culminated with a report in the Summer, 1981 issue (Vol 7, No 2) on two permaculture conferences.

The set of six Clover Project issues ($11.50 value) is available for $10 (plus 64¢ tax for Wash. residents) from **Tilth, 13217 Mattson Rd, Arlington, WA 98223.**

Kama Sutra

Masanobu Fukuoka, author of *The One-Straw Revolution*, first visited North America in 1979. When he travelled to a community in northern California he was surprised to learn that almost all their tools were designed for digging or plowing the earth. Because of their problem with annual weedy grasses, such as foxtail and wild oats, he suggested that they find a way to farm without turning the soil. Cultivation, he said, brings weed seeds to the surface, allowing them to germinate year after year.

To control weeds, Fukuoka suggested using the kama or Japanese sickle, one of the most commonly used tools of Japanese farmers. The best kamas are made like Japanese swords, with fine inlaid steel.

Japanese farmers use the short-handled kama for cutting weeds and for harvesting rice and winter grains. A kama with a four foot handle is used like our scythe for mowing in an orchard or for clearing brush. Kamas are becoming available in North America through **Smith & Hawken Tool Company (68 Homer, Palo Alto, CA 94301)**. Famous for their imported English tools, they will soon be importing a good selection of kamas and other high quality Japanese farming tools. Larry Korn is working with Smith & Hawken to evaluate each tool before it is imported.

George Ballis © 1980 NLP

Larry Korn demonstrated how to sharpen a kama at a Tilth workshop on natural farming. In addition to working with the kama, participants learned about Masanobu Fukuoka's no-till grain and vegetable techniques and about the role of trees in permanent agriculture. A complete report on the workshop was published in the Fall/Winter, 1980 issue of Tilth.

Physiographic Regions of the Pacific Northwest

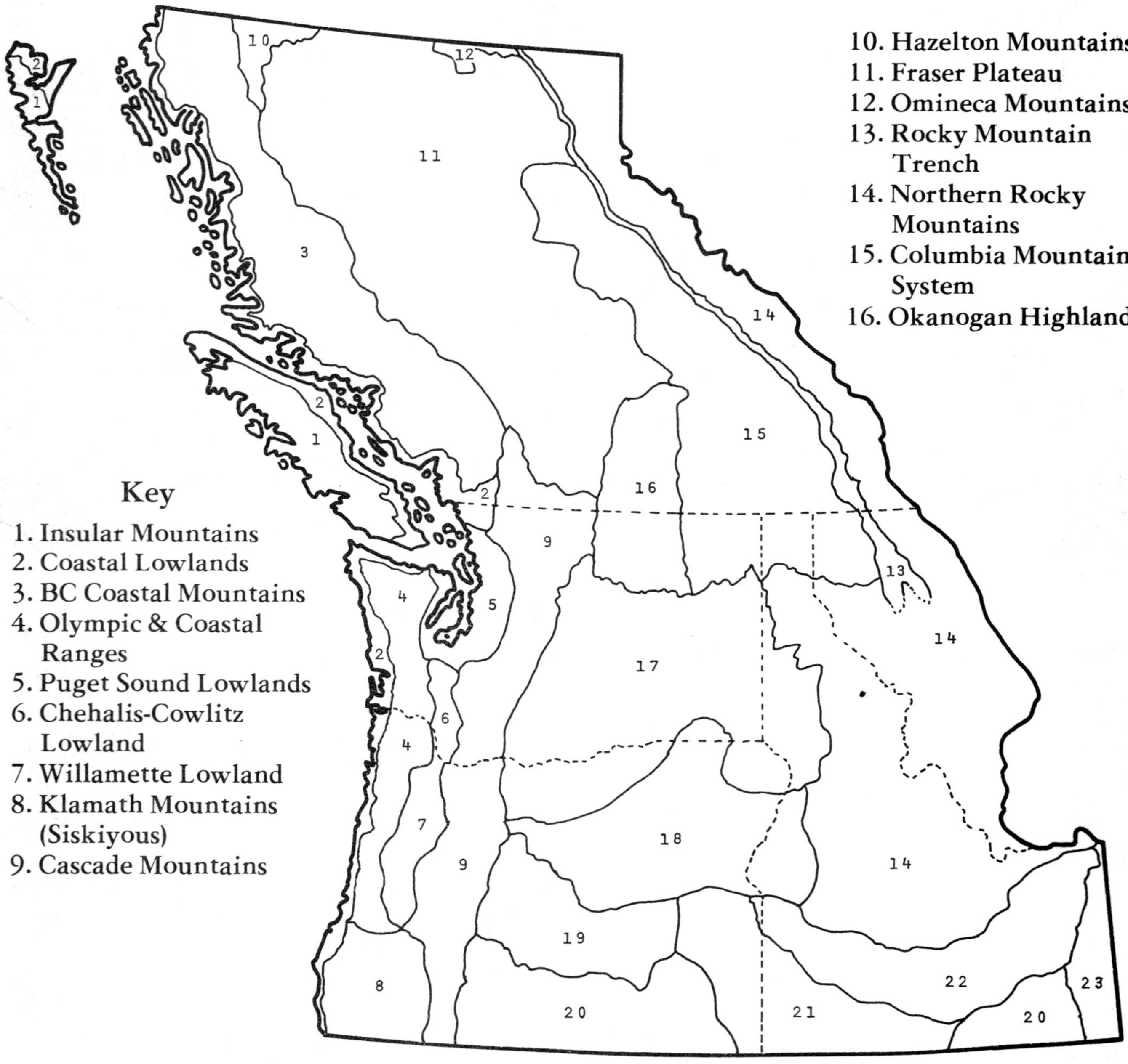

10. Hazelton Mountains
11. Fraser Plateau
12. Omineca Mountains
13. Rocky Mountain Trench
14. Northern Rocky Mountains
15. Columbia Mountain System
16. Okanogan Highlands

Key

1. Insular Mountains
2. Coastal Lowlands
3. BC Coastal Mountains
4. Olympic & Coastal Ranges
5. Puget Sound Lowlands
6. Chehalis-Cowlitz Lowland
7. Willamette Lowland
8. Klamath Mountains (Siskiyous)
9. Cascade Mountains

17. Columbia Basin
18. Central Mountains
19. Harney High Lava Plateau
20. Basin and Range
21. Malheur-Owyhee Uplands
22. Snake River Plains
23. Middle Rocky Mountains

Knowing Home

BLM

A bioregion, or life-region, is an area which has a unique interlocking web of life distinguishable from that of neighboring regions. A bioregion encompasses watersheds, plants and animals, peoples and tales, past and future. Because of this, a concept closely linked with bioregions is reinhabitation. In *Reinhabiting a Separate Country* (San Francisco: Planet Drum, 1978) Peter Berg explains that

> *Reinhabitation means learning to live-in-place in an area that has been disrupted and injured through past exploitation. It involves becoming native to a place through becoming aware of the particular ecological relationships that operate within and around it. It means understanding activities and evolving social behavior that will enrich the life of that place, restore its life-supporting systems, and establish an ecologically and socially sustainable pattern of existence within it....Useful information for reinhabitants can come from a wide range of sources.... Reinhabitants can apply this information toward shaping their own life patterns and establishing relationships with the land and life around them.*

The reason the concepts of reinhabitation and bioregionalism are relevant to this book is that in order to begin developing a sustainable agriculture, one needs to recognize and nurture this sense of place, this sense of rootedness and groundedness, this deep understanding of one's environment. And such an understanding is only possible when one's life has a rich, full-bodied context.

> *Hold in your mind the memory of the land as it is when you take it. And with all your strength, all your might and with all your heart, preserve it for your children.*
>
> Chief Sealth, 1855

CoEvolution Quarterly

One of the best ways to keep informed about our rapidly evolving planet is *CoEvolution Quarterly*, a unique journal created by the same folks who publish the *Whole Earth Catalog*.

The Winter, 1981 issue of *CoEvolution Quarterly* was devoted to the concept and philosophy of bioregions, with major essays on the theoretical and political implications of living and working in place.

A subscription to *CoEvolution Quarterly* is $14 per year. The bioregions issue is available for $4 from *CoEvolution Quarterly*, PO Box 428, Sausalito, CA 94966.

THE NORTHWEST EXPERIENCE

The Pacific Northwest is a land bounded by rainy winter seas and densely forested mountain ranges passable only for a matter of weeks during the summer. The indigenous peoples who inhabited these coasts and river valleys, over centuries and in many cases millenia, wove together the strands of a richly textured mythic and spiritual heritage with roots reaching back through the past ice age. These are the original myths of this region, sprung out of its communion with the first *homo sapiens* to settle here....

We who see ourselves as reinhabitants of this, or any region, inherit this body of myth as part of the water we drink, the air we breathe. But we also bring with us our own mythic heritage, trappings of the civilization that produced us: a skeptical rational intelligence, certain scientific assumptions, ecological insights, anthropological knowledge of other cultures, other religions; politics, poetry....All these shuffled and cross-bred into a cosmology which must, at this point in civilization, find its justification within the dictates of the biosphere itself. We must become true inhabitants of the planet, as in the Latin *habite* (to dwell, as within a home) if we are to stay here at all.

Tim McNulty, from "Myths and a Sense of Place" in *The Northwest Experience, Volume 1.*

There are now two volumes of *The Northwest Experience*. They contain essays, interviews, reports and comics which blend to provide historical, ecological, spiritual and futuristic perspectives on life in the Pacific Northwest.

As editor Lane Morgan says in her introduction to Volume 2:

"A regional perspective begins with individual experience. We need to know who we are and why we are here before we can talk sensibly about what we want to become. *The Northwest Experience* is a place to voice these individual Northwests."

The Northwest Experience, Volumes 1 and 2. Lane Morgan, Ed. Seattle: Madrona Publishers, 1980 and 1981.

To help further the development of such a context, this first chapter provides an overview of the past and present of the Pacific Northwest.

BOUNDARIES OF THE NORTHWEST

For the purposes of this book, the northern boundary of the Pacific Northwest is the 55th parallel of latitude (north), which is just north of Prince Rupert on the British Columbia coast. The region extends through British Columbia and western Montana, with the eastern boundary being the crest of the Rockies. The southern boundary is defined by the borders of Oregon and Idaho. Lastly, the western boundary runs along the edge of the continental shelf, for many of the region's resources are derived from the coastal and offshore waters. Within these boundaries are two sub-regions, the Maritime and the Interior, divided by the Cascade Mountains.

The temperate climate found in the region west of the Cascades, referred to in this book as the Maritime, is unique in North America. It bears more similarity to the climate of parts of Britain and western Europe than it does to most of continental North America, as Maritime gardeners quickly learn. For example, the Olympics and the mountains of Vancouver Island receive the highest amount of precipitation of any area in North America--over 200 inches annually. However, precipitation can vary widely within a small area, as is demonstrated by the rainshadow cast by the high Olympics; precipitation is as low as fifteen inches a year in the San Juan Islands.

Amount of precipitation gradually diminishes between the coastal mountains and the Cascades. The western side of the Cascade mountain belt still experiences relatively high precipitation, but the eastern slopes are much drier. Most of the area from the crest of the Cascades to the crest of the Rockies, referred to as the Interior, receives only ten to twenty inches of precipitation annually, with some areas receiving as few as five inches.

The length and intensity of the growing season varies greatly in the Pacific Northwest. Some areas can experience frost every month of the year, while others do so only rarely. Extremely hot summer temperatures in the Columbia and Snake River valleys (up to 115° F) give these areas a higher accumulation of degree-days than any other part of the region. This makes possible the ripening of some crops which cannot ripen in maritime areas, even though the latter have more frost-free days.

The Rogue River's interior basin has an especially good growing season, situated as it is between the Siskiyous and the southern Cascades. It has hotter and drier summers than any other area west of the Cascades.

Although the general climate patterns of the Pacific Northwest can be helpful in understanding the range in distribution of soil types and vegetation, factors such as variations in local topography may result in variable microclimates within even a small area. It is important to take these local variations into account when designing agriculture landscapes.

Reinhabitation

GARY SNYDER

In Wendell Berry's essay "The Unsettling of America" he points out that the way the economic system works now, you're penalized if you try to stay in one spot and do anything well. It's not just that the integrity of Native American land is threatened, or National Forests and Parks; it's *all* land that's under the gun, and any person or group of people who tries to stay there and do some one thing well, long enough, to be able to say, "I really love and know this place," stands to be penalized. The economics of it works so, that anyone who jumps at the chance for quick profit is rewarded—doing proper agriculture means *not* to jump at the most profitable chance—proper forest management or game management means doing things with the far future in mind—and the future is unable to pay us for it right now. Doing things right means living as though your grandchildren would also be alive, in this land, carrying on the work we're doing right now, with deepening delight ....

Re-inhabitory refers to the tiny number of persons who come out of the industrial societies (having collected or squandered the fruits of 8000 years of civilization) and then start to turn back to the land, to place. This comes for some with the rational and scientific realization of inter-connectedness, and planetary limits. But the actual demands of a life committed to a place, and living somewhat by the sunshine green plant energy that is concentrating in that spot, are so physically and intellectually intense, that it is a moral and spiritual choice as well.

Thomas H. Pohrt, *The Old Ways*

"Mankind has a rendezvous with destiny in Outer Space." Some say. We are already travelling in space. This is the galaxy, right here. The wisdom and skill of those who studied the universe first hand, by direct knowledge and experience, for millennia, both inside and outside themselves, is what we might call the Old Ways. Those who envision a possible future planet on which we continue that study, and where we live by the Green and the Sun, have no choice but to bring whatever science, imagination, strength, and political finesse they have to the support of the inhabitory people— natives and peasants of the world. Entering such paths, we begin to learn a little of the Old Ways, which are outside of history, and forever new.

Much of the conceptualization and language of bioregions and reinhabitation has come from a group of folks in northern California. Peter Berg and friends at the Planet/Drum Foundation coined the term "reinhabitation" and joined it with "bioregion" to introduce the idea of "living-in-place." The point is that living-in-place is an energizing pre-requisite to living and working on a scale that *matters*.

Planet Drum has published a three-part series of pamphlets on bioregional design: Toward a Bioregional Model, by George Tukel, Figures of Regulation, by Peter Berg, and Reinhabiting Cities and Towns, Designing for Sustainability, by John Todd. The complete Eco-Decentralist Design series is available for $10 ppd. Planet Drum also publishes Raise the Stakes, a thrice-yearly report on the worldwide movement toward bioregionalism and decentralization.

For more information, contact the Planet Drum Foundation, Box 31251, San Francisco, CA 94131.

In *The Old Ways*, Gary Snyder richly explains the context and importance of reinhabitation in his talk excerpted at left. In his newest book, *The Real Work: Interviews and Talks 1964-1979*, he discusses more fully how and why reinhabitation is such a difficult task, and also why it is such a necessary and fulfilling one. Gary's poetry and talks blend to provide a calm, enriching and heartfelt vision of the beauty and strength of a world inhabited by farmers, gardeners and other reinhabitants.

Turtle Island. New York: W.W. Norton & Co, 1974.

The Old Ways: Six Essays. San Francisco: City Lights Books, 1977.

The Real Work: Interviews and Talks 1964-1979. New York: W.W. Norton & Co, 1980.

Knowing Home

RAIN Staff. Knowing Home, Studies for a Possible Portland. 1981. RAIN, 2270 NW Irving, Portland, OR 97210.

Introduction to the setting, people, and potential of a city that people choose to call home.

Knowing Home: Studies for a Possible Portland is an excellent model of a way to develop a community's "sense of place." Built around a set of essays, including "Knowing Home: The Basics of Bioregions" and "Sustainable Portland: What We Need Is a City That Can Carry Us into the Next Century, Knowing Home develops the theme of Portland's past, present and future. Shorter essays on Portland's life support—Energy, Food/Agriculture, Arts, Waste Recycling, and others—succinctly describe Portland's existing reality, how this affects Portland's relation to the rest of the region and the world, and how the transition to a more sustainable self-reliant community might come about. Interspersed throughout are photos, interviews and quotes which increase the book's textural depth and effectiveness. (See excerpts below.)

The last section of the book contains an annotated guide to Portland-area community and national self-help groups and activities, and a bibliography designed to help the reader take the step from understanding to action.

Knowing Home was written and edited by the staff of RAIN magazine, a national bi-monthly which features appropriate technology articles and access information. This book was in part an attempt to define the staff's own sense of place, to ground their regional and national work in the local community. To this end they have also developed the Rain Community Resource Center, which develops programs, resources and publications to help people work together to build self-reliant communities.

Mark Musick

Bob Benson is a citizen-activist, working to preserve historical sites and endangered ecosystems. Knowing Home includes a biographical sketch of Bob by Richard Plagge. Below is an excerpt.

Bob's knowledge of the area he grew up in is uncanny. He sees things on several levels at once so that you sometimes feel you are in a car with some kind of X-ray machine that is equipped with a time-shift module. 'We are now passing over a latitude line,' he says. Then moments later, 'This road used to climb the grade up towards that farmhouse, but in '48 when this new highway went through, the state managed to finagle an easement through here. Or: 'That field is probably an erratic which floated over here in a chunk of ice during one of the post-glacial floods.

Why is this elusive quality, a sense of place, so valuable? Because it is a reference point--both a perspective on the larger world around us and a platform for local action. Held rightly, a sense of place is a tool for framing those bigger-than-we-care-to-imagine problems and bringing them back down to local scale. It helps us to focus our awareness on who we are, where we are headed, and what our next steps might be. In so doing, we often discover that the best solutions are those that can be found in our own back yards.

Steven Ames

In order to understand how to continue to live in this region, or any region we must look at how the region functions. We are, as poet Gary Snyder has said, extremely deficient in regional knowledge. There was a while, before Oregon became a part of the franchised and centralized economic structure, when citizens knew about the land, the seasons, and how to carve out a living. We cannot return to this past era, but we can change our direction and look at what kind of life our region is able to sustain.

Steve Johnson

In May of 1981 a survey of planners, architects and university professors on the future of American cities was published in *Next* Magazine. The conclusion, as usual, left little doubts as to how the experts view Portland. Among the 55 largest cities in the nation, Portland was ranked second for its overall future prospects. It was also judged most attractive city for its size and finished close to the top among trend-setting cities, best-managed cities and best cities for retirement ....

Why Portland? What makes us so prone to succeed? One could argue endlessly--and many people do--as to whether it is dynamic leadership or an active citizenry or any number of other factors. But this line of argument altogether bypasses some of the more indigenous qualities at work. One such quality, so pervasive and yet so subtle, is rarely if ever acknowledged: Portland is a successful place because it is a distinct *place*....

Steven Ames

Bibliography

ATLASES

Albright, R., et al. Coastal Zone Atlas of Washington. Olympia: Washington State Department of Ecology, 1980.
Detailed information on the interactions of living organisms from tidewater to coastal forest. Also useful for Oregon and B.C. seacoasts.

Farley, A. L., ed. The Atlas of British Columbia: People, Environment and Resource Use. Vancouver: University of British Columbia Press, 1979.
Up-to-date information with many maps and tables.

Highsmith, Richard, and Kimerling, A. Jon. Atlas of the Pacific Northwest. Corvallis: Oregon State University Press, 1979.
Each of the six editions has special features. The fourth is very good; the sixth is most up to date.

Loy, William G. Atlas of Oregon. Eugene: University of Oregon, 1976.
A beautiful book and useful research tool.

FLORA AND FAUNA

Floras contain the most detailed listing of native plants of their respective region. They are geared to the scientific botanist rather than the amateur. They are invaluable to the serious student who has learned, or will take the time to learn, the fine points of botanical keying out of plants.

Arno, Stephen F., and Hammerly, Ramona P. Northwest Trees. Seattle: The Mountaineers, 1979.
Nice imagery of the forests of the Northwest, with beautiful line drawings and descriptions of each tree.

Booth, W. E. Flora of Montana: Part 1. Bozeman, MT: Foundation of Montana State College, 1950.

____, and Wright, J. C. Flora of Montana: Part 2. Bozeman, MT: Montana State University, 1966.

Canada Department of Forestry. Native Trees of Canada. Ottawa: Canada Department of Forestry, 1979.

Clark, Lewis J. Wildflowers of the Pacific Northwest: From Alaska to Northern California. Sidney, B.C.: Gray's Publishing, 1976.
This large (604 pages) volume is a glossy, beautiful book meant to be used in the home rather than as a field guide. A collector's item, by one of the Northwest's wildflower masters.

Cronquist, Arthur; Holmgren, Arthur H.; Holmgren, Noel H.; and Reveal, James L. Vascular Plants of the Intermountain West, U.S.A. Vol. 1. New York: Mafner Press, 1972.

——, and Holmgren, Patricia K. Vascular Plants of the Intermountain West, U.S.A. Vol. 6.
These volumes are beautifully illustrated with many photographs. The paleobotany of the region is described in great detail.

Davis, Ray J. Flora of Idaho. Provo, UT: Brigham Young University Press, 1952.

Franklin, Jerry F., and Dyrness, C. T. Natural Vegetation of Oregon and Washington. Portland: U.S. Forest Service, 1973.
The authoritative text on the vegetation patterns of Oregon and Washington. Detailed information on which plants share habitats. Many photographs.

Gilkey, Helen M., and Dennis, La Rea J. Handbook of Northwest Plants. Corvallis: Oregon State University Bookstore, 1980.
This comprehensive key to flowering plants includes only keying out information and brief habitat descriptions. It covers the area from the summit of the Cascades to the coastline and from the Umpqua Divide to northern Washington.

Haskins, Leslie. Wildflowers of the Pacific Coast. New York: Dover, 1977.
When this book was first published in 1934 it was the first popular flora of the region, and it is still a useful book today. Includes an identification key, habitat, blooming period and native American uses, with black and white photos.

Hitchcock, C. Leo, and Cronquist, Arthur. Flora of the Pacific Northwest. Seattle: University of Washington Press, 1976.
The ultimate botanical key for Oregon, Washington and Idaho. It lists every native plant species classified as to its most recent revision. Useful for keying out plants and finding their range.

Hitchcock, C. L.; Cronquist, Arthur; Ownbey, Marion; and Thompson, J. W. Vascular Plants of the Pacific Northwest, 5 Vols. Seattle: University of Washington Press, 1955-1969.

Kozloff, Eugene, N. Plants and Animals of the Pacific Northwest. Seattle: University of Washington Press, 1976.
Plant and animal species arranged by habitat, with line drawings and color photos of plants in flower.

Krajna, V. J., and Brooke, R. C. Ecology of Western North America. Vancouver: University of British Columbia, 1970.
The definitive work on the vegetation patterns of British Columbia.

Maser, Chris; Mate, Bruce R; Franklin, Jerry F; and Dyrness, C. T. Natural History of Oregon Coast Mammals. USDA Forestry Service General Technical Report PNW-133. Portland: Pacific Northwest Forest and Range Experiment Station, 1981.
This thick book presents detailed information on the biology, habitats, and life histories of the 96 species of mammals of the Oregon Coast. Soils, geology, and vegetation are described and related to wildlife habitats for the 65 terrestrial and 31 marine species.

Ornduff, Robert. An Introduction to California Plant Life. Berkeley: University of California Press, 1974.
A fine overview of the vegetation patterns of California, including a chapter on the role of introduced species in the ecosystems. An affordable, readable book.

Peck, Morton E. A Manual of the Higher Plants of Oregon. Portland: Binfords & Mort, 1961.
This manual gives a nice synopsis of the vegetation patterns of Oregon.

Proctor, Charles M., et al. An Ecological Characterization of the Pacific Northwest Coastal Region. Biological Services Program FWS/OBS-79/11. Portland: U.S. Fish and Wildlife Service, 1980.
This five volume report contains an enormous amount of information. Plant, animal, fish and human interactions from the crest of the Coast range to the edge of the continental shelf. Covers northern California, Oregon and Washington.

Taylor, Roy L., and MacBryde, Bruce. Vascular Plants of British Columbia. Vancouver: The University of British Columbia Press, 1977.

Gathering What the Great Nature Provided

Food Traditions of the Gitksan

This is a singularly beautiful book, in which the Gitksan people, members of the Tsimshian tribe, describe the traditional methods and recipes used in the area of north central British Columbia around what is now called the Skeena River. Throughout, the cultural importance of food is emphasized as the group's elders share their knowledge simply, often with humor.

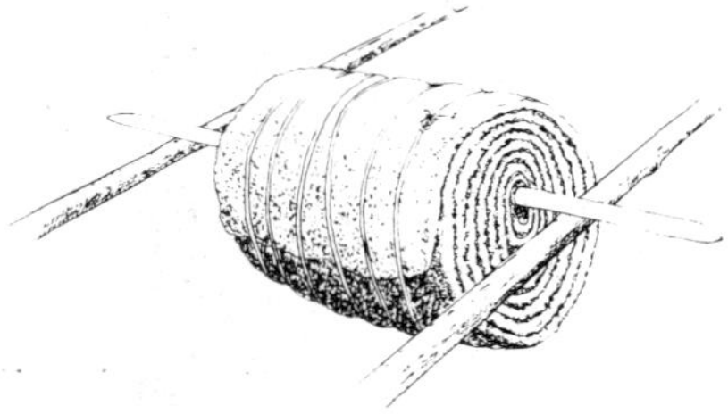

Carefully rolled dried berries.

Here are instructions on how to fillet, smoke, boil, toast, age, barbecue and bake fish; how to pick and preserve berries; how to use evergreen trees for food; how to preserve and cook moose, beaver, rabbit, porcupine and grouse.

Gathering What the Great Nature Provided. By the People of 'Ksan. Seattle: University of Washington Press, 1980.

Native People

The Native Americans saw the entire landscape as the source of their food. In addition to the abundance of fish, which was the staple of their diet, Northwest Indians harvested acorns, hazel nuts and pine nuts where available, and gathered the many vegetables, herbs and fruits which grew wild throughout the region.

One of the best introductions to the indigenous uses of native plants is provided by Nancy Turner in her two-volume set, *Food Plants of British Columbia Indians*, and in her companion volume, *Plants in British Columbian Indian Technology*. These books span the traditional uses of plants in both the Maritime and Interior regions.

Then as now, eating sometimes took on an added, social significance. As Nancy Turner describes it in *Food Plants of the British Columbia Indians, Volume 2, Interior Peoples*:

> Every year in late spring, when the "Indian potato" and yellow avalanche lily *(Erythronium grandiflorum)* corms and other "roots" were ready to be dug, families gathered at Botanie from all over the surrounding countryside— Thompson, Lillooet, and even Shuswap peoples, the former acting as hosts since they owned the valley— to partake in the harvest. The valley assumed the atmosphere of a country fair, with games and competitions, visiting between old friends, reuniting of relatives, and undoubtedly courting of sweethearts.

Turner, Nancy. *Food Plants of British Columbia Indians.* 2 Vols. Victoria: British Columbia Provincial Museum, 1975.

——. *Plants in British Columbia Indian Technology.* Victoria: British Columbia Provincial Museum, 1979.

Bibliography

Balls, Edward K. Early Uses of California Plants. Berkeley: University of California Press, 1962.
Useful California ethnobotany.

Densmore, Frances. How Indians Used Wild Plants for Food, Medicine and Crafts. New York: Dover, 1974.
Originally published in 1926.

Gilmore, Melvin R. Uses of Plants by the Indians of the Missouri River Region. Lincoln: University of Nebraska Press, 1977.
First published in 1919, many of the plants discussed are found in parts of the western Rocky Mountains.

Gunther, Erna. Indian Life on the Northwest Coast of North America. Chicago: University of Chicago Press, 1972.
Details the culture of Northwest peoples when they were first contacted by Europeans.

——. Ethnobotany of Western Washington. Seattle: University of Washington Press, 1981.
Culinary and medicinal uses of wild plants by Native Americans of the Maritime region.

Harrington, H. D. Edible Native Plants of the Rocky Mountains. University of New Mexico Press, 1968.
Useful guide, large bibliography.

Mabey, Richard. Plantcraft: A Guide to the Everyday Use of Wild Plants. New York: Universe Book, 1978.
Everyday use of wild plants by Native Americans and pioneers.

Murphy, Edith Van Allen. Indian Uses of Native Plants. Palm Desert, CA: Desert Printers, 1959.
Ethnobotany for the Indians of eastern Oregon and Utah.

Robinson, Peggy. Profiles of Northwest Plants: Food Uses, Medicinal Uses, Legends. 2nd ed. Portland: Far West Book Service, 1979.
Alphabetical list of common Northwest plants.

Sauer, Carl. Man in Nature: America Before the Days of the White Man. Berkeley: Turtle Island Foundation, 1975.
Accurate portraits of American Indian tribes and their ways of life.

Stuhr, E. T. Manual of Pacific Coast Drug Plants. Lancaster, PA: Science Printing Co., 1933.
Scientific and common names, uses and range of 1,100 species used by Pacific Coast Indians or settlers for medicinal purposes.

FOR MORE INFORMATION

Bear Tribe Medicine Society
P.O. Box 9167
Spokane, WA 99209
Modern-day medicine society; publishes a magazine and catalog of Native American books and conducts seminars on healing.

Native Plants

A native plant preserve is an area in which native plant communities are protected so that their natural interactions can be observed with as little interference as possible from non-native plants, grazing animals or people. Plant preserves maintain diverse gene pools, and give people an opportunity to study and observe the plant forms which have naturally evolved in that location.

In 1927, the Forest Service joined other federal land management agencies in establishing a system of preserves for major native plant ecosystems. Since then, other public agencies and private organizations have joined the effort to preserve natural areas. The Pacific Northwest is fortunate to have one of the most well-developed systems of natural areas in the nation. The 1979 *Atlas of the Pacific Northwest* (6th ed.) lists 105 Natural Areas established in Oregon, Washington and Idaho. The atlas contains a map showing the location of each of these natural areas and its acreage, as well as the agency or foundation which manages it. Permission to visit these preserves must be obtained from the managing agency.

Franklin, Jerry F.; Hall, Frederick C.; Dyrness, C.T.; and Maser, Chris. Federal Research Natural Areas in Oregon and Washington: A Guidebook for Scientists and Educators. 1972. Available from USGPO, Superintendent of Documents, Washington, DC 20402.

Physical, biological and historical features of the 48 Research Natural Areas of the Pacific Northwest.

Franklin, J.F. "Scientific Reserves in the Pacific Northwest and their Significance for Ecological Research." In Proceedings of the Symposium on Terrestrial and Aquatic Ecological Studies of the Northwest, March 26-27, 1976. Cheney, WA: Eastern Washington State College Press, 1976.

Kruckeberg, Arthur R. Gardening with Native Plants of the Pacific Northwest. Seattle: University of Washington Press, 1982.

A beautiful introduction to the diverse and colorful world of native plants that are welcome additions to the garden. Included are valuable notes on plant propagation, design considerations, and plant sources.

Spurr, Joy. Wild Shrubs, Finding and Growing Your Own. Seattle: Pacific Search Press, 1978.

Guide to the selection, propagation, and culture of forty ornamental native shrubs. Includes a Landscaping Guide with notations on color, habitat, and edibility of fruit for humans and wildlife.

NATIVE PLANT SOCIETIES

Native plant societies work with professionals and amateurs to preserve endangered species of native flora. Their objective is to increase the knowledge of their members and the public in identification and conservation of these native plants.

Washington Native Plant Society, c/o Dr A.R. Kruckeberg, Dept. of Botany, University of Washington, Seattle, WA 98195.

Native Plant Society of Oregon, 1920 Engle Ave NW, Salem, OR 97304.

Idaho Native Plant Society, Herbarium, College of Idaho, Caldwell, ID 83605.

California Native Plant Society, 2380 Ellsworth, Suite D, Berkeley, CA 94704.

THE MARIPOSA FOUNDATION

The Mariposa Foundation is an Idaho-based non-profit organization which is developing a conservation program for the revival of valuable plants, especially wild lilies. They are concentrating on the two most important lilies, the camas and the sego. These lilies were once widespread in the native prairie ecosystem. They provided an important part of the food supply for native Americans and later for trappers and pioneers, and are highly prized by wild animals as food.

Lilies of B.C.

The camas and sego lilies have been greatly reduced in numbers and in their natural range. In some cases subspecies have become extinct, and many more are endangered. If lost, the genetic base of the lily family as a whole would be weakened, thus limiting the ability of lilies to survive changing environmental conditions.

For more information, contact The Mariposa Foundation, 415 Residence St., Moscow, ID 83843.

OPERATION WILDFLOWER

"Operation Wildflower" is a National Council of State Garden Clubs effort dedicated to the the beautification of our roadsides and other approved sites and to the education of the public in the appreciation, preservation and propagation of our native wildflowers and grasses.

The Council has published a book entitled Resources on Wildflower Propagation. It is a comprehensive directory published in 1981 to encourage greater appreciation and use of native plants. It includes extensive bibliographies and access to researchers across the nation. The address is 4401 Magnolia Avenue, St. Louis, MO 63110.

Botanical Gardens
University of British Columbia
Vancouver, B.C. V6T 1W5
110 acres; native, medicinal and imported plants.

Van Dusen Botanical Gardens
5251 Oak Street
Vancouver, B.C. V6M 4H1
55 acres of demonstration gardens; 3,000 species grouped into geographic sections.

Butchart Gardens
Victoria, B.C.
The most beautiful private garden in the Northwest.

Berry Botanic Garden
11505 SW Summerville Avenue
Portland, OR 97219
Alpine plants, primulas, natives, rhododendrons, lillies and more.

Hoyt Arboretum
4000 SW Fairview Blvd.
Portland, OR 97221
214 acres; conifers, native shrubs, flowers and wildflowers.

Peavy Arboretum
Oregon State University
School of Forestry
Corvallis, OR 97331
80 acre site adjoining 11,500 acre University forest. Contains native and non-native species.

Finch Arboretum
W 3404 Woodland Avenue
Spokane, WA 99204
Greenhouses closed due to budget cuts. Grounds are still open.

Hiram Chittenden Locks Garden
Market St. at 26th Ave., NW
Seattle, WA 98107
Extensive, colorful worldwide selection of plant material.

University of Washington Arboretum
University of Washington
Seattle, WA 98185
Created in 1894, the 200 acre site is considered among the best in the U.S. Classes, periodicals and publications are offered.

Strybing Arboretum
San Francisco, CA 94122
1,000 acres; one of the finest in the U.S. Rare subtropical plants, natives and cold climate plants.

U.S. National Arboretum
24th and R Street NE
Washington, DC 20002
National botanical resource and research center.

Public Gardens

Human immigrants have carried seeds of their familiar, useful plants with them wherever they have gone, introducing them into plant communities often very different from their original communities. This mingling of gene pools has caused the extinction of many plant species, and the near extinction of countless others. In recognition of the importance of species diversity, arboreta, botanical gardens and related horticultural research centers around the world have done a tremendous amount of work preserving plant species for display, research and planting.

The Pacific Northwest is well-endowed with arboreta and botanical gardens. Gardening columnist Ed Hume's book *Northwest Gardens to Visit* (see below) lists about 30 public plantings of interest for the Northwest.

Mary Morgan

If you are so lucky as to live near one of these arboreta, take the time to visit there and learn from direct observation about the growing habits of different plants. Most of the gardens are planned for beauty, so they make a fine outing for the whole family.

Public support is needed to see that the arboreta receive the financial support they need to carry on their work.

Wind River Arboretum
U. S. Forest Service
Carson, WA 98610
Preserve for the study of Northwest forest ecology.

ORGANIZATIONS

American Horticultural Society
Mount Vernon, VA 22121
Publishes a newsletter and magazine bi-monthly. Special offers on books, gifts and rare seeds.

The Washington Nature Conservancy
618 Smith Tower
Seattle, WA 98104
Has helped preserve 1.7 million acres of islands, marsh, forest, streams, lakes and prairie throughout U.S., Canada, Latin America and the Caribbean.

BIBLIOGRAPHY

Northwest Gardens to Visit, Ed Hume, PO Box 99505, Tacoma, WA 98499.
Listing of some 30 arboreta and botanical gardens in the Northwest.

American Gardens-Traveler's Guide, Brooklyn Botanic Gardens, 1000 Washington Ave., Brooklyn, NY 11225.
Describes several hundred famous gardens and arboreta around the U.S. and Canada.

Logan, Harry Britton. A Travelers Guide to North American Gardens. New York: Charles Scribner's Sons, 1974.
Listing of North American public gardens. Many photos.

Garden in the Desert

On a once-barren bluff overlooking the Wenatchee Valley one family's vision has transformed the desert into a bit of paradise.

Surrounded by vast treeless hills, the green alpine forest known as Ohme Gardens has grown over a period of fifty years to stand as a beacon to human vision and determination, and to the power of plants and water to bring life to the land.

In 1929 the bluff was simply part of a 40 acre parcel of land on which Herman and Ruth Ohme planned to start an orchard. Born in Illinois, Herman Ohme was fascinated by the rugged mountains of the West. Impressed by the bluff's spectacular views of the Columbia River and the Cascade Mountains, the Ohme's set out to begin transforming the dry, rocky hill into an alpine forest. That vision became a dream to which they devoted the rest of their lives, and the dream has been carried on by their children and grandchildren.

Working by hand, the sagebrush was grubbed from the dry slopes to make room for trees and shrubs transplanted from the Cascades. Large boulders were moved and the first ponds were dug with the help of a mule. By 1939 the results of their efforts were already becoming apparent to people in the valley below. Visitors persuaded the Ohme's to open their garden to the public. Today the cool green garden, set against spectacular sandstone formations, covers nearly ten acres of former desert and attracts more than 30,000 visitors a year.

Where hot, dry winds once swept unobstructed, the bluff is now protected by more than a thousand ponderosa pine, Douglas-fir, grand fir, western red cedar, mountain hemlock, and alpine fir. Hand-hewn stone paths and steps connect sparkling pools, streams and waterfalls which flow among meadows of thyme, dianthus, phlox, sedum, vinca minor, mosses and ferns.

The Ohme's goal was to re-create a natural high mountain forest overlooking the arid valley, and Ohme garden *is* a delight. It is, however, far from being a self-sustaining landscape. Water, sometimes as much as 100,000 gallons an evening, is pumped from the Columbia River to the ponds, and from there through the irrigation system which makes the emerald forest and beautiful flowers possible.

The Ohme family achieved its goal of bringing the forest to one piece of the earth. The costs in terms of energy and water were not a primary consideration, as they are today. But by adding drought-hardy plants, a similar accomplishment is possible on a sustainable basis by using drip irrigation and other water conserving techniques. The Ohme family's achievement, however, still serves as an inspiration to people working to restore green to arid landscapes.

Mary Morgan

Ohme Garden is located three miles north of Wenatchee. Visitors are welcome to tour the grounds from April through October for a modest admission fee. For more information contact Ohme Garden, 3327 Ohme Road, Wenatchee, WA 98801.

Dalichow, Fritz. Agricultural Geography of British Columbia. Vancouver: Versatile, 1972.
Well-researched overview of the B.C. agricultural industry.

Daubenmire, Rex. Plant Geography. New York: Academic Press, 1978.
A fascinating look at plant geography with special emphasis on North America, by a person who understands the vegetation of the Pacific Northwest.

—. Structure and Ecology of Coniferous Forests of the Northern Rocky Mountains. Missoula: University of Montana Foundation, 1969.
Guide to the habitat zones of the Northern Rockies.

Dorcey, H. J. The Uncertain Future of the Lower Fraser. Vancouver, B.C.: Westwater Research Centre at the University of British Columbia, 1976.
This book is the result of years of research into the ecology of the lower Fraser River and the pollution and degradation to which the river is being increasingly subjected. The summary points out that the river is still relatively clean and explains steps which can be taken to reverse the degradation processes and maintain the river's natural productivity and health.

Heilman, Paul; Anderson, Harry W.; and Baumgartner, David M. Forest Soils of the Douglas-Fir Region. Pullman: Washington State University Press, 1979.
An in-depth look at the soils and vegetation of the Maritime Northwest. Concise, readable account of the region's geological history.

Hunt, Charles B. Natural Regions of the United States and Canada. San Francisco: W. H. Freeman, 1974.
Illustrated guide to bioregions. Detailed and authoritative.

Jones, R. L. Biogeography: Structure, Pattern and Change Within the Biosphere. Amersham, Buckinghamshire: Hulton Educational Publications, 1980.
A survey of the world's ecosystems. Primarily a study of the movement of world land masses and plant species from Precambrian times to the present.

McKee, Bates. Cascadia, The Geologic Evolution of the Northwest. New York: McGraw, 1972.
Detailed but readable text on how the Northwest came into being. Includes British Columbia. Many aerial photographs of Northwest mountainscapes.

Sauer, Carl O. Land & Life. Berkeley: University of California Press, 1963.
Collection of works by one of North America's most far-sighted cultural geographers.

Valentine, K. W. G., et al, eds. The Soil Landscapes of British Columbia. Victoria: Resources Analysis Branch,
Contains useful charts, excellent fold-out maps, and color photos of the major soil types. A good way to get to know British Columbia from the ground down.

PERIODICALS

Capital Press
PO Box 2048
Salem, OR 97308
This excellent weekly newspaper provides personal and professional insights into the state of Northwest farming, forestry and fisheries. Many of the stories have a pro-agribusiness slant, but for current information on the Northwest's agricultural situation, the Capital Press can't be beat. The large classified section is an excellent place to find used farm equipment at good prices.

Clinton St. Quarterly
2522 SE Clinton
Portland, OR 97202
This is a fun paper. Each issue contains an invigorating blend of articles on local and regional politics, theater and art; travel and perspectives on foreign affairs; sports, jazz and rock and roll; and lots more. Even the ads are good reading. CSQ is free in the Portland area, $5/yr by mail.

(Lewisia Tweedii)

Common Sense: Examining Man's Role in Nature
Kootenay Environmental Institute, Ltd.
Box 2, Galena Bay
Via Revelstoke, BC V0E 1X0
This informal bimonthly magazine contains reprints from many other environmental journals. Philosophical and factual articles are interspersed with inspirational quotes.

Earthwatch Oregon
Oregon Environmental Council
2637 SW Water Avenue
Portland, OR 97201
Broad coverage of environmental issues, ranging from bicycle commuting to land use planning.

Okanogan Natural News
Box 1000
Oroville, WA 98844
This eclectic newspaper comes out a few times a year from the highlands of northeastern Washington. Each issue is filled to overflowing with poems, reprints and articles on world affairs, agriculture, community, spirituality, diet, health and more.

Pacific Northwest
222 Dexter Ave North
Seattle, WA 98109
This magazine, published ten times a year, contains a wonderful variety of essays, articles and editorials on the many aspects of life in the Pacific Northwest. Recent articles included a highly controversial piece on the town of Forks, Washington, a spring Restaurant Guide, and a report on a Tacoma psychologist who is doing studies on the interactions between color, nutrition, environment and human behavior. Published by Pacific Search Publications, whose book-publishing division, Pacific Search Press, has produced such books as Common Seaweeds of the Pacific Coast and Fire and Ice: The Cascade Volcanoes.

The Smallholder
Argenta, BC V0G 1B0
Published bimonthly, The Smallholder is an excellent information network serving rural people. Readers send in their questions on agriculture, forestry, aquaculture, shelter, food storage, and cottage industries. Answers or comments by other readers are published in subsequent issues. Also lists upcoming meetings and events.

Syesis
c/o British Columbia Provincial Museum
Victoria, BC V8W 1A1
Syesis is an Indian word meaning a story about an actual event in one's history. Syesis contains well-researched articles on the natural history of British Columbia, and specifically the ecology of the plant and animal communities found there.

Kay Hoffman, after a photograph by Jan Kepley

Sustainable Agriculture

An index of the health of a rural community—and, of course, of the urban community, its blood kin—might be found in the relative acreages of field crops and tree crops. By tree crops I mean not just those orchard trees of comparatively early bearing and short life, but also the fruit and nut and timber trees that bear late and live long. It is characteristic of an unsettled and anxious farm population—a population that feels itself, because of economic threat or the degradation of cultural value, to be ephemeral—that it farms almost exclusively with field crops, within economic and biological cycles that are complete in one year. This has been the dominant pattern of American agriculture. Stable, settled populations, assured both of an economic sufficiency in return for their work and of the cultural value of their work, tend to have methods and attitudes of a much longer range. Though they have generally also farmed with field crops, established farm populations have always been planters of trees.

Wendell Berry. A Continuous Harmony; Essays Cultural and Agricultural. New York: Harcourt Brace Jovanovich, 1970.

Agriculture and forestry are cultural and ecological enterprises. A number of nations are developing approaches to the production of farm and forest resources which rely less on industrial processes and more on ecological principles.

The **International Federation of Organic Agriculture Movements** *(IFOAM) is a global network linking people who are actively creating a new agriculture. In mid-October, 1977 nearly 200 members from around the world gathered in Ebenrain, Switzerland for an international research conference. Below is an excerpt from the introduction to the conference in which Colin Fisher of England's Pye Research Center defined the components of a sustainable agriculture. It is reprinted from the conference proceedings,* Towards a Sustainable Agriculture, *Besson & Vogtmann, eds.*

COLIN FISHER

What do we mean by "a sustainable agriculture"? There are, in my view, eight basic components which are a prerequisite for the achievement of "sustainability" in an agricultural system:

1. Paradoxically, I would nominate "dynamism" in the system as the first of these. In biological systems that which is static is rarely sustainable. A sustainable agriculture is, therefore, of a kind which improves rather than deteriorates in response to the demands made on it and the methods employed in the achievement of optimum production.

2. Even more obviously, a "sustainable system of agriculture" must be one that achieves the production of crops and livestock and the management of the farm's resources in a way that harmonises rather than conflicts with natural systems. This does not mean that we must always despise to use those resources which are devised by man *(sic)* rather than the bounty of nature. To do so would, in my view, be to treat with contempt the very special gift of intelligence which distinguishes us from other species. Nor need a "sustainable" system be entirely self-sufficient. Too dogmatic a pre-occupation with self-sufficiency can

result in a waste of resources. The efficient use of resources (be they natural or devised by man) is the key to the dynamism of constant improvement, and the test of the validity of the resource input should be that it contributes to what I will call "dynamic sustainability".

3. Thirdly, I believe that the system must be diverse and that we have yet to achieve sufficient understanding of the way we must plan and design for diversity in the pursuit of optimum production.

4. Next, our "sustainable agriculture" must be one which relies primarily on renewable resources for the achievement and maintenance of the basic fertility essential to optimum production.

5. The system must be one in which the input of thought, ingenuity, care and personal involvement can be judged to be more significant than the inputs of technology, although I believe our requirement for technology will increase rather than diminish in response to our closer understanding of biological systems. It is, however, unlikely that the sort of technology I have in mind would resemble the crude technologies, both mechanical and chemical, which characterise the sort of agriculture with which all of us from industrialised countries are familiar.

6. A "sustainable agriculture" should be one which recognizes the contribution of good nutrition to the health of populations and accepts that the farmer or producer has a special responsibility to ensure, not only that he or she eliminates hazards of toxicity, but enhances (within the limits of what can be understood) the potential for discriminating between the mediocre, the good and the excellent in aiming always for the optimum nutritive value of staple foods.

7. "Sustainability" must embrace more than crops, livestock and the soils which support their production. It must also include the people who work or live on the holding, and the relationship of the holding to the rest of the rural community in which it is situated. "Sustainability" is about sociology as well as biology.

8. I will complete my list by urging that a farm system is more likely to be "sustainable" if it is aesthetically pleasing to those who work on or live near it, and if it enhances rather than scars (as is often the case) the landscape of which it forms a part. Individually and collectively we are more likely to help "sustain" that which pleases rather than disturbs us.

It would not be difficult to add to my list and, in particular, to discuss the economic (I use the term in its broad not its purely fiscal sense) implications of "sustainability."

Synthesised to its finest a "sustainable agriculture" is about the pursuit of excellence. There can be few areas of human activity where excellence proves itself so exclusive as in agriculture. But to strive for less than excellence is to betray that very special responsibility that is both the burden and the privilege of those of us to whom the ownership or management of a farm is entrusted.

BIBLIOGRAPHY

Baker, Richard St Barbe. My Life, My Trees. 1979. Findhorn Foundation, The Park, Forres, Scotland IV36 0TZ.
Autobiography of the Father of the Trees. A personal account of St Barbe's campaign to save native forests and halt advancing deserts. (See also p. 147).

Berry, Wendell. The Gift of Good Land. San Francisco: North Point Press, 1981.
Essays on people and places where the potential still exists for a sustainable agricuture. A down-to-earth sequel to The Unsettling of America.

——. Recollected Essays, 1965-1980. San Francisco: North Point Press, 1981.
Eleven essays on preserving a sense of place, the health of the earth and human communities.

——. The Unsettling of America: Culture and Agriculture. San Francisco: Sierra Club Books, 1978.
The connections between people and land, and the attitudes and technologies that disrupt them. From hard-hitting critique to an inspired vision of re-settlement.

Besson, Jean-Marc, and Vogtmann, Hardy, eds. Towards a Sustainable Agriculture. 1978. IFOAM Secretariat, c/o chez Les Quatre Saisons, 6 rue Saulnier, F75009 Paris, France.
Proceedings of the 1977 international research conference on sustainable agriculture. Includes 25 papers on current research around the world.

Boeringa, R., ed. Alternative Methods of Agriculture. Amsterdam: Elsevier Scientific Publishing, 1980.
English translation of "The Dutch Report on Alternative Agriculture." It is an in-depth review of alternative agriculture methods and philosophies currently in practice in Europe.

Committee on Foreign Affairs. Background Papers for Innovative Biological Technologies for Lesser Developed Countries. Washington, DC: USGPO, 1981.
A series of papers presented by the Office of Technology Assessment at a special workshop for the Committee on Foreign Affairs. Government policy may change slowly, but these papers provide a wealth of stimulating ideas and information on new crops, biological nitrogen fixation, multiple cropping systems, utilization of native plants, and mycorrhiza technology.

Douglas, J. Sholto, and Hart, Robert A. de J. Forest Farming. Emmaus, PA: Rodale Press, 1978.
Holistic look at the role tree crops can play in building a sustainable agriculture worldwide.

Eckholm, Eric. Planting for the Future: Forestry for Human Needs. Worldwatch Paper 26. Washington D.C.: Worldwatch Institute, 1979.
An overview of world forest trends, the forces behind deforestation, and the need for community-based forestry.

Friends of the Earth. Progress as if Survival Mattered. San Francisco: Friends of the Earth, 1981.
Comprehensive guide to environmental issues, with agriculture seen in the broad context of population and energy.

Howard, Sir Albert. An Agricultural Testament. Emmaus, PA: Rodale Press, 1973.
Pioneering statement of the principles of biological agriculture.

Jackson, Wes. New Roots for Agriculture. San Francisco: Friends of the Earth, 1980.
Incisive analysis of the crisis in conventional agriculture. The solution? Stop tillage and plant perennials, with the potential of perennial grains as an alternative for the future.

King, F.H. Farmers of Forty Centuries: Permanent Agriculture in China, Korea and Japan. Emmaus, PA: Rodale Press, 1973. First published in 1911.
F.H. King's agricultural knowledge and keen observation reveal the agricultural traditions of China, Japan and Korea.

Merrill, Richard, ed. Radical Agriculture. New York: Harper & Row, 1976.
A collection of essays that helped lay the groundwork for alternative agriculture. Outlines major problems and suggests solutions.

Pratt, Martha M. Proceedings of the P.E.I. Conference on Ecological Agriculture. 1979. Institute of Man and Resources, Box 2008, Charlottetown, P.E.I., Canada C1A1A4.
Transcripts of a dozen presentations on the potential for ecological approaches to agriculture. Includes discussions of forest farming, soil testing, farm-scale composting, IPM, and agriculture in Europe and China.

INTERNATIONAL FEDERATION of ORGANIC AGRICULTURE MOVEMENTS

IFOAM Bulletin, c/o Rodale Research Center, RD 1, Box 323, Kutztown, PA 19530.
Quarterly review of new developments in ecological agriculture. The Rodale Research Center is publishing the English edition of the IFOAM Bulletin in support of the International Federation of Organic Agriculture Movements. It includes research reports, journal abstracts, literature reviews, meeting announcements, and IFOAM committee reports. The Bulletin also serves as a forum for the diverse opinions and activities of IFOAM members.

Quinney, John; Barnhart, Earle; and Massion, Maia. Farming in the Image of the Forest. The New Alchemy Institute, 237 Hatchville Road, East Falmouth, MA 02536. Forthcoming.
The ecological basis for a new agriculture, with an emphasis on permanent crops.

Science Council of Canada. Canadian Food and Agriculture: Sustainability and Self-Reliance. 1979. The Science Council of Canada, 100 Metcalfe Street, Ottawa, Ontario, Canada K1P 5M1
This discussion paper is the report of a committee set up to develop recommendations on "Canada's Scientific and Technological Contribution to World Food Supply." While the main concern of the committee was gaining an understanding of the international food system, committee members soon realized that this question, as it related to Canada, was closely tied to Canada's approach to her own national development, the long-term stewardship of her renewable and non-renewable resources, and national and regional self-reliance. As their conclusion notes, "A sustainable food system implies frugality—the willingness to live within physical and social limits for the long-term benefit of all. Current economic reasoning and policy, however, are generally premised on the permissibility of overcoming those limits, and it is precisely this approach which must now change. Industrial thinking now needs reform."

Smith, J. Russell. Tree Crops: A Permanent Agriculture. New York: Devin Adair Co., 1950.
The classic book on tree crops. Most of the text consists of detailed descriptions of the use of tree crops around the world as a solution to the destruction of soils for the production of annual crops. Readable and inspiring.

Todd, John, and Todd, Nancy Jack, eds. The Village As Solar Ecology. East Falmouth, MA: The New Alchemy Institute, 1980.
Inspired by Margaret Mead's suggestion that New Alchemy begin designing for larger communities, this is the first look at integrated life-support systems on a village scale. This beautifully produced book emphasizes the spiritual as well as the technical aspects of creating new communities.

——, with Tukel, George. Reinhabiting Cities and Towns: Designing for Sustainability. San Francisco: Planet Drum, 1981.
Part of Planet Drum's Eco-Decentralist Design series, this is a thorough introduction to biological design principles. Included is a discussion of patterns in nature, and the application of bioregional design concepts to community water supply, shelter, food production, and energy systems.

Todd, Nancy Jack, ed. The Book of the New Alchemists. New York: E.P. Dutton, 1977.
The first synthesis of New Alchemy's work in aquaculture and bioshelter design. Also included are essays on women and ecology, and spirituality.

USDA. Report and Recommendations on Organic Farming. Washington, DC: USGPO, 1980.
The first official response of the USDA to the potential of organic farming as an alternative to soil erosion and the decline of modern agriculture.

Weiner, Michael. Plant a Tree: A Working Guide to Regreening America. New York: Macmillan, 1975.
Readable information on planting trees; well illustrated and full of enthusiasm. The beautiful photos of mature trees for each species described is enough to inspire any homeowner or tree-planter to action.

Ecology and the Farm

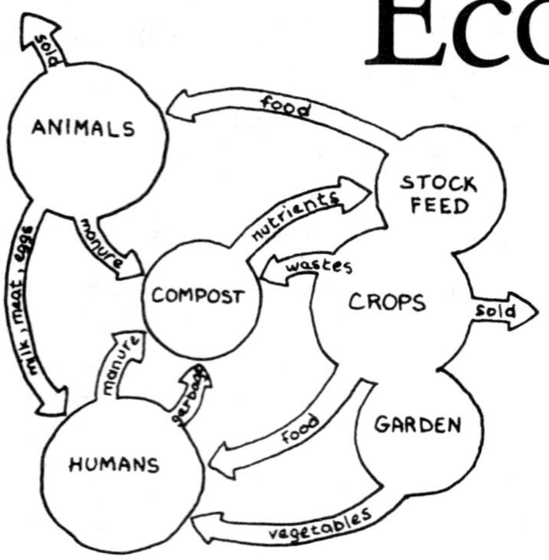

EARTHCARE

Ecological principles relevant to farm practices include:

- the idea that everything affects everything else; an eco-system (including a "farm eco-system") is an interrelated, interdependent whole. At the same time, to achieve its unity, the system maintains a diversity of plants and animals in an intricate system of checks and balances to both foster growth and regulate populations of the individual species within the system.
- the idea that eco-systems are sustainable through internal re-cycling of substances. The system is energized by the sun, a renewable resource, and depends on the cycle of continuous growth and decay to supply nutrients.
- the idea that nothing is lost within the system, and also that its resources and capabilities are limited. Pollutants must be prevented from poisoning the system through the cleansing ability of microorganisms. The capability of the system to absorb toxic materials, as well as to provide nutrients without the complementary process of recycling is finite.

With these and other ecological principles we begin to shape some idea of how our farms—which are eventually accountable to these principles—must be developed. An ecological farm models itself on a natural eco-system as closely as possible. Sustainability, the ability to continue permanently, depends on the extent to which a farm can model itself after the balanced ecological pattern.

THE FARM ORGANISM

By following ecological principles within the borders of the farm, that is, treating the farm as a living organism, community, or eco-system, we are achieving the objective of working in harmony with natural processes; when a farm is an integrated system it contributes to the stability of the larger ecological surroundings of which it is a part. A farm should be based in a healthy environment which is, in turn, composed of healthy farm units—one cannot function in a healthy way without the other.

Ecological systems are interrelated, but they are also self-supporting—the needs of its members are met within the system. Rudolf Steiner, one of the founders of ecological agriculture, said that a farm should try to follow this idea: "A farm is true to its essential nature, in the best sense of the word, if it is conceived as a kind of individual entity in itself—a self-contained individuality. Every farm should approximate to this condition. This ideal cannot be

The article at right is reprinted from Earthcare, a new book which provides an excellent synthesis of ideas and information on ecological agriculture. Written for people farming the prairies of Saskatchewan, Earthcare is the most comprehensive and detailed text available on the principles of organic agriculture. Emphasis is on viewing each farm as a living organism, with details on agricultural ecology, soil management, crop rotations, legumes, weed control, shelterbelts, pest management, the role of livestock, and wildlife enhancement.

Edited by Paul Hanley, Earthcare, Ecological Agriculture in Saskatchewan (1980), is published by the Earthcare Information Centre, Box 1048, Wynyard, Saskatchewan S0A 4T0.

BIBLIOGRAPHY

Cox, George W, and Atkins, Michael D. Agricultural Ecology, An Analysis of World Food Production Systems. San Francisco: W.H. Freeman, 1979.
Detailed text on agricultural ecosystems. Includes sections on crop and livestock agriculture, soil ecology, and integrated pest management.

Hill, S.B., et al. Reading References on Ecological Agriculture: Eight Annotated Bibliographies. 1976. University of Regina, Department of Extension, Regina, Saskatchewan, Canada.
Annotated bibliographies on ecological agriculture, soil biology, fertilization and human waste. Covers books, journals and reports.

absolutely attained, but it should be observed as far as possible. Whatever you need for agricultural production, you should try to possess within the farm itself...."

Stated simply the first principle of ecological agriculture is that a farm should be developed similarly to a natural eco-system:

- It should include a diverse community of plants and animals (including a portion of the land in natural habitats, such as bush or marsh).
- A farm should attempt to contain within it all that it needs to maintain itself, starting with feed, a personal food supply, a portion of its energy requirement and its fertilizer requirements (manure, green manures, etc.)

THE DIFFERENCE BETWEEN A FARM AND A NATURAL ECOSYSTEM

While ecological farming aims at managing a farm as if it were a "micro-ecosystem," a farm will always be different than a completely natural habitat. In a natural system, practically all substances that are produced are eventually recycled within the system; on a farm, the objective is to produce food, not only for the residents, but mainly for export into urban areas. The main difference, then, between a farm and a natural habitat, is the export of production, which is never recycled. We could assume that nutrients will continue to be depleted from the system until its productivity is seriously depleted; however, in reality a properly managed farm can maintain its productivity for several reasons.

Compared to the natural habitat, on a farm the substance cycles are intensified and qualitatively changed, that is, the turnover of organic materials is accelerated. So even though crops are exported from the farm, greater production can be achieved and fertility maintained if adequate recycling of organic matter takes place and if sufficient nitrogen is brought into the soil by nitrogen-fixing bacteria (a process stimulated by good husbandry).

A farm becomes sustainable when:

- Tillage practices improve the rooting zone, soil life, release of nutrients and turnover of organic matter.
- A proper ratio of livestock is achieved and manures are properly used to provide humus and enhance soil life, while providing greater amounts of nitrogen and other elements.
- Appropriate crop rotations increase the depth of the rooting zone; stimulate beneficial organisms; accumulate minerals in the top soil from the subsoil; increase water retention and aeration.
- The ratio of land in soil restoring crops, windbreaks, tree crops, etc. is improved.

Lotan, James E., et al. Effects of Fire on Flora: A State-of-Knowledge Review. USDA Forest Service Gen Tech Rep WO-16. Washington, DC: USGPO, 1981.
The scope of this report includes all main vegetation types in the US. Dozens of literature citations.

Lovelock, James E. Gaia, A New Look at Life on Earth. Fair Lawn, NJ: Oxford University Press, 1979.
Scientific understanding of the Earth as a living organism.

Odum, E.P. Fundamentals of Ecology. New York: Holt, Rinehart & Winston, 1971.
A popular introductory ecology text, with a strong emphasis on human ecology, systems ecology, ecological principles, and environmental protection.

Ricklefs, R.E. Ecology. Concord, MA: Chiron Press, 1979.
An unusually lucid style makes this exhaustive text easy for the beginner to read. Its broad scope, complete references and many examples also make it useful to the advanced reader. A strong evolutionary approach is used.

VITA. Environmentally Sound Small-Scale Agriculture Projects, Guidelines for Planning. 1979. Volunteers in Technical Assistance, 3706 Rhode Island Avenue, Mt. Rainier, MD 20822.
Basic introduction to agricultural ecology. Clear illustrations and definitions of ecosystems, food webs, diversity, succession, and limiting factors.

PERIODICAL

Agro-Ecosystems
Associated Scientific Publishers
PO Box 211
Amsterdam, The Netherlands
An international scientific journal on the ecology of agriculture and forestry. Emphasis is on ecological interactions within and between managed agro-ecosystems, including coverage of crops, soils, livestock, and wildlife.

If your aim is to create a self-sustaining ecosystem, consider carefully the crops you will rely on for market. From an ecological point of view, the best goal would be to export only the net increase gained from water, air and sunshine. The agricultural products which come closest to realizing this goal are fruits, berries, oils, biomass fuel (ethanol, methanol, methane), honey, and flowers. In essence, the objective is the export of pure sunshine.

MARK MUSICK

Green Manure— A Technique for Renewed Fertility

Don Lambert grows dryland wheat and peas on 800 acres in the Palouse region of eastern Washington. His land was originally purchased by his father in 1909 and, with the exception of one trial application back in the 1960s, no commercial fertilizers have been used on the farm since it was acquired. Instead, Lambert relies on green manures to ensure the health and fertility of his soils, and as a result he achieves yields nearly equal to his neighbors' at less cost.

Green manures are crops grown strictly for soil conservation and improvement. Crops such as oats, Austrian winter peas, annual rye, vetch, lupine, sweetclover, and buckwheat may be used. Legumes are especially useful because they have the capacity to fix atmospheric nitrogen in the soil. Green manures draw nutrients from the subsoil and, in the process, they improve soil structure, stimulate biological activity, provide organic matter, and reduce nutrient leaching, while suppressing weeds and diseases.

Wheat and peas are Lambert's commercial crops, but he balances his exports with the periodic return of green manures to continually renew his soil. In Lambert's crop rotation, winter wheat is followed by green peas. After pea harvest his fields are chisel plowed and then sown to Austrian winter peas. The next July the winter peas are plowed down and his fields are fallowed to conserve moisture for a crop of spring wheat the following year. Alfalfa is used periodically to retire the soil from cultivation for two years. Alfalfa is a nitrogen-fixing legume with deep, vigorous roots which greatly enhance soil fertility.

A lifetime of such farming practices has ensured Lambert of biologically active soils which are in excellent tilth and have fertility levels similar to conventional farms in his area. Because Lambert's practices are a unique exception in a region which switched to chemical farming a generation ago, his farm has become the subject of several research projects directed by Robert Papendick, a USDA soil scientist stationed at Washington State University in Pullman.[1]

Papendick, who headed the USDA's Organic Farming Study Team a few years ago, has found that Lambert's fields compare very favorably with those of his neighbors. Not only has Lambert virtually eliminated soil erosion, but the structure of his soil is superior to that of his neighbors. And Andrea Patten, a Master's candidate working under Papendick's direction, has shown that the organic matter, total nitrogen, and extractable phosphorus and potassium tend to be higher in Lambert's soil than in neighboring fields.[2] As a result, Lambert's farm is highly productive, and is likely to remain so while his neighbors face higher costs for petroleum-based fertilizers.

[1] Ramsey, E.W. "Green Manure Crops Maintaining Fertility." Washington Farmer-Stockman, January 7, 1982, p. 24.

[2] Patten, Andrea G. Weilgart. "Comparison of Nitrogen and Phosphorus Flows on an Organic and Conventional Farm." Master's thesis, Washington State University, 1982.
Report on the results of a two year study on Don Lambert's farm. Copies of the thesis are available from the Washington State University Library, Pullman, WA 99164.

ORGANIZATIONS

Ecological Agriculture Project, PO Box 225, MacDonald Campus, Ste-Ane-de-Bellevue, Ontario, Canada H9X 1C0.
Leading Canadian center for information on new approaches to agriculture. Extensive library. Several bibliographies and research papers available.

Frank Porter Graham Center, Rt. 3, Box 95, Wadesboro, NC 28170.
Seeking new agricultural alternatives for the South. Experimenting with natural farming techniques at their demonstration farm.

Friends of the Farm, Rt. 1, Box 32, Dalton City, IL 61925.
Non-profit educational organization promoting biological agriculture. Services include tours of demonstration farms, advice on agricultural questions, and a newsletter.

International Institute of Biological Husbandry, 9 Station Approach, Needham Market, Ipswich, Suffolk, IP6 8AT, England.
An institute devoted to the research and development of biological husbandry and its application worldwide.

Maine Organic Farmers and Gardeners Association, PO Box 188, Hallowell, ME 04347.
Highly effective membership organization serving gardeners and small farmers in the state of Maine. Their annual fair and their newspaper are both significant community services.

Natural Organic Farmers Association, 43 State Street, Montpelier, VT 05602.
An annual conference, farmers' market organizing, bulk purchases, and a newspaper all serve to promote local agriculture and regional self-reliance.

Northwest Bio-Dynamic Agricultural Society, PO Box 97, Chilliwack, BC V2P 6H7.
Dedicated, friendly group of gardeners and farmers interested in putting Rudolph Steiner's Bio-Dynamic principles into practice. They host speakers from around the world, sponsor conferences, and distribute Bio-Dynamic literature and preparations. Send a self-addressed stamped envelope for information.

Oregon Grange, 1313 SE 12th, Portland, OR 97214
State branch of the nation's oldest farm organization.

Oregon-Washington Farmers Union, 10453 Fourth Plain Blvd. NE, Vancouver, WA 98662.
Activist organization which promotes producers' cooperatives and campaigns for political issues vital to family farmers.

Tilth, 13217 Mattson Road, Arlington, WA 98223.
Regional association serving agriculture and forestry in the Northwest. Membership organization, with active chapters throughout the region.

Washington State Grange, 3104 Western Avenue, Seattle, WA 98121.
The largest state Grange in the country, with ongoing social and political action programs.

Women in Farm Economics (WIFE), c/o Ruth Conarro, Star Rt, Mesa, WA 99343
Education and lobbying organization working for the economic survival of the family farm.

Insectary Plants

KATE BURROUGHS

Most ecologists agree that the practice of monoculture has led to an increase in pest and disease problems. When the same crop stretches for miles, unbroken by other plant species, it is much easier for herbivores and wind-borne plant disease spores to spread. Native plants are removed in the process of clearing the land for agriculture, and native insects have to switch to crop plants for their survival. Introduced pests which are adapted to agricultural crops have compounded the problems.

Increasing the diversity of plants within and surrounding the agricultural ecosystem is one strategy for controlling pests using biological restraints. Careful consideration must be given, however, to the particular plant species to be added, because simply increasing the number of different plants may have the undesired result of *increasing* plant pests. Therefore, it is important to observe and understand the biology and ecology of potential crop pests and of their natural enemies. Adding the appropriate species, even in small numbers, is the key to establishing stability.

Crop and non-crop plant species affect the colonization of pests by their visual appearance and smell, or by acting as a physical barrier or diversionary host. Some introduced insect pests, such as the cabbageworm *(Pieris rapae)* and the cabbage aphid *(Brevicoryne brassicae)* naturally exploited the temporary habitats of early succession plants on disturbed ground. These pests are especially attracted by the visual appearance of open stands of green plants silhouetted against brown soil, similar to the conditions created by plowed-field agriculture. Other species, such as the corn borer (*Ostrinia furnicalis)* and the fruit fly (*Oscinella frit)*, which formerly exploited dense stands of wild grasses, are less attracted to widely-spaced crops.

The dispersal of insect pests, nematodes and diseases can be impeded in several ways when host and non-host plants are grown together. Because insects often locate their host plants by smell, most companion planting schemes are designed to mask the smell of crop plants by interplanting them with other strong-smelling plants. For example, cabbages are sometimes protected against flea beetles by intercropping with tomatoes. Companion planting also decreases plant-to-plant colonization by the relatively immobile pests living in the soil.
Non-susceptible plants help prevent the dissemination of wind-borne plant disease spores, such as powdery mildew. Intercropping can also provide a physical barrier to protect crops. For example, the cabbage root fly *(Erioischia brassicae)* can be impeded from laying its eggs in fields which have a cover crop of clover.

Another way in which increased plant diversity helps decrease the presence of insect pests is by supporting

WHAT IS IPM?

Integrated pest management (IPM) is the management of pest populations below levels that cause economic damage by using a compatible balance of biological, cultural, chemical, genetic or other control methods. Control may be aimed at one or more pests depending upon the scope and complexity of the management system.

The IPM concept comes from the realization that any disruption of a pest will tend to affect other pests and beneficials in the crop complex. Integrated pest management attempts to develop and use techniques to manage pests, not to eradicate them.

Dan Meyer and Dennis Johnson. Concepts of Integrated Pest Management in Washington (EB 0753). Pullman, WA: Extension Service, 1980.

PERIODICAL

IPM Practitioner
Rt. 2, Box 28A
Winters, CA 95694

The best single source of current information on Integrated Pest Management. The monthly newsletter includes reports on current research, reviews of literature, products, and services, and features on control strategies for specific farm and household pests.

BIBLIOGRAPHY

Baker, Kenneth F., and Cook, R. James. Biological Control of Plant Pathogens. San Francisco: W.H. Freeman, 1974.
Special emphasis on the role of soil ecology in biological control.

"Beneficial Predators and Parasites Found on Washington Crops." Extension Bulletin 640. Pullman, WA: Cooperative Extension Service, 1975.
Profusely illustrated identification guide to a dozen insects you'll want to keep around.

Callahan, Philip S. Tuning into Nature. Old Greenwich, CT: Devin-Adair, 1975.
Fascinating investigations of the secret lives of plants and insects which will lead to new, ecologically sound methods of pest management.

Carr, Anna. Rodale's Color Handbook of Garden Insects. Emmaus, PA: Rodale Press, 1979.
Over 300 color photographs of garden insects, including range maps, life cycles, host plants and natural controls.

Davis, D.W., et al. Biological Control and Insect Pest Management. Davis, CA: Division of Agricultural Science, Univ. of California, 1979.
Good general text. Topics include sampling and monitoring, economic injury levels, cultural control, and the development of pest management programs.

DeBach, Paul, ed. Biological Control of Insect Pests and Weeds. NY: Reinhold, 1964.
Definitive text on biological pest management. Technical and thorough. Topics include basic insect population dynamics and methods for establishing and evaluating control programs. Includes methods for enhancing natural controls by protecting and encouraging beneficial insects.

——. Biological Control by Natural Enemies. London: Cambridge University Press, 1974.
Details the ecology of pest/predator interactions, with case histories of successful biological control of pest outbreaks from around the world.

Doane, C.C., and McManus, M.L. The Gypsy Moth: Research Toward Integrated Pest Management. FS-SEA Tech Bulletin 1584. Washington, DC: USGPO, 1981.
Essential reference work for everyone involved in gypsy moth programs. This is a comprehensive review of the history, ecology, and alternative management systems for control of this insect which is becoming an increasing threat in the Northwest.

Flint, Mary Louise, and Van den Bosch, Robert. Introduction to Integrated Pest Management. New York: Plenum Press, 1981.
Current information on IPM strategies and programs.

Frank, J. Ray, ed. "Symposium on Integrated Pest Management." In HortScience, Vol. 16(4), August 1981.
This issue of HortScience features six papers which give an overview of the current status of integrated pest management by some of the leaders in the field. They cover urban pest management, weed control practices, pests in horticultural cropping systems, regional planning and coordination of IPM programs.

beneficial insects which prey on pests. At some point in their life cycle, most beneficials need nectar, pollen, alternate hosts or other food sources. Proper nutrition of beneficials is essential because it affects their ability to survive and lay eggs. Again, plant species added for this purpose must be selected carefully because they may also cause an increase in the population of pests.

Plantings can be diversified in several ways. Depending on the situation, one or more of the following systems can be used.

1) Hedgerows: A mixture of plant species, from annual herbaceous plants to perennial shrubs and trees, can be planted on the edges of crop fields, along fencerows, or in woodland clearings.

2) Polyculture or Companion Planting: Two or more crops can be interplanted, or one crop can be sown into another before the first crop is harvested. These strategies are especially useful for annual crops, and also help keep troublesome weeds under control.

3) Encouraging beneficial weeds and/or insectary plants: Specific species can be planted or native volunteers allowed to grow. Some insectary plants are also valuable as cover crops and can be used to enrich the soil as well.

HEDGEROWS

Hedgerows can increase the number of beneficial insects by providing shelter and alternative sources of food. They provide nectar and pollen for beneficials such as ladybird beetles (ladybugs) and lacewings when crop aphid populations are low. Most of the parasitic wasps also require nectar, pollen, and/or honeydew during their adult stage. Wild plants and hedgerows growing on uncultivated land also provide alternate sources of food for beneficial insects when prey populations are low.

Hedgerow plants leaf out in the spring, well before most crop plants. They support insect populations which provide food for the beneficial insects and allow the beneficials to produce an early generation. This early generation of natural enemies will then move into the crop fields as pest populations begin to build. In the same manner, a diverse community of insects in the hedgerow increases the diversity of insects in the adjacent fields. The closer the hedgerow is to the field, the greater the beneficial influence.

Upstanding hedges interfere with air currents, catching weakly flying insects and depositing them on both lee and windward sides of the hedge. More strongly flying insects, such as syrphid flies and wasps, tend to congregate on the lee side. Hedgerows also facilitate insect migrations after hibernation and from spring to summer habitats, so their beneficial influence may extend far beyond their immediate area.

POLYCULTURE OR COMPANION PLANTING

The practice of growing many crops together in the same field has been used for millennia throughout the world, and it is still being practiced in many places. One combination which has been used extensively by native people in North America is corn, squash and beans. Polyculture increases crop protection by copying the structure and function of natural ecosystems, and maintains diversity in crop rotations. Pest infestations in polyculture agroecosystems are significantly lower than those of monoculture systems. This effect is due to a greater number and more effective dispersal of natural enemies, or the disruption of colonization and dispersal of pests, or both.

The population of the diamondback moth *(Plutella xylostella)*, a pest of tomatoes, has been regulated by a chemical repellency, or masking, of the tomatoes by intercropping them with cabbage. White and red clover intercropped with cabbage has been shown to reduce the native cabbage aphid and exotic cabbageworm populations by interfering with the colonization of the pests and increasing the number of predator ground beetles.

BENEFICIAL WEEDS AND INSECTARY PLANTS

Nectar and pollen are important for the survival of beneficial insects when hosts are scarce. Access to nectar and pollen sources increases the longevity and egg laying ability of the beneficial insects and thus their effectiveness in controlling pests. For many predators, hosts are scarce during midsummer. When designing the system it is important to determine when pest populations are likely to be at their peak so the main bloom of insectary plants will attract beneficial insects at the same time. A good strategy when using native weeds or planting insectary plants is to establish successions of various useful plants (in the case of annuals or biennials), or those with overlapping bloom periods (in the case of perennials) to provide flowers throughout critical times.

Secondary plants growing with the crops in the field often attract beneficial insects or repel insect pests. Certain predators of aphids, for example, are able to colonize brussel sprout plants grown with weeds more easily than brussel sprout plants grown on bare ground. Ground beetles *(Lebia* spp.) are associated with certain weed species such as *Solanum* and *Amaranthus*. Removing these weeds also removes the beneficial ground beetles.

The presence of an alternate host for beneficials on insectary plants should not coincide with the periods critical for control of insect pests. If alternate hosts are not controlled, the beneficials could be satisfied by feeding on them instead of the pests.

Beneficial insects living on weeds or insectary plants will usually disperse to crop plants by themselves, but sometimes the weedy plants may need manipulation to force movement. One strategy is to allow weeds to grow and concentrate insect numbers, then cut the weeds to force the insects to move into crop fields or up into orchard trees. The timing of these manipulations should take into consideration the biology, and particularly the life cycle, of the beneficial insects.

"Garden Insect Control Without Synthetic Insecticides" (EM 3757). Pullman, WA: Cooperative Extension Service, 1973.
Geared for the backyard gardener, this leaflet explains the use of insecticides such as rotenone and nicotine. More than half the information, however, focuses on attracting and keeping birds, with a table of birdhouse dimensions for 27 species.

Jordan, William H. Windowsill Ecology. Emmaus, PA: Rodale Press, 1977.
Greenhouse and house plant biological control techniques.

McLeod, Edwin. Snail and Slug Biology and Management: A Review. Organic Agriculture Research Institute, Box 475, Graton, CA 95444.
Brief literature survey. Fourteen references.

Ridgway, R.L., and Vinson, S.B. Biological Control by Augmentation of Natural Enemies. New York: Plenum Press, 1977.
This in-depth text deals with principles, methodology, and the economic feasibility of mass rearing and release of natural enemies.

FOR MORE INFORMATION

Beneficial BioSystems
1603 63rd Street
Emeryville, CA 94608
Supply Pteromalids, tiny parasites which lay eggs in the pupae of manure-breeding flies, and fly traps with a money-back guarantee. Geared especially for livestock owners.

Center for the Integration of the Applied Sciences
1016 Grayson Street
Berkeley, CA 94710
Formed by entomologists William and Helga Olkowski, the CIAS works as an interface between humans and the species humans regard as pests, helping cities, parks and institutions develop conscious, practical pest control strategies. Their catalog lists the many publications they have available on common pest problems. Include SASE with catalog request.

Organic Pest Control Naturally
David C. Mirgon
19920 Forest Park Drive NE
Seattle, WA 98155
Wholesale and retail preconditioned ladybugs, praying mantis eggs, and other predators and parasites.

Insectary Plant Index

KATE BURROUGHS

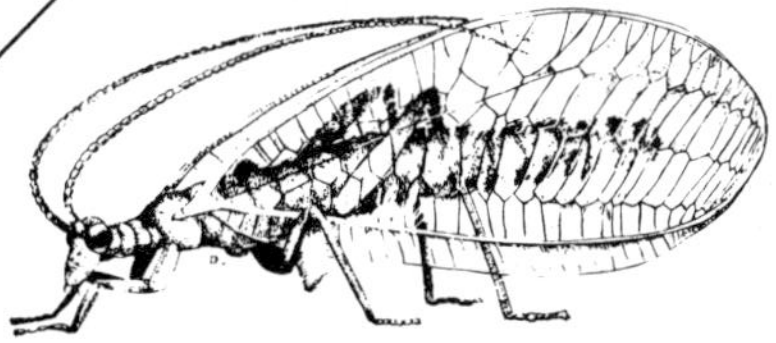

INTRODUCTION

This chart is a guide to attracting beneficial insects through the use of insectary plants. It is intended as a reference only, and must be adapted to local conditions.

Beneficial insects listed across the top are keyed to perennial and annual plants listed down the left side. Where they are known, special benefits and potential problems are noted in the comments column. See the references in the bibliography below for additional information.

Beneficial Insects Attracted — INSECT ORDER	COLEOPTERA	DIPTERA (Flies)			HEMIPTERA	HYMENOPTERA (Bees and Wasps)									NEUROPTERA		
INSECT FAMILY / **Plant**	Ladybugs	Chloropid Flies	Flower Flies	Tachina Flies	Minute Pirate Bugs	Aphelinid Parasitic Wasps	Bees	Braconid Parasitic Wasps	Chalcid Parasitic Wasps	Potter Wasp	Ichneumon Parasitic Wasps	Mud Daubers, Sand Wasps	Trichogramma Parasitic Wasps	Hornets and Yellow Jackets	Lacewings	**BLOOM PERIOD**	**COMMENTS**
I. PERENNIALS																*Bloom period based on Northern California. Adjust for local area.*	
A. GROUND COVERS																	
Melilotus alba, White Sweet Clover				●			●					●		●		May to September	
Trifolium pratense, Red Clover							●									April to October	
Trifolium repens, White Clover						●	●									April to December	Benefit : Supports parasites of wooly apple aphid.
B. HERBACEOUS PLANTS																	
Achillea spp., Yarrow	●						●	●								Depends on species	
Eryngium spp., Coyote Thistle						●							●			Depends on species	
Foeniculum vulgare, Common Fennel			●	●			●	●			●	●		●		June to October	
Ruta graveolens, Rue										●	●	●				May to June	

C. SHRUBS																	
Atriplex spp., Saltbush			●							●		●				Late May to late June	Dioecious species: blossoms on female plants support beneficials.
Baccharis pilularis consanguinea, Coyote Brush		●	●	●	●		●	●	●	●	●	●		●		September, October and later	Benefit : Orange tortrix prefers to wine grape. Hazard : Attracts flies, corn maggot, false chinch bug.
Eriogonum spp., Wild Buckwheats		●	●	●	●		●					●		●		Depends on species	May increase harmful lygus bug.
Euonymous japonica, Evergreen Euonymous	●	●	●	●			●	●	●	●	●	●			●	Late May through June	
Rhamnus californica, Coffeeberry	●		●	●							●	●				Mid-April to late June	Attracts many parasites, predators. Hazard: Host of grape whitefly.
Symphoricarpos spp., Snowberry			●	●										●		Depends on species	
D. VINES																	
Hedera spp., Ivy			●	●			●	●		●		●		●		September and October	
Polygonum aubertii, Silver Lace Vine			●	●			●									October	
E. TREES																	
Ailanthus altissima, Tree of Heaven		●	●	●			●	●		●	●				●	Early May through June	
II. ANNUALS OR BIENNIALS																	
A. GROUND COVERS																	
Brassica hirta, White Mustard								●			●					March to August	Benefit : Supports parasite of lepidopterous larvae & pea aphid.
Fagopyrum esculentum, Domestic Buckwheat						●		●			●						Benefit : Supports parasites of European pine shoot moth.
B. HERBACEOUS PLANTS																	
Angelica spp., Angelica	●									●		●			●	Depends on species	
Coriandrum sativum, Coriander			●				●									May to July	Benefit: supports parasites of pea aphid.
Daucus carota, Wild Carrot			●	●	●		●	●	●		●			●	●	May to September and later	

BIBLIOGRAPHY

Altieri, M.A., and Whitcomb, W.H. "The Potential Use of Weeds in the Manipulation of Beneficial Insects." *Hort Science* 14 (1979): 12-18.
Excellent overview of how weeds can contribute to pest control by hosting beneficial insects and alternate prey. Includes 75 references.

International Tree Crops Institute, Box 666, Winters, CA 95694. 1981.
Unpublished sheet on perennial insectary plants compiled by Robert Bugg. Much of the information on the chart above came from this list.

Leius, K. "Influence of Wild Flowers on Parasitism of Tent Caterpillar and Codling Moth." *Canad. Entomol.* 99 (1967): 444-446.
Details rates of parasitization versus types of flowering plants. Includes 21 literature references.

Van Emden, H.F. "The Role of Uncultivated Land in the Biology of Crop Pests and Beneficial Insects." *Sci. Hort.* 17 (1965): 121-136.
A classic article on the biological and ecological relationships between hedgerows, uncultivated land and insects. Includes 81 references.

William, R.D. "Complementary Interactions Between Weeds, Weed Control Practices, and Pests in Horticultural Cropping Systems." *Hort. Science* 16 (1981).
Overview of interaction between weeds and insects, and how to manage weeds in crops. Includes 161 references.

Nitrogen-Fixing Plants

PHIL ANDRUS

Seventy-eight percent of the air we breathe is nitrogen. Unlike oxygen, the other important constituent of air, nitrogen is not highly reactive. Nitrogen serves as the atmospheric medium in which the more active gases are carried, kept separated, kept quiet. We cannot do without nitrogen, but we do not do much with it.

Plants, however, do a lot with nitrogen. It is an essential component of all plant life, being one of the basic elements in the chemical structure of proteins. However, nitrogen in its gaseous state cannot be used by plants. Rather, nitrogen must first be converted into various nitrate compounds for it to sustain plants. This conversion, called fixation, occurs in many ways, both naturally and artificially: in fertilizer factories, by lightning storms, by certain plants, and by some special soil bacteria, yeasts and fungi. In addition, nitrogen compounds are concentrated when consumed by fish, birds and animals, all of whom secrete this concentrated nitrogen in their manure.

In spite of all the nitrogen in the atmosphere, and all the ways in which nitrogen can be fixed into compounds usable to plants, it is still difficult to provide plants with the nitrogen they need. The use of industrially-produced nitrogen for agricultural use has increased from 500,000 tons annually to nine *million* tons during the last thirty years, at an annual cost of billions of cubic feet of natural gas. That is not a sustainable agriculture. Agriculture that is to survive beyond the last puff of natural gas must get its nitrogen from processes not dependent on fossil fuels. Biological nitrogen-fixation processes are many, and they are effective.

NITROGEN-FIXING PLANTS

Two groups of herbaceous and woody plants have the ability to fix atmospheric nitrogen in the soil, legumes and actinomycete-nodulated angiosperms. Together in these two groups are at least 1,350 species of plants capable of nitrogen fixation, although only about 25 are extensively used today in agriculture and forestry. Included among the legumes are beans, peas, clover and alfalfa. Certain legumes fix nitrogen because of their symbiotic relationship with *Rhizobium* bacteria, which form nodules in the roots of their host plants. The plants provide the bacteria with carbohydrates for energy and a stable environment for growth, while the bacteria give the plants usable nitrogen and other essential nutrients. The rest of the soil community and neighboring and succeeding plants benefit from the nitrogen and other rich compounds exuded from the nodules, and from the recycling of the nitrogen as the

Peas, Please

Humans and plants have much in common. Both enjoy sunlight, music and tender loving care. And, despite outward appearances, similarities extend to the structure of our cells.

With the exception of one atom, chlorophyll (the center for photosynthesis in plants) and hemoglobin (the vehicle for respiration in animals) are structurally identical. At the center of each chlorophyll molecule is an atom of magnesium which gives plants their characteristic green color. At the center of each hemoglobin molecule, on the other hand, is an atom of iron which causes blood to be red.

Legumes take this similarity one step futher. If you carefully remove a pea or clover plant from the soil and slice open one of the nodules formed by rhizobia bacteria on its roots you will see a pink or red color. This color indicates the presence of hemoglobin—the same compound which is found in the blood of humans and other mammals. This compound exists in neither the free-living bacteria nor the host plant, but is formed in the union of plant roots and bacteria.

So, the next time you extend your fork for a mouthful of peas remember that—for the sake of one atom....

LAURA STUCHINSKY

plant drops its leaves or decays. *Rhizobium* bacteria usually are annual in nature, developing in the spring and decaying in the autumn.

A second, less familiar form of nitrogen fixation occurs in plants that are nodulated by actinomycete fungi of the genus *Frankia*. Although the basic nitrogen fixation process is the same with these plants as with legumes, most of their nitrogen contribution is in the form of falling leaves and decaying litter. Their nodules are perennial rather than annual. Examples of this group of plants are alder, ceanothus and Russian olive.

The planting of nitrogen fixers such as alfalfa and soybeans as green manure crops is a long-established practice. Trees and shrubs can be used in a similar fashion. For example, alder has been reported as being included in the rotation of Asian rice fields, as a companion plant in Guatemalan corn fields, and as a companion plant for apples in the Netherlands. Masanobu Fukuoka interplants acacia trees in his Mandarin orange groves in Japan, and autumn olive is being used successfully as a nurse crop for walnuts in the American Midwest. In the coniferous forests of the Pacific Northwest, some foresters are studying the use of red alder as a nitrogen-fixing companion for Douglas-fir.

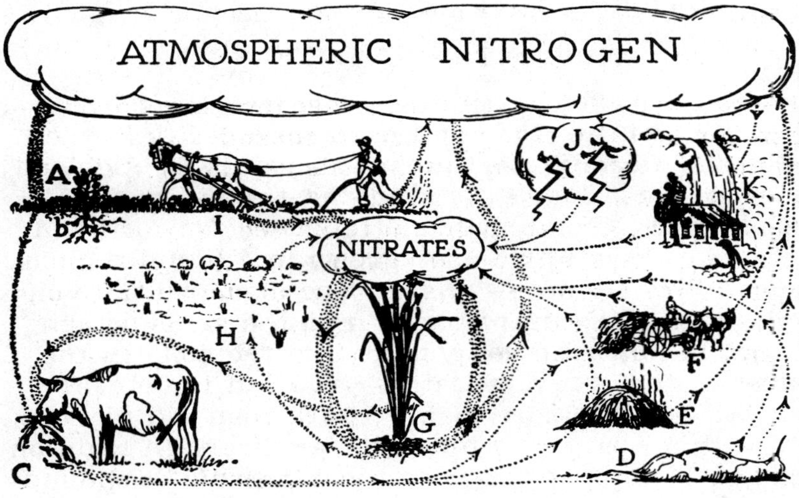

Reprinted from *The Small Farmer's Journal*, Spring, 1979

Note the courses nitrogen may take; through the root nodules (b) of legumes (A) and into the bodies of animals (C); through their decay (D) into the soil (where it is changed to nitrates) or back into the air; through manure piles (E) and freshly spread manure (F) both of which deliver some to the soil and lose some into the air; through the plowed-under cover crop (I); through the action of thunderstorms (J) and the efforts of humans in making nitrogen fertilizers (K); both of which methods change the atmospheric form directly into nitrates. Note also how plants (G) taking the nitrates, may either decay and give the nitrogen back to the air and to soil bacteria, or may be eaten as fodder (C) and start the cycle anew; or as stubble plowed under (H), may again deliver their nitrogen to be remade into nitrates and started on another journey of service. The farmer's task is to promote all those changes which hold the nitrogen in the soil and to check those which result in the loss of nitrogen into the air.

Several factors enhance nitrogen fixation. For maximum production, the soil must be low in nitrogen but otherwise in good condition, with good structure and aeration and adequate levels of calcium, phosphorus and potassium. The nitrogen production of legumes often is enhanced by inoculation with a healthy strain of *Rhizobium* especially suited to crop and location.

AZOLLA

Azolla is a genus of small fern that lives naturally in lakes, swamps and streams. Its agricultural

BIBLIOGRAPHY

Allen, O.N., and Allen, Ethel K. The Leguminosae, A Source Book of Characteristics, Uses and Nodulation. Madison: University of Wisconsin, 1981.
This compendium of information on the legumes is the result of 45 years of research by the Allens. They give detailed information on each of the 750 known genera of Leguminosae, plus details on many of the 20,000 known species.

Clark, Wilson. "China's Green Manure Revolution." In Science '80, September/October, 1980.
Review of subtropic green manures which features azolla culture, including descriptions of harvesting and re-establishment of azolla "seed."

Del Tredici, Peter. "Legumes Aren't the Only Nitrogen-Fixers." In Horticulture, March, 1980.
Review of current research.

Delver, P., and Post, A. "Influence of Alder Hedges on the Nitrogen Nutrition of Apple Trees." In Plant and Soil, 28:2 (1968), pp 325-336.
Report on controlled study of an apple orchard in the Netherlands interplanted with hedges of European black alder (*Alnus glutinosa*). "Trees 5 m or less from the hedge had higher leaf nitrogen contents, a 37 percent higher yield and made better shoot growth."

FAO. China: Azolla Propagation and Small-Scale Biogas Technology. Soils Bulletin #41, 1979. UNIPUB, 345 Park Avenue S, New York, NY 10010.
Uses of azolla for fertilizer and as a feedstock for methane fuel production.

Funk, David T.; Schlesinger, Richard C.; and Ponder, Felix, Jr. "Autumn-Olive as a Nurse Plant for Black Walnut." Botany Gazette 140:110-114, 1979.
In tests at five locations in Missouri, Illinois and Indiana, planting autumn-olive as a nurse crop stimulated growth of black walnut. The mixed plantings have greater soil nitrogen and (during the growing season) lower soil and air temperatures than the pure walnut plantings.

Gordon, J.C.; Wheeler, C.T.; and Perry, D.A. Symbiotic Nitrogen Fixation in the Management of Temperate Forests. 1979. Forest Research Laboratory, Oregon State University, Corvallis, OR 97331.
Proceedings of a workshop surveying the biology and potential of employing nitrogen-fixing plants in silviculture. One of the best available collections of information on the topic.

importance lies in its symbiotic relationship with the blue-green algae *Anabaena*, a relationship very similar to those described above for legumes and actinorhizal plants. Azolla provides nutrients and protection for *Anabaena*, while the algae produces more than enough nitrogen for both itself and its host. Under the right conditions, the azolla can double its weight every three to five days, and can fix nitrogen at a higher rate than most legumes. Azolla is used most extensively in China and Vietnam. For example, it is cultivated as a green manure crop on about 60% of the rice area in Vietnam's Red River Delta. Azolla plants are described by the Vietnamese as "indestructible fertilizer factories," for they were the only nitrogen "factories" which continued to produce even during the worst of the Vietnam war.

In addition to producing nitrogen for crops, azolla absorbs from the water nutrients that might otherwise be lost, storing them with the nitrogen until the azolla mat is incorporated into the soil. Often the stored nutrients include much of the phosphorus applied by the farmer to stimulate azolla growth and nitrogen fixation, thus eliminating the need to apply phosphorus directly to the crop. Another benefit of azolla is that its thick mat on the water's surface curtails the growth of many aquatic weeds, both by blocking their sunlight and by physically inhibiting their emergence.

OTHER NITROGEN-FIXERS

Two new areas of agricultural research involve nitrogen fixation by algae and bacteria directly in the soil. Battelle's Pacific Northwest Laboratory in Richland, Washington, for example, is studying the use of two species of blue-green algae as a fertilizer for corn and tomatoes. In this study algae, grown in a liquid medium, is sprayed directly onto the soil. The algae multiply on the soil surface, fixing nitrogen from the air and then excreting the nitrogen compounds into the soil. In the first year of the study, tomatoes grown in the algae-treated soil had 45% more plant growth than did plants treated with a commercial nitrogen fertilizer. In addition to nitrogen, the algae provides other organic compounds that help build soils and bind micronutrients, making them more available to plants.

Robert H. Faust, an agronomist and ecological agriculture consultant in Twin Falls, Idaho, has been studying the effectiveness of certain bacteria as inoculants for non-leguminous plants. The nitrogen-fixing bacteria *Azotobacter* has been used for some years in the Soviet Union, but only recently in the U.S. In 1980, Faust conducted test plantings on a farm outside Twin Falls. Wheat seed treated with *Azotobacter* yielded 11.6% more bushels per acre than did either untreated or chemically-treated seed. In order for *Azotobacter* to be effective, soil pH must be between 6.8 and 8, and optimally between 7.2 and 7.8. Faust markets *Azotobacter* as 'Azobac,' available from **Faust Bio-Agricultural Services, PO Box 1150, Twin Falls, Idaho 83301.** The Faust Services catalog is available for $1.

Haines, Sharon G., and DeBell, Dean S. "Use of Nitrogen-Fixing Plants to Improve and Maintain Productivity of Forest Soils." In Impact of Intensive Harvesting on Forest Nutrient Cycling. 1979. College of Environmental Science and Forestry, State University of New York, Syracuse, NY 13208.
State-of-the-art survey of potential uses of nitrogen-fixing plants in forest management.

Leonard, Lewis T. "Nitrogen-Fixing Bacteria and Legumes." In The Farmstead Book 1, Paul Hammond, ed. Seattle: Cloudburst, 1978.
Good introduction to nitrogen-fixation and the use of legumes in agriculture.

Lumpkin, Thomas A., and Plucknett, Donald L. "Azolla, a Low-Cost Aquatic Green Manure for Agricultural Crops." In Background Papers for Innovative Biological Technologies for Lesser Developed Countries. U.S., Congress, House, Committee on Foreign Affairs, 97th Congress, 1st Session, 1981, pp. 311-348.

McGuire, W.S.; Dawson, M.D.; and Crofts, F.C. Effective Nodulation and Production of Subterranean Clover with Pelleted and Small Amounts of Lime. Bulletin 633. 1978. Ag. Exper. Station, Oregon State University, Corvallis, OR 97331.
Report on the value of pelletizing legume seed with lime and bacterial inoculant for effective nodulation. Tests conducted in Douglas County, OR, demonstrate the effectiveness of this technique in renovating pastures on harsh sites.

Musick, Mark. Nitrogen-Fixing Trees and Shrubs for the Pacific Northwest. Arlington, WA: Tilth, 1980.
Brief survey of 21 species of N-fixing trees and shrubs native or adapted to this region, with notes on uses, list of sources and bibliography.

Taylor, T.M.C. The Pea Family (Leguminosae) of British Columbia. Handbook No. 32. Victoria, BC: British Columbia Provincial Museum, 1974.
Botanical guide to over 150 species of legumes either native or introduced to British Columbia. Precise botanical descriptions, diagnostic illustrations, and distribution maps are provided for each species.

Whyte, R.O. et al. Legumes in Agriculture. New York: Food and Agriculture Organization of the United Nations, 1973.
World-wide survey of legume research. First published in 1953.

BOOK SOURCES

The companies listed below are sources of books related to agriculture and forestry.

Ag-Ho Service Company
306 North Alamo Drive
Vacaville, CA 95688
Catalog includes more than 3,000 agriculture-related titles from 250 publishers.

Earthbooks Lending Library
Rt 1, Box 346
Mountainburg, AR 72946
Mail-order rental library specializing in self-sufficiency information.

Horticultural Books, Inc
PO Box 107
Stuart, FL 33494
Wide range of foreign and domestic books.

Hurley Books
Westmoreland, NH 03467
Source of rare agricultural textbooks.

International Scholarly Book Service
Box 1632
Beaverton, OR 97075
Exclusive North American distributor for many Australian, European and Japanese horticultural publishers.

Israel Program for Scientific Translations
Kiryat Moshe
PO Box 7145
Jerusalem, Israel
Publishers of a wide range of books, including the wealth of USSR research in agriculture and forestry.

Powell Bookstore
1005 West Burnside
Portland, OR 97209
The largest stock of new and used books on agriculture, forestry and ecology in the Northwest. No catalog, but the studious plant enthusiast visiting Portland will be well rewarded by a personal visit.

Pomona Book Exchange
Highway 52
Rockton, Ontario L0S 1X0
Large catalog of old and new books.

UNIPUB
345 Park Ave South
New York, NY 10010
Distributor of publications by United Nations agencies, including the Food and Agriculture Organization (FAO).

Whole Earth Household Store
Fort Mason Center, Bldg F
San Francisco, CA 94123
Distributor of books listed in The Next Whole Earth Catalog, which is a comprehensive source of information on all aspects of the transition to a sustainable society.

Farming with Computers

Agriculture On-Line Access (AGRICOLA)
National Agriculture Library, USDA
Beltsville, MD 20705
Computer access to the 1.3 million files in the National Agricultural Library, plus 140,000 new entries made each year from over 6,500 journals published around the world. AGRICOLA can be accessed online through several computer brokerages.
Other computer-based services of the National Agricultural Library include the Current Awareness Literature Service (CALS), the Current Research Information System (CRIS), and the Food and Nutrition Information Center (FNIC).

A.T. Times

Brunner, Jay. "Pest Management Via Computer Models is Becoming a Reality." Dec 1, 1980, pp 41-44. The Goodfruit Grower, Washington State Fruit Commission, 1005 Titon Drive, Yakima, WA 98909.

"Good computer models of insect development using long-term weather forecasts can predict the occurrence of events (e.g. codling moth egg hatch) within 1 to 2 days of their actual occurrence in the field....

"Population dynamics models, for example predator-prey interaction models, will predict the outcome of interactions between pests and their natural enemies. The availability of a model coupled with information on the economic injury level of a pest will provide an orchard manager with a decision making tool to determine not only the optimum timing of control sprays but if indeed they are even required."

Computerized Plant Germplasm Inventory System
National Seed Storage Laboratory
Colorado State University
Fort Collins, CO 80523
Data on over 95,000 seed samples stored at the National Seed Storage Laboratory. Searches for specific germ plasm provided on request. Seeds are available to plant scientists only.

Control Data Corporation. "Bringing New Opportunities Down on the Farm." In Contact, Vol. 7, No. 4 (March/April, 1981). CDC, Box 0, Minneapolis, MN 55440
The first of a series of articles on CDC's Rural Venture Project. Rural Venture is an innovative (and controversial) demonstration of the application of computer technology to small farm production and management.

Grassroots
Videotex Systems for Agriculture
122 St Patrick Street
Toronto, Ontario M5T 2X8
Marketing information, research assistance, and satellite weather forecasting available through a two-way computer network which utilizes telephone lines and home TVs. Developed in Canada, Grassroots is now expanding to serve U.S. farmers.

Hinman, Herbert R, and Willett, Gayle S. Farming With Computers: What Are the Alternatives? EB 0953. Pullman, WA: WSU Extension Service, 1981.
Discussion of programmable calculators, computer networks, and microcomputers, with a listing of programs.

Johnson, Steve. Information and Communications Technology for the Community. 2nd ed, 1982. Center for Urban Education, 0245 SW Bancroft, Portland, OR 97201. $5.00.
Thorough introduction to the development and potential of computers, with an emphasis on their use in meeting human needs. Includes an extensive bibliography and listing of community groups using computers.

Landscape Electronics
9 North May Street
Joliet, IL 60435
Developers of computer programs for designing landscapes, preparing bids, monitoring costs, and selecting plants.

Wright, M. Anthony, and Kendrick, James G. AGNET: Management Information for Agriculture. EM 4645. Pullman, WA: WSU Extension, 1981.
Introduction to the AGNET system, a computer network serving agriculturists in 30 states and Canada. Over 200 programs are available (with names like FEEDMIX, IRRIGATE, and WOODLOT) to aid farmers in planning and management. A messaging system also allows for two-way networking. AGNET terminals are available in many County Extension offices. Inquire about access.

Design

MAIA MASSION

Design is the process of creating patterns which unite elements in an integrated whole. "Permaculture" is a word coined by Bill Mollison and the Tagari group in Australia to describe the conscious design of self-sustaining agricultural landscapes. Rather than dominating the environment, permaculture seeks to enhance it. The goal of permaculture design is to integrate livestock, food production, and human shelters in an agriculture which embodies the stability and resilience of natural ecosystems.

Below is an excerpt from a report by Maia Massion on permaculture design principles. Her complete article was published in the Fall, 1981 issue of **Tilth**.

The most important skill a designer or steward can develop is the ability to see clearly what is occurring in the landscape. Keen observation and careful notation of even the most minute or mundane natural phenomena is important. Essential information about the ecology, history and potential productivity is revealed, for example by rocks, insects, trees and shifting air currents. All are indicators of the flow of life and energy across the landscape. Awareness of the ecology and patterns created by these flows is fundamental in assessing the resources, limitations and potential of the site, and is the basis for developing designs for sustainable agriculture.

Site strategy and tactics are important aspects of the overall permaculture design process. The "strategy" for a site refers to planning its development and management through time. For example, strategizing for proper woodlot managment involves planning a diverse succession of species suitable to fuel, timber and animal fodder production over the years.

Design "tactics" are simply the techniques of management or production. Coppicing is an example of a tactic or technique that can be implemented for sustained woodlot management.

Site zonation and sector planning help determine the placement of elements in the landscape. "Zones" emanate outward in concentric circles from the center of greatest activity on site. This is usually the main dwelling, designated as **Zone 1**. **Zone 2** would be the area immediately surrounding the house. **Zone 3** would begin beyond the back yard, and so on. Larger acreages may even encompass a **Zone 5**, the domain of forest woodlands and wildlife.

Elements are placed in zones according to the maintenance and harvest activity they require. Chicken sheds and greenhouses, needing daily attention, would be located in Zone 2. Sporadic and seasonal visitations to the woodlot indicate that it could be located in Zone 4.

BIBLIOGRAPHY

Alexander, Christopher, et al. A Pattern Language. New York: Oxford University Press, 1977.

This is a unique and extraordinarily useful book. The authors provide a "language" of 253 patterns which distill centuries of wisdom about the relationship of people to natural and built environments. Each pattern is described in detail, illustrated with drawings and photographs, and linked with other patterns to create a comprehensive whole. While a companion volume needs to be created for landscapes and agriculture, this is an excellent tool for all designers.

Britz, Richard, et al. The Edible City Resource Manual. Los Altos, CA: William Kaufman, 1981.

Strategies for bringing agriculture into the city. Stimulating ideas and information for creating the self-sustaining city of the future.

Creasy, Rosalind. The Complete Book of Edible Landscaping. San Francisco: Sierra Club, 1982.

Handbook for gardeners and landscape architects integrating edible trees, shrubs, flowers and vegetables into beautiful, food-producing landscapes. Sierra Club went all out with this book. A design manual, edible plant encyclopedia, and sourcebook all in one.

Planet Drum. Eco-Decentralist Design (See page 17).

Smyser, Carol A., et al. Nature's Design: A Practical Guide to Natural Landscaping. Emmaus, PA: Rodale Press, 1982.

Useful introduction to site analysis and landscape planning, with details on do-it-yourself construction and plant propagation.

ZONES

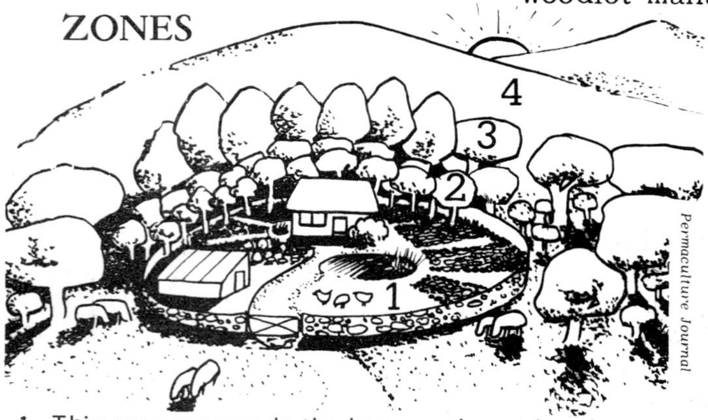

1. This zone surrounds the house and contains the greenhouse, intensive vegetable garden, and herbs.
2. Dense planting of fruit and nut trees, chicken forage, and herbs that still require some care.
3. Hardy plant structures including thickets, wind breaks, and agro-forestry.
4. Low intensity self forage system for large animals.

"Sectors" are wedges of the zonation circle. They are delineated according to site orientation and relations to sun angles, wind directions and fire patterns. Thus the optimal location for a water reservoir needed for agricultural and/or human use would be in the sector from which fires are most likely to enter and threaten the site. The reservoir could then also function as fire prevention and control. Woodlots as effective windbreaks could be established in the sector defined by the direction of cold winter winds.

IN COMPLEXITY LIES STABILITY

The stability of a system is its ability to withstand stress and continue functioning in a productive manner. Diversity is the number of different elements in the system, and complexity is the number of functional relationships between those elements. Diverse polyculture systems of mixed plant and animal species are more productive than monocrop systems because energy is captured, stored and utilized in so many different ways.

Natural ecosystems and permaculture designs tend toward maximizing diversity and complexity. Stability is insured if every function of the system is fulfilled by several different elements, and if every element fulfills several functions. For example, if there are several sources of chicken feed (mulberries, acorns, grains, greens, insects, kitchen garbage), then the function of feeding the chickens will still be fulfilled if a few crops fail. Meanwhile, the chicken is functioning as a mulberry- or acorn-into-egg transformer, biological pest controller and "fertilizer tractor" for the garden.

Successful permaculture design requires a subtle and profound evolution in thinking which shifts the emphasis from the object to its function. Even our language is object-oriented. We use the word "tree," while some other, more functionally oriented culture might refer to the same being as "giver of shade and fruit."

As people learn more about permaculture they have many real and very challenging questions about the specific applications of its principles and methods. Permaculture experimentation involves uncertain economic parameters and potential financial risk.

A lot of the information people desire concerning permaculture economics and applied ecology does not yet exist for the Pacific Northwest. Each bioregion will need to develop its own specific permaculture techniques and sustainable agriculture strategies according to its own people, economics, climate and appropriate plant and animal species. The exchange of information and mutual support, especially through bioregional networks, will be essential in facilitating the development of Pacific Northwest permaculture and in the evolution of a sustainable, ecological agriculture with which we can heal the earth and support ourselves.

PERIODICAL

Naturescape
PO Box 514
Wheaton, IL 50187

Written for professional landscape architects, but valuable for anyone interested in urban planning or landscape design and maintenance. Emphasis on low maintenance perennials, conservation, wildflowers, and the use of native plants.

MAPPING

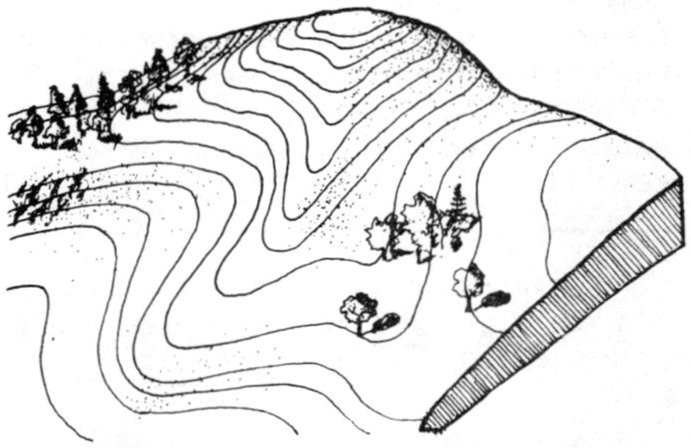
The Owner-Built Homestead

Maps are tools for planning and record keeping. They are useful for recording basic data such as soil types, vegetation, topography, or any other physical or cultural features.

Barbara and Ken Kern provide a concise, step-by-step introduction to map making in their book, The Owner-Built Homestead (New York: Charles Scribner's Sons, 1977).

Design With Climate

Franklin, T. Bradford. Climates in Miniature: A Study of the Micro-Climate and Environment. Westport, CT: Greenwood Press, 1972.

Delightful introduction to research on the effect of micro-climates in the garden, on the farm, in woodlots, and in the home.

Geiger, Rudolph. The Climate Near the Ground. Cambridge, MA: Harvard University Press, 1965.

Textbook on micro-climates. Full of useful material for anyone seeking a better understanding of ground-level ecology.

Olgyay, Victor. Design With Climate: Bioclimatic Approach to Architectural Regionalism. Princeton, NJ: Princeton University Press, 1962.

Olgyay draws on the architecture of cultures from around the world, the building patterns of birds and other animals, and the sciences of meteorology and engineering to come up with one of the best guides for siting and designing climatically sensitive buildings. Aspects covered include design of shading devices, shading effects of trees and vegetation, flow patterns inside buildings, thermal effects of materials, and micro-climate effects.

Agricultural Research
Science and Education Administration
US Department of Agriculture
Washington, DC 20250
Monthly review of topics related to agricultural research, including crops, livestock, engineering, weed management, and insects. The journal reflects the policy of the USDA, with increased emphasis on biological pest management and on the research needs of small farms.

Levitan, Lois. Improve Your Gardening With Backyard Research. Emmaus, PA: Rodale Press, 1979.
An introduction to working in the living laboratory of the garden. Detailed information on tools and techniques for setting up controlled experiments, and on how to observe, record, and compare data.

Research

For the past generation the needs of people concerned with environmentally sound approaches to agriculture and forestry have largely been ignored by the research establishment. Now, however, a profound shift is underway, with a greater emphasis in public and private research centers on more sensitive, resource-conserving approaches to the production of food and timber.

Several centers publish research reports which are available to the general public. The notes on the work of each center provide only a general introduction. Contact the centers directly that are doing work related to your interests for a list of available publications.

Research Centers

BRITISH COLUMBIA

Abbotsford Small Fruits Substation
510 Clearbrook Road
Abbotsford, BC
Strawberries and raspberries.

Agassiz Research Station
Box 1000
Agassiz, BC V0M 1A0
General agricultural research for western British Columbia. Projects in overwintering crops and year-round tunnel crops. Animal emphasis on dairy cattle and poultry genetics.

Forest Entomology and Pathology Laboratory
Canada Department of Forestry
409 Federal Building
Victoria, BC
Information on forest insects and control measures.

Kamloops Research Station
3015 Ord Rd.
Kamloops, BC V2B 8A9
Primary British Columbia research station for rangeland management.

Prince George Experimental Farm
RR #1
Prince George, BC V2N 2H8
Pasture management, forage utilization and potato research.

Saanichton Research Station
RR #1
Saanichton, BC
Weed and insect control, vegetable and root crops.

Sidney Research Station
8801 E. Saanich Rd.
Sidney, BC V8L 1H3
Greenhouse crops, soilless culture, weed control, ornamentals.

Summerland Research Station
Summerland, BC V0H 1Z0
Includes a 7-acre isolation orchard for research in viral tree diseases, general tree fruit research and hardiness trials.

Vancouver Research Station
6660 N.W. Marine Drive
Vancouver, BC V6T 1X2
Main center for plant virus research in Canada. Provides information on insects and plant diseases for coastal British Columbia.

IDAHO

Entomology Research Laboratory
PO Box 67
Twin Falls, ID 83301
Biological, physiological and ecological studies to control insects affecting vegetables in arid areas.

Forestry Sciences Laboratory
PO Box 469
Moscow, ID 83843
Timber conifers, bark beetles, western white pine disease, streamflow in northern forests, insect control.

Idaho Agricultural Experiment Station
University of Idaho
Moscow, ID 83843
Primary center for agricultural research in Idaho. Emphasis on plant diseases and potato culture.

Northwest Watershed Research Center
PO Box 2700
Boise, ID 83701
Influences of vegetation, climate, and land management on the movement of water and sediment.

Southwest Idaho Research and Extension Center
Parma, ID 83660
Tree and vegetable crops, onion diseases.

United States Sheep Experiment Station
Dubois, ID 83423
National breeding and research center for range sheep. Developers of three new sheep breeds: Columbia, Targhee and Polypay. Currently testing guard dogs for predator control.

Water Resources Research Institute
University of Idaho
Moscow, ID 83843
Answers queries and makes referrals on pesticides, pollution, fisheries and wild and scenic rivers.

University of Idaho Research and Extension Center
Aberdeen, ID 83210
Research on rangeland forage plants, including the new Delar small burnet, which has proven superior for winter forage in the intermountain West. Also work on potatoes and crop diseases.

MONTANA

Forestry Sciences Laboratory
Montana State University
Bozeman, MT 59715
A field unit of the Intermountain Forest and Range Experiment Station. Lodgepole pine silviculture, mountain grasslands and cutover forest ranges, engineering systems for steep mountain forestry.

Forestry Sciences Laboratory
Drawer 7, University of Montana
Missoula, MT 59801
Forest management in north central Rocky Mountain area, silviculture of western larch and Engelmann spruce.

Montana Agricultural Experiment Station
Montana State University
Bozeman, MT 59715
Diseases of potatoes and other field crops. Specializes in soils and weeds, genetics, and environmental horticulture. Staff apiarist and horticulturist.

Montana Forest and Conservation Experiment Station
School of Forestry
University of Montana
Missoula, MT 59801
Soil and water conservation, wildlife, hydrology.

Northern Montana Agricultural Research Center
Star Route 36, Box 43
Havre, MT 59501
Dryland shelterbelts, dryland fruit culture and variety trials.

Northwestern Montana Agricultural Research Center
1570 Montana 35
Kalispell, MT 59901
Weed control, plant breeding, potato varieties.

Rangeland Insect Laboratory
South 11th Ave
Bozeman, MT 59777
Pioneers in biological control of grasshoppers utilizing *Nosema locustae*.

Western Montana Agricultural Research Center
531 N.E. Quest Lane
Corvallis, MT 59828
Insect biology and control, plant nutrition, variety trials.

OREGON

Eastern Oregon Agricultural Experiment Station
PO Box 833
Burns, OR 97720
Semi-arid crop culture and grassland production. Researching use of fire for range and habitat improvement on the Malheur National Wildlife Refuge near Burns.

Forest Sciences Laboratory
3200 Jefferson Way
Oregon State University
Corvallis, OR 97331
Forest genetics, ecology, silviculture, reforestation systems, mycorrhizal associations, ecosystems management.

Intermountain Experiment Station
PO Box 389
Roseburg, OR 97470
Range rehabilitation.

Malheur Agricultural Experiment Station
Rt. 1, Box 620
Ontario, OR 97914
Livestock feeding, irrigation, soil management, variety trials of forage crops, potatoes, onions and tree fruits.

Mid-Columbia Agricultural Experiment Station
Hood River, OR 97031
Orchard pests and diseases, soil fertility, winter hardiness, pollination.

North Willamette Agricultural Experiment Station
Aurora, OR 97002
Vegetables, including year-round production, nursery crops, berries, small fruits, study of 2,4-D trace effects on vegetables, municipal wastes as vegetable fertilizers, testing 54 varieties of grapes (cuttings available).

> Because of the mild weather west of the Cascades, it is possible to grow peas through most of the summer. Peas have traditionally been grown only in early spring, however, because of destructive viruses which are transmitted by aphids in warm weather.
>
> Thanks to field tests at Washington State University's Research and Extension Center in Puyallup, five varieties of peas were selected which demonstrate virus resistance: Corvallis, Knight, Trident, Aspen, and Oregon Sugar Pod. Use of these varieties will make it possible to extend the pea season months longer than ever before.

Northwest Plant Germplasm Repository
Oregon State University
Corvallis, OR 97331
National collection of living plants of all small fruits, pears, filberts, hops and mint. Available to plant breeders, plant physiologists and the nursery industry.

Oregon Agricultural Station
Oregon State University
Corvallis, OR 97331
Wide range of research programs, including the development of meadowfoam, an oil crop with commercial potential for the Willamette Valley.

Pacific Northwest Forest and Range Experiment Station
PO Box 3141
Portland, OR 97208
Coordinates research at laboratories around the region. Primary emphases are timber management, watersheds, recreation, range and wildlife habitat, forest protection, economics. Produces annotated lists of publications and The Forester's Almanac 1977 (General Technical Report PNW-62), an encyclopedia of forest and range research in the Northwest. Address information and publication requests to Informational Services Division.

Rabbit Research Center
Oregon State University
Animal Sciences Department
Corvallis, OR 97331
The only research center in North America devoted to rabbits. Efforts directed at increasing commercial production on small farms. Newsletter on current research available.

Range and Wildlife Habitat Laboratory
Rt. 2, Box 2315
LaGrande, OR 97850
Forest and range grazing strategies.

Silviculture Laboratory
1027 NW Trenton Ave.
Bend, OR 97701
Fire and timber management for eastern Washington and Oregon commercial forestland.

Southern Oregon Agricultural Experiment Station
569 Hanley Rd.
Medford, OR 97502
Small test orchard of English cider apples, with a limited amount of scion wood available for Porters Perfection, Bramley, Sweet Alford and Cox's Orange Pippin. New varieties of pears, peaches, apples and sweet corn, pear breeding and rootstock research, pest control, cultural practices.

Umatilla Agricultural Experiment Station
Hermiston, OR 97838
Crop research for the Hermiston Irrigation District: variety trials, weed, insect and disease control, winter hardiness.

Water Resources Research Institute
Oregon State University
Corvallis, OR 97331
Annual Water Research Summary and seminar proceedings and reports on soil, water and erosion, aquaculture, inter-regional water diversion, pesticides and toxic substances in water.

WASHINGTON

Coastal Washington Research and Extension Unit
Rt. 1, Box 570
Long Beach, WA 98631
Cranberries.

Entomology Research Laboratory
3706 Nob Hill Rd.
Yakima, WA 98901
Chemical and biological control of vegetable and fruit insect pests.

Forest Hydrology Laboratory
1133 N. Western Ave.
Wenatchee, WA 98801
Watershed research.

Forestry Sciences Laboratory
3625 93rd Ave. SW
Olympia, WA 98502
Intensive culture of Douglas-fir and assocated species, old-growth forests and wildlife habitats. Leading center for research on alder utilization.

Irrigated Agriculture Research and Extension Station
Prosser, WA 99350
All major irrigated crops, except apples and pears.

Northwest Washington Research and Extension Unit
1468 Memorial Highway
Mt. Vernon, WA 98273
Western Washington vegetables and small fruits including seed crops, grapes, ornamentals and Christmas trees. Some fava bean trials.

Pacific Northwest Forest and Range Experiment Station
Forestry Sciences Laboratory
4507 University Way NE
Seattle, WA 98105
Logging systems for fragile mountain terrain.

Soil Erosion Research Laboratory
215 Johnson Hall
Washington State University
Pullman, WA 99164
Dryland soil erosion and runoff, deep plowing, green manure, nitrogen fertilizer use, alfalfa and grass cropping effects.

Southwestern Washington Research Unit
1918 NE 78th St.
Vancouver, WA 98665
Small fruits, beans, fruit trees, Christmas trees.

Tree Fruit Research Center
1100 N. Western Ave.
Wenatchee, WA 98801
Biological control of tree fruit pests, computer-based codling moth model, virus diseases, soil fertility and nutrition, rootstock evaluation, pruning, orchard management.

Washington Agricultural Experiment Station
Washington State University
Pullman, WA 99163
Pollination and bee poisoning, greenhouse crops, urban development and parks, fruit tree rootstocks and nursery stock.

Western Washington Research and Extension Center
West Pioneer
Puyallup, WA 98371
Commercial and garden vegetables, stone fruits, garden pest control.

NORTH AMERICA

Beltsville Agricultural Research Center
Beltsville, MD 20705
A 7,000-acre complex with eight institutes and a staff of thousands, one of the biggest in the world. For a description of research programs write for The Beltsville Agricultural Research Center, (#729-888/2154), from USDA, Office of Governmental and Public Affairs, Room 507A, 14th and Independence Ave, Washington, DC 20250.

Kusa Research Foundation
PO Box 761
Ojai, CA 93023
A new organization, "designed to serve as a public gene bank" for ancient cereal grains and North American pioneer wheat. Plans for a seed nursery, storage unit, and publication. Include $1 for postage and handling with information requests.

Livingston and Associates
Box 2910
Santa Cruz, CA 95063
Leading center for research and dissemination of information on the use of earthworms in the composting of municipal wastes.

Organic Agricultural Research Institute
Box 475,
Graton, CA 95444
Provides research and education for farmers to promote organic methods. Publications available on weed management, goats, and slug and snail control.

Rodale Research Center
Box 323, RD 1
Kutztown, PA 19530
Major North American center for research on biological agriculture. Current projects include studies on problems commercial farms encounter in the transition to organic practices, the development of amaranth as a new grain crop, and the development of small scale aquaculture systems.
An annotated list of research reports is available without charge. Individual reports, available for $1 each, include detailed studies of intercropping corn and vegetables in legume and grass sod, aquaculture techniques, insectary plants, grain and vegetable amaranth, and winter vegetable production.
The Rodale Reader Service Dept. (33 E Minor, Emmaus, PA 18049) is also available to respond to requests for agriculture-related information, such as sources of seed, parts or equipment, or information on crop rotations, pest control methods, or farm finance.

Soil and Health Society
33 E Minor St
Emmaus, PA 18049
Non-profit membership organization headed by Robert Rodale. Sponsors research on food production, nutrition and health. Offers tests of drinking water, human hair, and farm or garden soil to detect heavy metal contamination.

Woods End Agriculture Institute
Orchard Hill Rd.
Temple, MA 04984
Publishes translations of European research in biological agriculture.

INTERNATIONAL

International Council for Research in Agroforestry
PO Box 30677
Nairobi, Kenya
The objective of ICRAF is to improve the nutritional, economic and social well being of people in developing countries through agroforestry. A free newsletter, available upon request, reports on developments in the tree crops movement worldwide.

Henry Doubleday Research Association
Convent Lane
Bocking, Braintree
Essex, England
Largest organic gardening group in England. Members do extensive research, conducting experiments coordinated by the organization. Vegetable gene bank and sanctuary gardens for old varieties. Publishes quarterly newsletter and frequent reports. Membership is $10, U.S. and Canadian.

Commonwealth Scientific and Industrial Research Organization, GPO Box 350, Adelaide, South Australia 5001
The CSIRO is an information network for agricultural and forestry research throughout the British Commonwealth. They publish a Research Index which lists participating agricultural research stations and their main research topics.

International Research Center on Environment and Development (CIRED)
54 Boulevard Raspail, Room 311
75270 Paris CEDEX 06
France
Publishes Ecodevelopment News, a French and English language newsletter on grass-roots projects in rural cultures around the world.

Tools for Agriculture

Borremans, Valentina. Guide to Convivial Tools. Ann Arbor, MI: R.R. Bowker, 1979.
Comprehensive guide to publications and organizations creating alternatives to industrial society.

Boyd, John. Tools for Agriculture, A Buyer's Guide to Low-Cost Agricultural Implements. 1976. Intermediate Technology Development Group, Box 337, Croton-on-Hudson, NY 10520.
Guide to over 300 manufacturers of simple agricultural tools from around the world. Illustrations provide ideas for how to build many tools yourself.

Branch, Diana, ed. Tools for Home-Steaders, Gardeners and Small-Scale Farmers. Emmaus, PA: Rodale Press, 1978.
Access to over 700 hard-to-find tools and implements from around the world. Now out of print. Check it out at the library.

Clark, Wilson. Energy for Survival. Garden City, NY: Doubleday, 1975.
A good introduction to net energy analysis and the many alternatives available for creating a sustainable society.

Edlin, Herbert L. Woodland Crafts in Britain. North Pomfret, VT: David & Charles, 1973.
One of the best sources of information available on the many aspects of woodcraft. Herbert Edlin takes us to the heart of a culture which was founded on the use of local resources. Subjects covered include: alder coppicing, ash and beechwood crafts, chair making, birches and besoms, hawthorns and hedgecraft, hazel hurdles, hoops and crates, the cleaving of oak, willow weaving and basketry, millwright and wheelwright, charcoal burning, herbs, dyes, resins, and food from the forest. The detailed text is accompanied by 150 photos of people at work in forest and workshop.

Harter, Kimberly. Northwest Alternative Energy Directory. 1980. Bonneville Power Administration, Box 3621, Portland, OR 97208.
Comprehensive directory of public agencies and private organizations active in all aspects of alternative energy development. Includes sources of information on alcohol fuels, methane, small scale hydro, photovoltaics, solar energy, and wind power. Periodic updates are sent to people who order the directory.

Illich, Ivan. Tools for Conviviality. New York: Harper & Row, 1973.
Guidelines for how we can regain control of our tools for the creation of meaningful work and a sustainable society.

Lawrence, James, ed. The Harrowsmith Sourcebook. Camden East, Ontario: Camden House, 1979.
Sources of tools, breeding stock and alternative technologies in Canada. Lively and useful.

Merrill, Richard, and Gage, Tom. Energy Primer: Solar, Water, Wind and Biofuels. Pine Brook, NJ: Delta Books, 1978.
Comprehensive review of alternative energy and food production strategies. The reviews of agriculture, biomass fuels, aquaculture and wood production are thorough, with extensive bibliographic references.

Partridge, M.O. Farm Tools Through the Ages. Boston: New York Graphic Society, 1973.
Covers the evolution of farm tools to the early 1900's. Many illustrations and clear explanations of simple hand and animal powered tools.

TOOL SOURCES

A.M. Leonard, Inc.
6665 Spiker Road
Piqua, OH 45356
Full range of horticultural tools and supplies.

Alfred Teufel Nursery, Inc.
Main Office:
12345 NW Barnes Road
Portland, OR 97229
and
Branch Office:
666 134th Street SW
Everett, WA 98204
Wholesale distribution of greenhouse and horticultural supplies. Largest and most complete selection in the region.

Small Farm Equipment Co.
205 Industrial Drive
De Kalb, IL 60115
At last—North American versions of the small walking and cultivating tractors used in China and Japan. Moderately priced, high value machines for small farms and orchards.

Northwest Fur & Trading Company
2807 Tide Court
Salem, OR 97303
Distributor of Aermotor and Wind-charger windmills, hydraulic rams, submersible pumps, and solar collectors, all at discount prices. Catalog $2.00 (refundable with purchase).

By Hand & Foot, Ltd.
PO Box 611
Brattleboro, VT 05301
Expanding source for "tools dependent on human energy." Advanced scythes, carts and wood-splitting tools currently available.

Garden Way Living Center
Parkway Plaza
17740 South Center Parkway
Tukwila, WA 98188
and
1802 Jantzen Beach Center
Portland, OR 97217
Complete line of gardening tools and supplies, including Todd Speedling trays and Troy-Bilt equipment.

Nasco Farm & Ranch Catalog
Box 3837
Modesto, CA 95352
Wide selection of livestock, fencing, horticultural, forestry, and general farm equipment.

Orchard Equipment & Supply Co.
Route 116
Conway, MA 01341
Specializes in commercial cider-making equipment.

Sears Farm & Ranch Catalog
Distribution Center
Seattle, WA 98184
Supplies for livestock, beekeeping, farming and gardening. Sears distributes two small-scale seed and grain cleaners.

Smith & Hawken Tool Co.
68 Homer
Palo Alto, CA 94301
Best source of garden and woodlot tools of the highest quality.

Tradewinds, Inc., PO Box 1191, Tacoma, WA 98401.
Distributors of Humus Retracting Cultivators, which are ideal for mechanical cultivation in orchards and vineyards.

FORESTRY TOOLS

Ben Meadows Co.
3589 Broad Street
Atlanta, GA 30366

Forestry Suppliers
205 W. Rankin Street
PO Box 8397
Jackson, MS 39204

International Reforestation Suppliers
PO Box 5547
Eugene, OR 97405

Nancy J Turner. Plants in British Columbian Indian Technology. Handbook No. 38. British Columbia Provincial Museum, Victoria, B.C. V8V1X4

Sensitive and thorough introduction to the many uses of plants by the native people of the Northwest. Dozens of plants are described, along with detailed descriptions of their use by tribes both east and west of the Cascades. Like all of Nancy Turner's work, this is an outstanding piece of scholarship and communication.

In British Columbia, where winters are long and cold, where most foods must be cooked to make them palatable, where waterproof clothing and watertight shelters are indispensable, where the most efficient means of transportation in the early days was by water, where the native hunting/fishing/gathering economy required a myriad of special tools, and where metal of any type was relatively rare and pottery virtually unheard of, plant materials were not only useful but absolutely essential for survival.

Heat, shelter, transportation, clothing, implements, nets, ropes, and containers—necessities of life in the Pacific Northwest environment—were all afforded by the great abundance and variety of vegetation in the area.

Edward S. Curtis

Indian hemp (Apocynum cannabinum) *was without doubt the most important source of plant fibre for native people in the southern Interior....*

The Okanagan and other Interior groups, including the Thompson, Lillooet, Shuswap, and Kootenay, used Indian hemp twine for fishing lines and nets—because it keeps its strength under water and will not shrink—and for deer nets, slings, bowstrings, bridle ropes, nooses for game birds, hide stretchers, for binding implements, for sewing moccasins, clothing, baskets, birch-bark canoes, and cat-tail mats, and for weaving tumplines, garments, baby bedding, and bags. It was often woven with other plant fibers, such as silverberry, willow, sagebrush, and tule, and in making garmets it was sometimes spun with deer hair....

Raw or spun, the fibre was a common trading product, not only amongst the various Indian groups of the southern Interior, but between them and the Coastal people, as far west as Vancouver Island. In the early days of contact with the white man, a large bundle of prepared twine was worth as much as a horse in the Okanagan area. On the Coast, the twine was a prized material for net-making, superior even to stinging nettle twine.

Compost Power

Biothermal Energy Center
PO Box 3112
Portland, ME 04101

Non-profit organization promoting the use of compost as a source of energy as well as a means of recycling wastes and building soil fertility. The center distributes information, publishes a quarterly newsletter, and serves as a communications link between researchers and projects around the world. They are developing and testing designs for composting greenhouses and are disseminating information on Jean Pain's methods of producing heat, fuel and compost (see below).

Pain, Ida, and Pain, Jean. The Methods of Jean Pain. 1980. Biothermal Energy Center, PO Box 3112, Portland, ME 04101.

Report on techniques and equipment for generating heat, methane and compost from chipped brush. Jean Pain, developer of the system, used the heat to warm his house, the methane to fuel his trucks, and the compost to fertilize his farm. His work is an important model for the integration of agriculture and forestry.

"Composting in solar greenhouses has shown encouraging results. In composting greenhouses in Illinois and Oregon, research pioneer Dan Knapp found that compost dramatically influences plant growth, vitality and frost resistance. Little or no watering is required due to the large amounts of water vaporized by the heat of the compost. The moist air prevents soil from drying out. Frost protection is provided as the vapor condenses and crystalizes on the plant surfaces, forming an insulating layer that permits greenhouse temperatures to drop to 15° F with no damage to the plants....

"The combined benefits of a low cost heat, CO_2, nitrogen and frost protection while composting year round make the development of commercial scale composting greenhouses attractive and economically feasible. With the wealth of knowledge available in composting, CO_2 enrichment, solar greenhouse design and farm oriented engineering, composting greenhouses could soon become a major contributor to a successful local agriculture and economy."

Bruce Fulford
BioThermal Energy Center

FOR MORE INFORMATION

Miles Solar Systems
Box 350
Philomath, OR 97370

Inventors of heat exchange modules which extract energy from compost to produce heat day and night, all year round. They claim that the "Miles Compost Module is the only commercially available device to extract heat from compost."

Adams, Robin Billau; Averill, Jon; and Daniels, Ida. Compost Toilets, A Guide for Owner-Builders. 1979. National Center for Appropriate Technology, PO Box 3838, Butte, MT 59701.

Detailed information on compost toilet alternatives. Practical and useful.

BioCycle, Journal of Waste Recycling
Box 351
Emmaus, PA 18049

Bimonthly technical journal on large-scale composting systems for municipal, industrial, and agricultural wastes. Emphasis is on recycling nutrients for the benefit of agriculture and forestry.

Fiber Crafts

Dempsey, John M. Fiber Crops. Gainesville, FL: Univ. of Florida, 1975.
Dempsey covers only nine fiber plants, but each one is covered in very great detail. Flax, hemp, ramie, and an *Abutilon* species are included. Their adaptations, culture, diseases, insects, processing methods, and uses are discussed, with additional references provided.

Dodge, Charles R. A Report on the Uncultivated Bast Fibers of the United States. USDA, 1894. Copies available for $2.50 from Friends of the Trees, Star Rt Box 74, Oroville, WA 98844.
Included in this report is rare information on little-known fiber plants, plus the early history of experiments with native and non-native fiber plants.

Hoppe, Elizabeth, and Edberg, Ragnar. Carding, Spinning and Dyeing: an introduction to the traditional wool and flax crafts. NY: Van Nostrand Reinhold, 1975.
Translated from Swedish, this book gives very good coverage of working with wool and flax.

Hunter, Dard. Papermaking, the History and Techniques of an Ancient Craft. NY: Dover, 1943.
Dard Hunter spent decades making paper from dozens of plant materials. Excellent introduction to papermaking as a cottage industry. Fortunately, it's still in print.

Kirby, R.H. Vegetable Fibers. NY: Interscience, 1963.
One of the most comprehensive texts on fiber crops. All of the major fiber crops (and many of the minor ones) are covered in detail. Information for each fiber crop includes varieties, sowing, culture, labor requirements, harvesting, processing, fiber extraction, and uses.

The Needlework Guide to the Pacific Northwest, 4326 Dyes Inlet Rd NW, Bremerton, WA 98312.
Directory of textile farmers, crafters, teachers, and merchants.

Straw Into Gold, 3006 San Pablo Ave, Berkeley, CA 94702. Catalog $1.
A friendly company which carries a wide assortment of fibers, spinner's supplies, basketry materials, yarns, and natural dyes, plus hundreds of how-to books on fiber crafts, papermaking, and basketry.

Greenhouses

Chapel, Paige. The City Greenhouse Book. 1980. Center for Neighborhood Technology, 570 W. Randolph Street, Chicago, IL 60606.
Guide for organizers and planners developing urban greenhouse projects. Covers potential conflicts and legal barriers in addition to details on greenhouse designs and management. Extensive bibliographies and source lists.

Ecotope Group. A Solar Greenhouse Guide for the Pacific Northwest. 1979. Ecotope Group, 2332 E. Madison, Seattle, WA 98112.
Step-by-step instructions for building attached solar greenhouse. Includes solar radiation information for British Columbia, Washington and Oregon.

Fisher, Rick, and Yanda, William. The Food and Heat Producing Solar Greenhouse. Santa Fe, NM: John Muir, 1976.
Comprehensive source of information on the design, construction and management of solar greenhouses.

McCullagh, James C. The Solar Greenhouse Book. Emmaus, PA: Rodale Press, 1978.
Details on solar greenhouse construction and management. Covers both urban and rural greenhouses.

NCAT. Biological Management of Passive Solar Greenhouses. 1979. National Center for Appropriate Technology, Box 3838, Butte, MT 59702.
Annotated bibliography, with references to sources of information on all aspects of greenhouse production.

Nearing, Scott, and Nearing, Helen. Building and Using Our Sun-Heated Greenhouse. Charlotte, VT: Garden Way, 1979.
Introduction to year-round food production in an owner-built greenhouse. Especially good for cold-climate areas.

Smith, Shane. The Bountiful Solar Greenhouse. Santa Fe, NM: John Muir, 1982.
Detailed guide to solar greenhouse gardening, including design considerations, propagation techniques and insect control, plus common and exotic crops for year-round vegetable and fruit production. Well written, this book provides a wealth of information.

PERIODICALS

Alternative Sources of Energy
107 S Central Avenue
Milaca, MN 56353
Lively exchange of information on all aspects of energy alternatives—wind, solar, hydro and biomass. ASE also publishes technical papers on topics ranging from solar grain driers and alcohol fuels to wood heat and windmills.

Journal of the New Alchemists
New Alchemy Institute
237 Hatchville Road
East Falmouth, MA 02536
Annual report on New Alchemy's garden, aquaculture, and village technology research. They also publish a quarterly newsletter for members. An information brochure on their projects and publications is $2.

Mother Earth News
PO Box 70
Hendersonville, NC 28791
Jam packed journal. Most articles written by its readers. Emphasis on alternative energy, ecology, and doing more with less.

RAIN
2270 NW Irving
Portland, OR 97210
Access guide to appropriate tools for building self-sufficient, just and ecologically sound communities.

Small Farm Energy
PO Box 736
Hartington, NE 68739
Bimonthly newsletter which communicates information generated from the Small Farm Energy Project, a research and demonstration project designed to install energy-saving devices on small farms. Emphasis is on practical alternatives that can be built on the farm, including solar heating and grain drying systems, and the production of alcohol and vegetable oil fuels.

TRANET
Box 567
Rangeley, ME 04970
Quarterly review of appropriate technology resources, projects and programs around the world. Excellent for people seeking a global perspective on social and technological change.

Undercurrents
27 Clerkenwall Close
London ECIR OAT, England
British sister publication to RAIN magazine, this is a bimonthly review of "soft technology and hard politics."

A.T. Organizations

Alternative Energy Resources Organization (AERO)
Stapleton Bldg, 424
Billings, MT 59101
Statewide organization promoting conservation and renewable energy.

Amity Foundation
Box 11048
Eugene, OR 97440
Non-profit organization which conducts research and provides consulting on solar greenhouse design and management, small-scale aquaculture, small farm water management, urban agriculture, and educational programs.

Citizens for a Solar Washington
1932 First Avenue, #917
Seattle, WA 98101
Statewide organization of local groups and individuals promoting conservation and energy alternatives. CSW's bimonthly magazine, Solar Washington, provides thorough coverage of activities in the state. The January/February and March/April 1982 issues featured special reports on appropriate technology, including a valuable directory of agriculture, alcohol, biomass, hydro, and solar projects. The special issues are available for $2 each.

Carlotta Collette

Cynthia Stull of Dallas, Oregon, is one of dozens of experimenters awarded research grants through the U.S. Department of Energy's Appropriate Technology program. Working with her husband, William, Cynthia's project was to study the economic feasibility of a small on-farm alcohol still.

Funding for the AT program has been discontinued, but a great deal was learned while it was in existence. Reports on the Stull's alcohol experiment and a wide range of other innovative projects are available in each state.

In addition to several fuel alcohol experiments, the Oregon program included methane, low-head hydro, solar greenhouse, and intensive food production projects. Information on the program in that state is available from **Oregon Appropriate Technology, Box 1525, Eugene, OR 97440.**

Reports on the Washington AT program, which included projects on solar grain drying, aquaculture, small-scale cheese manufacturing, use of modern horse-drawn tillage implements, and Integrated Pest Management, can be obtained from **Citizens for a Solar Washington, 1932 1st Ave, #917, Seattle, WA 98101.** For information on results of the program in other states, contact the state energy office.

Aprovecho Institute
442 Monroe Street
Eugene, OR 97402
Research and demonstration network devoted to the application of appropriate technologies throughout the world. Major projects include windmills, low-cost housing, solar water heaters, and development of new stove technologies to conserve firewood. They sponsor research on bamboo production and use, and publish "Cookstove News," an international newsletter on cookstove research.

Canadian Society Promoting Environmental Conservation
2150 Maple Street
Vancouver, BC V6J 3T3
Center for energy and conservation related information.

Domestic Technology Institute
PO Box 2043
Evergreen, CO 80439
Non-profit research, education and service organization devoted to development and dissemination of information on appropriate technologies for food production, processing and storage; alternative energy sources, including solar greenhouses, ethanol, methane, water pumping; and building construction technologies.

Ecotope Group
2332 E. Madison
Seattle, WA 98112
Involved with the design and construction of solar homes and greenhouses, methane systems, and other energy alternatives.

Farallones Institute
Integral Urban House
1516 5th Street
Berkeley, CA 94710
and
Rural Center
15290 Coleman Valley Road
Occidental, CA 95465.
Research and demonstration centers for applying appropriate technologies in urban and rural environments.

Intermediate Technology Development Group of North America
Box 337
Croton-on-Hudson, NY 10520
Source of information on international approaches to appropriate technology. Distributes Appropriate Technology, an excellent quarterly journal produced by ITDG/London, and a wide range of publications on agriculture, energy, rural development, economics, and water supply and sanitation.

National Center for Appropriate Technology, Box 3838, Butte, MT 59702.
Extensive listings of general and technical publications on energy conservation, food production, and alternative energy sources.

Northeast Washington Appropriate & Creative Technology (NEW ACT), Box 385, Republic, WA 99166.
Workshops, book service, resource center, and group purchases of photovoltaic cells.

Oregon Appropriate Technology, Box 1525, Eugene, OR 97440.
Consulting and design for solar greenhouses, energy conservation, recycling, and energy planning.

Responsible Urban Neighborhood Technology (RUNT), 3116 N. Williams Ave, Portland, OR 97227.
Active neighborhood group which is developing the Eliot Energy House as a community resource and education center.

Solar Greenhouse Society, 3535 NE 27th, Portland, OR 97212.
Local group of greenhouse builders and owners which meets to share skills and techniques.

Volunteers for International Technical Assistance
1815 N Lynn Street, Suite 200
Arlington, VA 22209
Non-profit organization which promotes appropriate technologies in developing countries. VITA publishes over 140 books, bulletins and manuals on agriculture and solar, wind and water systems. They also publish VITA News, an outstanding review of international appropriate technology activities.

The Farm

JIM BAUERMEISTER

The people who homesteaded Franklin County, Washington were latecomers to the Columbia Plateau. They had heard of the rich grain growing region called the Palouse, and were on the tail end of a thirty year land rush. They settled under land laws written for a fictitious west and bought land from the railroads at very reasonable rates. As the Columbia Plain filled in from the north and east, the timbered river bottoms were taken first, then the rolling hills close to the forests.

By 1900 the more recent homesteaders were miles from water and days from trees. But they believed the old adage, "where wheatgrass grows, wheat will grow," and a newer corollary, "where sagebrush grows big, wheat will do well." They brought their sodbuster plows from the Great Plains and turned under the bluebunch wheatgrass and lichens, wildflowers and sage. Many broke their health and destroyed their families trying to farm 160 acres in an area that got less than ten inches of rain a year. Ten years after the landrush ended, half the settlers were gone.

The survivors learned how to grow grain in the arid land. They replaced their moldboard plow with the disc and the chisel plow. They let half the land lie fallow each year to collect two seasons' rains. But they didn't create a sustainable agriculture. One diesel cat replaced thirty-six horses, but allowed farmers to disc up steeper, more erosive slopes. Dust storms still darken the skies, and a sudden cloudburst or spring thaw sends tons of topsoil to the bottoms of coulees.

Because the land in the Columbia Basin is too arid to grow cover crops for nitrogen and the fields are too

PUBLICATIONS

A.S.A. Organic Farming: Current Technology and Its Role in Sustainable Agriculture. Forthcoming (1983). American Society of Agronomy, 677 S Segoe Road, Madison, WI 53711.

Proceedings of a major symposium on biological agriculture, with papers on soil and crop research in Europe and North America. Over a dozen papers, including presentations by Vogtmann, Harwood, Lockeretz, Papendick, Parr, Eggert, Cook, Pimentel, and Youngberg.

FAO. China: Recycling of Organic Wastes in Agriculture. FAO Soils Bulletin 40. Rome, Italy: Food and Agriculture Organization of the U.N., 1978 (Distributed by UNIPUB, New York).

Based on a month-long tour of China, this is a report on the numerous techniques the Chinese have developed for turning wastes into wealth. Included is a survey of methods for composting animal, human and crop wastes, a chapter on the use of azolla and blue-green algae, details on the use of legumes for green manuring, and a thorough discussion of methane digesters for producing high quality fertilizer and fuel.

Harmond, Paul, ed. The Farmstead Book 1. Mayne Island, BC: Cloudburst, 1978.

Reprints of turn-of-the-century bulletins on farming and woodlot management. Valuable vintage information on legumes, crop rotations, and green manuring.

Kern, Barbara, and Kern, Ken. The Owner-Built Homestead, A How-To-Do-It Book. New York: Charles Scribner's Sons, 1977.
One of the first contemporary efforts to integrate tree crops, livestock and plants into a whole system. Advocates no-till approach, with emphasis on mulch gardening and ley farming.

Koepf, Herbert H.; Petersson, Bo D.; and Schaumann, Wolfgang. Bio-Dynamic Agriculture, An Introduction. Spring Valley, NY: The Anthroposophic Press, 1976.
Bio-Dynamics views the farm as a living organism set within the larger organism of the landscape. This is the most comprehensive introduction to the Bio-Dynamic school of agriculture.

Logsdon, Gene. Small Scale Grain Raising. Emmaus, PA: Rodale Press, 1977.
Useful for growing grain crops on a small scale using organic methods. Includes discussion of crop rotations using clover and other legumes.

Papendick, R.I., et al., eds. Multiple Cropping. ASA Pub. 27:1, 1976. American Society of Agronomy, 667 S Segoe Road, Madison, WI 53711.
Technical review of multiple cropping systems around the world.

Schmid, Otto, and Klay, Ruedi. Green Manuring, Principles and Practice. 1980. Woods End Agricultural Insititute, Orchard Hill Road, Temple, ME 04984.
Review of European green manure research and techniques. Special section on undersowing crops with legumes.

big to spread manure on, the introduction of chemical fertlizers in the 1950's was a godsend to the farmers. Herbicides were welcomed to the sixty year battle with Russian thistle and cheatgrass.

In addition to increased erosion, mechanization, and dependence on chemicals, farms have become larger and more centralized. My family's farm near Connell in north Franklin County is a conglomeration of a half-dozen different homesteads. The nearest neighbor is three miles away. Most of the land adjacent to my folks' farm is farmed by people who live as far as twenty miles away. Nearby communities, such as Kahlotus, Washtucna, Hatton and Lind have withered away because there isn't the population to support stores, newspapers, movie theatres or banks.

The lack of diversity and of neighbors has done nothing to encourage young people to stay in the community. There is a certain inbred feeling of inferiority and fatalism from being what one government commission described as "the people left behind."

The wheat growing region of eastern Washington is linked to a global food and energy system subject to all the whims of international politics and intrigue. In 1978 the price of wheat dropped abruptly, and farmers hoarded gas and diesel and worried whether they would obtain enough fuel to get through harvest—all because of a revolution in Iran. Iran had oil and used it to buy wheat. The downfall of a despotic shah was felt in Connell. If the link to the food and energy trade were broken abruptly (and farming methods did not adapt) the farming community could not survive in Franklin County. If there is a gradual transition to a sustainable agriculture, the community will prosper.

Achieving a sustainable agriculture, however, will require a concerted public and private effort. Many

A Perennial Grain Available Now

In New Roots for Agriculture, Wes Jackson suggests that perennial grains may be a long term solution to erosion and energy problems. Sixty years ago American scientists began experimenting with a hybrid cross of wheat *(Triticum)* and wheatgrass *(Agropyron)*. The result was a new plant species, "agrotriticum," a perennial grain which bred true to its parent. Agrotriticum was found to be very resistant to fungus diseases, lodging and insect damage. The strong root system made it tolerant of drought.

Russian researchers developed an agrotriticum in the 1950's which had a protein content of 25%. This compares to 15% for the best red winter wheats and 20% for wheatgrass. During the same period a high protein agrotriticum was developed at the University of California (Davis) which had superior baking characteristics. First year production rivaled wheat in yields. An article in the USDA's Agriculture Research announced that commercial varieties of agrotriticum would be available "in a few years."

That was 25 years ago. In the interim, few efforts have been made to improve agrotriticum. Generally, American researchers gave up on the idea of perennial grain. They discovered that agrotriticum would only produce well for three or four years, with production declining after the first year. Controlling weeds in the agrotriticum was also perceived as a problem. Researchers instead focused on borrowing chromosomes from agropyron to increase rust- and mosaic-resistance in annual wheat.

The Russian researchers found that nearly every variety of wheat crossed relatively well with four varieties of agropyron: *Agropyron elongatum* (tall wheatgrass), a late maturing bunchgrass adapted to alkaline soils in climates of moderate rainfall; *A. intermedium* (intermediate wheatgrass), a sod-forming grass well adapted to the intermountain prairies; *A. trachophorum* (pubescent wheatgrass), a sod-former similar to *A. intermedium*; and *A. repens* decisions will have to be made far from Connell. Land laws will need to be amended to encourage the creation of new farms. Government farm programs will need to be reoriented to benefit low and moderate income farmers. Research will have to focus on ways to increase soil fertility without depleting other resources, and on reducing health and environmental risks from agriculture. Technologies that aid in the management and stewardship of land, and do not stifle human resourcefulness, will have to be adopted.

Yet most of the research and work will be done on the individual and community level. The first requisite of a sustainable agriculture is a sense of land. Land, as in Aldo Leopold's definition, is a whole system including climate, vegetation and animal and soil life. There needs to be an attitude that the care of land is a trust, an implicit contract not only between people and the land, but between present and future generations. Land can be used, but should not be abused. With a proper feeling for land, people can shape environments that will provide them with food, shelter, and commerce.

A sustainable agriculture is not a return to the past, though many ideas that worked in the past can be applied to the future. Seventy-some years ago a homesteader built a dike along one of the edges of what became my grandfather's farm. The homesteader hoped that he could impound enough rainfall to grow alfalfa. Even though the dike broke, the concept was sound; if not for alfalfa, then for a planting of trees.

The development of perennial grain varieties would be a major advance for this arid region. In Franklin County most of the rain falls in the winter, and is lost to shallow rooted annual grains. But the roots of perennials can tap moisture deep in the loess soil. Wes Jackson and the Land Institute of Salina, Kansas are working on developing agrotriticum, a perennial

Schwenke, Karl. Successful Small-Scale Farming. Charlotte, VT: Garden Way, 1979.
This clearly-written, thorough book introduces farming, from contour maps to gate styles, from soil types to planning square dances. Its text on plants, machinery, farm practices, cash and specialty crops, and "the whole farm" (roads, water, fencing and the farm shop) are directed to the intelligent, enthusiastic new farmer. Many useful charts and tables are in the back of the book. Livestock is not discussed.

Seymour, John. The Guide to Self-Sufficiency. New York: Popular Mechanics Books, 1976.
A comprehensive, delightful introduction to the art and craft of homesteading and small-scale farming. Beautifully illustrated, with concise introductions to everything from growing barley and making beer to gardening through the year.

——, and Seymour, Sally. Farming for Self-Sufficiency. New York: Schocken Books, 1976.
Beautifully illustrated homesteading handbook from a British perspective.

Thomas, Sherry, and Tetrault, Jeanne. Country Women, A Handbook for the New Farmer. Garden City, NY: Anchor Books, 1976.
A gift of knowledge and experiences shared between country women. This is a practical homesteading manual, with details on topics ranging from purchasing land, selecting tools, and establishing a garden to raising small livestock and butchering sheep, deer and goats.

(quackgrass), an aggressive, very persistent grass that has become a weed in most of the humid, temperate areas of the world. The Russians' success with these different crosses suggests the possibility of many potential agropyron-triticum crosses, which could produce perennial grains adapted to many regions of the world.

One group actively doing breeding and research work on agrotriticum is The Land Institute in Salina, Kansas, founded by Wes and Dana Jackson. Researcher Walter Pickett is working on improving agrotriticum's winter hardiness, and is experimenting with field burning and light tilling to increase second and third year production. He believes that production levels within five bushels of wheat will be economically acceptable, especially since the costs of production should be lower than for wheat because little or no cultivation is needed.

Pickett thinks that interplanting could eliminate agrotriticum's weed problem, and is interplanting it with perennial sunflowers and sorghum. The grains will be timed to ripen simultaneously, and can either be separated at harvest or used as a mixed feed. He believes that with concerted effort, commercial perennial grains could be available in ten years.

At present there is only one agrotriticum variety available in North America. It is called Agrotriticum W-21, and is grown by CalWest Seeds, a farmers' cooperative in Woodland, California. Dr. Don Smith, CalWest's research director, says that, in their area, W-21 yields 40 to 60% of what wheat does. The entire stand ripens simultaneously.

W-21 is marketed exclusively by Kester's Wild Game Food Nursery of Omro, Wisconsin, as a forage plant for wild geese, ducks and deer. They recommend planting agrotriticum in the fall, like winter wheat. If weeds threaten the crop they suggest mowing, or spraying with a broadleaf weed killer. Agrotriticum W-21 is winter hardy in the Maritime. In colder regions, a disking in the early fall of the following year will seed new plants and not kill the established stand.

If you would like to experiment with growing Agrotriticum W-21, it is available from Kester's, PO Box V, Omro, WI 54963.

USDA. Report and Recommendations on Organic Farming. Washington, DC: USGPO, 1980.
The first official USDA study on the potential of organic farming as an alternative to soil erosion and the decline of modern agriculture.

Walters, Charles, Jr., and Fenzau, C.J. An Acres USA Primer. Raytown, MO: Acres USA, 1979.
Useful introduction to the principles and practice of eco-agriculture.

Wolf, Ray, ed. Organic Farming: Yesterday's and Tomorrow's Agriculture. Emmaus, PA: Rodale Press, 1977.
This collection of essays mainly serves to make the point that organic farming is possible, desirable, and in fact essential for the continued survival of agriculture. Stories featuring successful organic farmers, economics comparisons, research results, and the basics of organic practices (organic matter, rotations, weed and pest control) educate the reader about the principles and techniques that really work, now.

wheat-wheatgrass cross which would be ideal for this region. It may also be possible to develop other alternative crops, such as nitrogen fixing grains or legumes (edible lupines, for example) for arid soils.

The revival of old technologies and the development of new ones will be essential to the farms of the future. Windpower was once a significant source of energy for the isolated ranches of Franklin County. Improved wind generation and photovoltaic systems could produce power more cheaply and more responsibly than the Washington Public Power Supply System. Today, mulching and drip irrigation make gardens and orchards possible even on farmsteads with low volume wells and windmills. Computers are becoming useful farm tools in communicating information and collecting and analyzing data on crop production, input costs, weather and research.

The methods Franklin County farmers develop to create a sustainable agriculture will, in many cases, be unique to their area, and will change with time. Creating a sustainable agriculture is an ongoing process. Climates change, and plants and animals continue to evolve. Human creativity and our concepts of farm ecosystems must also evolve. Creating a sustainable agriculture is a project for perpetuity.

PERIODICALS

Acres USA, Box 9547, Raytown, MO 64133.
The voice for ecological agriculture in the U.S.A., particularly in the Midwest. Valuable information for the commercial, organic, or transition farmer.

Bio-Dynamics, Box 253, Wyoming, RI 02898.
Primary source of information on the practice and philosophy of Bio-Dynamic agriculture. Scientific articles and an international perspective on the Bio-Dynamics movement.

Biological Agriculture and Horticulture. A B Academic Publishers, PO Box 97, Berkhamsted, Herts HP4 2PX, England.
Quarterly journal published by the International Institute of Biological Husbandry. Includes scientific studies on ecological approaches to agriculture and forestry, including soil fertility, pest management, energy utilization, appropriate technology, and the application of biological agriculture in developing countries.

Capital Press (see page 24).

Conservation Update Publication, PO Box 2356, Green Bay, WI 54306.
Monthly reports on the use of Basic-H in crop and livestock production, with formulations for numerous applications based on the experience of farmers and orchardists around the country.

Countryside, 312 Portland Road, Waterloo, WI 53594.
Practical information on all aspects of homesteading, with especially good coverage of livestock.

Farm Show, Box 704, Lakeville, MN 55044.
Bimonthly report on new machinery, livestock equipment, and alternative energy sources. Fascinating review of new developments in agricultural technology.

Farmer-Stockman, 926 W Sprague Avenue, Spokane, WA 99210.
Separate issues of the Farmer-Stockman are issued for each of the Northwest states.

Farmstead, Box 111, Freedom, ME 04941.
How-to for homesteaders, from green manure to tanning buckskin.

Harrowsmith, Camden House Publishing, Camden East, Queen Victoria Road, Ontario, Canada K0K 1J0.
A magazine for homesteaders and country people in Canada. Also looks at broad social issues.

The Land Report, The Land Institute, Rt. 3, Salina, KS 67401.
Reports on steps toward a new agriculture in the prairies, including Wes Jackson's work in the development of high-yielding perennial grain polycultures.

The New Farm, 33 E Minor Street, Emmaus, PA 18049.
Well-researched articles on soil and resource conserving techniques for commercial farmers either practicing or moving toward organic methods. Also good reports on government research and legislation.

Prairie Sentinel, Center for Rural Affairs, PO Box 405, Walthill, NE 68067.
A clear and lively bimonthly journal which documents the transition of agriculture in the heartland to more sustainable methods. Excellent features on crops and production methods.

Small Farmer's Journal, 3890 Stewart Road, Eugene, OR 97402.
The place for information on farming with draft horses. Also discusses draft oxen and mules. Includes general information on livestock and crops, much of it selected from old-time farming literature. Well-suited for the practical, small farmer.

Tilth, 13217 Mattson Road, Arlington, WA 98223.
Quarterly journal of biological agriculture in the Northwest, with practical information on farming and gardening techniques and reports on community activities around the region.

Soils Bibliography

Albrecht, W.A. The Albrecht Papers. Raytown, MO: Acres, USA, 1975.
Anthology of Albrecht's research on the relationship of organic matter and trace minerals in the soil to crop, livestock and human health.

Anon. "Selenium Deficiencies Documented in PNW." Washington Farmer-Stockman, 107:3 (Feb. 4, 1982), p 20.
"Certain volcanic soils, such as those found in the Pacific Northwest, are particularly deficient in selenium." As a result, severe selenium deficiencies have been shown in livestock.

Baker, Kenneth F., and Cook, R. James. Biological Control of Plant Pathogens. San Francisco: W.H. Freeman, 1975.
Primary text on plant pathogens and antibiotics in the soil, including an analysis of the value of green manuring in disease control.

Barrett, T.J. Harnessing the Earthworm. Ontario, CA: Bookworm, 1976.
A fascinating scientific account of the development of the "red wriggler" hybrid. Includes information on the culture and management of the earthworm, and the use of earthworms in agriculture.

Bennett, Hugh Hammond. Soil Conservation. New York: McGraw Hill, 1939.
Written just after the great Dust Bowl, this is a compendium of information on soil erosion and conservation by one of this century's leading conservationists. Still one of the best references on the subject of soil conservation.

Brady, Nyle C. The Nature and Properties of Soils. New York: MacMillan, 1974.
The standard text on soils. It is a basic, readable reference source for information on all aspects of soil science.

Darwin, Charles. Darwin on Humus and the Earthworm. London: Faber and Faber, 1966.
Classic study of humus formation and earthworm ecology.

Deavin, A. "The Rusch Concept of Soil Fertility and Method for its Micro-biological Evaluation." In Towards a Sustainable Agriculture, edited by Jean-Marc Besson and Hardy Vogtmann. Switzerland: Oberwil: IFOAM, 1978.
Introduction to new techniques for qualitative evaluation of fertility based on biological activity in the soil.

Dindal, Daniel L. Ecology of Compost. Office of Public Service, College of Environmental Science and Forestry, State University of New York, Syracuse, NY 13210.
Brief, well-illustrated introduction to the life of a compost pile.

To recognize the extent and the destructiveness of our "urban waste" is to recognize the shallowness of the notion that agriculture is only another form of technology to be turned over to a few specialists. The sewage and garbage problem of our cities suggests, rather, that a healthy agriculture is a cultural organism, not merely a universal necessity but a universal obligation as well. It suggests that, just as the cities exist within the ecology, they also exist within agriculture. It suggests that, like farmers, city-dwellers have agricultural responsibilities: to use no more than necessary, to waste nothing, to return organic residues to the soil.

Wendell Berry, *in* The New Food Chain, Jerome Goldstein, ed. Emmaus, PA: Rodale Press, 1973.

Farb, Peter. Living Earth. New York: Harper & Row, 1959.
An introduction to the ecology of the unnoticed world underfoot.

Griffen, David M. Ecology of Soil Fungi. Syracuse, NY: Syracuse University Press, 1972.
An advanced text of soil life and interactions.

Hill, S.B. et al. Ecological Soil Management. Ecological Agricultural Project, Box 225, MacDonald College, Ste. Anne de Bellevue, Quebec, Canada H9X 1C0.
Comprehensive bibliographic guide to the literature on soil science and ecological soil management. Thorough and useful.

Menge, John A. "Mycorrhiza Agriculture Technologies." In Background Papers for Innovative Biological Technologies for Lesser Developed Countries. U.S., Congress, House, Committee on Foreign Affairs, 97th Congress, 1st session, 1981, pp. 383-424.
Review of current information on mycorrhizal fungi and their potential for increased use in agriculture. 100 references.

Laboratories

Soil testing is a controversial subject. Some people say it is essential. Others say it's a waste of time and money. In their January, 1982 issue, The New Farm magazine began a five-part series comparing recommendations from 69 major soil labs for nitrogen, phosphorus, potassium, lime, and micronutrients. The results were very inconsistent, with nitrogen recommendations, for example, ranging from zero to 230 pounds per acre based on the same soil sample. What's the conclusion? If you do want your soils analyzed, find a lab you can trust, and then compare their findings with your own experience.

Government soil testing labs can be located through your local Extension agent or farm advisor. Below are private labs, several of which specialize in serving organic farmers.

Biosystem Consultants, Inc., PO Box 43, Lorane, OR 97451.
Mineral and organic matter tests designed for organic growers, with specific recommendations for both gardeners and commercial farmers.

Pacific Soil Analysis Inc., 1560 Rand Avenue, Vancouver, BC V6P 3G2.
Complete lab for testing soil fertility and structure, as well as plant tissues. Included are tests for total nitrogen, organic matter, sulphur, cation exchange capacity, micronutrients, and water retention.

Peaceful Valley Farm Supply, 11173 Peaceful Valley Road, Nevada City, CA 95959.
Comprehensive soil, leaf and petiole analysis, with detailed recommendations for organic farmers.

Soil and Health Society, 33 East Minor Street, Emmaus, PA 18049.
Non-profit research on soil and human health. Tests for heavy metal contamination in drinking water, soils, and human hair. Also soil nutrient tests with specific information for organic growers.

Soil and Plant Laboratory, Inc., PO Box 1648, Bellevue, WA 98009.
Soil nutrient, pesticide residue tests.

Stemilt Testing Lab, 4173-18 Joe Miller Road, Malaga, WA 98828.
Soil nutrient and leaf analysis, for both eastern and western Washington.

U.S. Testing Company, 2800 George Washington Way, Richland, WA 99352.
Soil nutrient, pesticide residue tests.

Woods End Laboratory, RFD Box 65, Temple, ME 04984.
Bio-Dynamic soil analysis, chromatograms.

Yakima Testing Lab, 1011 South Third Street, Yakima, WA 98901.
Leaf analysis (mostly for fruit trees.)

No-Till Farming

MARK MUSICK

Because of the twin crises of increased soil erosion and soaring petroleum costs, a new method called "no-till farming" has been in the spotlight in recent years. Farmers are seeking ways to minimize both the amount of tillage and number of tractor operations. In the Northwest no-till is especially being touted as a means of halting erosion in the dryland farming regions of the Interior. By this method, stubble and crop residues are left on the field, and herbicides are used to control re-growth and weed competition.

Although much research on chemical no-till is still underway, the practice has been slow to catch on. Planting into the remains of a previous crop does reduce soil erosion, but farmers are finding problems with competition from wild grasses and weeds not killed by the herbicides, rodents multiplying in the fields, and lower yields as the result of decreased humus and increased soil acidity. Pathogens in both the soil and crops are also becoming problems in chemical no-till, leading to the need for additional fungicides and pesticides.

A unique, biological approach to no-till farming has been proposed by Wes Jackson, director of The Land Institute in Salina, Kansas. In his book, *New Roots for Agriculture*, Jackson outlines a breeding program for the development of high yielding perennial grains. Jackson points out that all of our major grain crops are grown as annuals, including wheat, oats, corn, barley, rye, sorghum, rice, and millet. Yet, he says, these grains are all grasses closely related to perennials which have never been developed as food. In his book Jackson was quite conservative in predicting the advent of perennial grains, stating that their development might take fifty years or more. But recent work with agrotriticum, a cross between wheat and perennial wheatgrass, suggests that perennial grains suitable for commercial production could be developed within a decade. (See page 50).

Growing a permanent groundcover to act as a living mulch, and then planting right into it, is another no-till approach with potential for the Northwest. The most sophisticated example of this method is the grain rotation developed in Japan by Masanobu Fukuoka. In much the same way that Fukuoka learned to grow vegetables in a semi-wild fashion in his hillside orchard (see page 82), he also sought to devise a high yielding method for growing rice, wheat and barley that does not require turning the soil. After years of testing various legumes, Fukuoka concluded that low-growing, mat-forming white clovers were best for forcing out weeds and enriching the soil.

RODALE RESEARCH REPORTS

The Rodale Research Center is conducting an ongoing series of controlled experiments on minimum-tillage techniques for organic farmers. The following research reports are available for $1 each from the Rodale Research Center, RD 1, Box 323, Kutztown, PA 19530.

Ganser, Stephen. Vegetable Production in a Living Sod. Report 80-2, 1980.
Report on the planting of tomatoes, sweet corn and winter squash directly in legume and grass sods. "Crops in a mulched, unplowed grass sod gave consistently higher yields than did those planted in clean-cultivated fields."

Hofstetter, Robert. Legume Sod Interplanting. Report 80-10, 1980.
Report on first two years of experiments with corn and soybeans double row cut in an alfalfa sod.

Hofstetter, Robert, and Palada, Manuel C. Sod Interplanting with Corn and Soybean. Report 81-31, 1981.
This report includes a good overview of approaches to planting in a living sod, with 29 references for further research.

Palada, Manuel C. et al. Overseeding Legume Cover Crops on Corn and Soybean. Report 81-29, 1981.
First-year results with sowing red clover, white clover, crimson clover, and hairy vetch under established stands of corn and soybeans for soil conservation and improvement.

Volak, Barney, et. al. Productivity of Corn-Alfalfa Strip Intercropping System. Report 81-30, 1981.
Trials on different width tillage strips for sowing corn in alfalfa sod.

CORNELL RESEARCH

Anon. "The Concept of a Living Mulch." American Vegetable Grower, September, 1981, pp 38-46, 61.
Excellent overview of minimum-tillage research underway at Cornell.

Vrable, T.E.; Minotti, P.L.; and Sweet, R.D. Seeded Legumes as Living Mulches in Sweet Corn. Paper No. 769. Vegetable Crops Department, Cornell University, Ithaca, NY 14850.
Report on the effect of timing in the sowing of legumes under corn.

Using techniques described in *The One-Straw Revolution*, a succession of summer and winter grains is sown directly into the living clover mulch. After the rice and other grains are harvested, all of the straw is returned to the surface where it helps control weeds, retain moisture, build humus, and feed soil microorganisms. Ducks are occasionally allowed into the fields for weeding and manuring, and a little chicken manure is scattered on the fresh straw to aid in decomposition.

Following this pattern, Fukuoka has been able to achieve yields of rice, wheat and barley equal to the highest in Japan, all with hand labor in fields that have not been cultivated for thirty years, and without the use of chemical fertilizers or pesticides. Weeds are controlled and fertility maintained by crop rotations, returning straw to the field, and the living white clover ground cover. Only the grain is removed from the fields. All other nutrients are continually recycled.

In North America, controlled experiments on organic approaches to minimum tillage farming are currently being conducted by the Rodale Experimental Farm in Pennsylvania and by Cornell University in New York. Rodale has conducted trials on growing corn and soybeans in strips tilled into fields of alfalfa as well as trials on tomatoes, corn and winter squash seeded directly into grass and red clover sods. Preliminary trials showed that vegetables planted in holes cut into grass sod and mulched heavily with fresh alfalfa clippings consistently achieved higher yields than controls planted in clean-cultivated fields.

Similar work is being conducted at Cornell. Motivated by the urgent need to reduce soil erosion, improve soil structure, reduce leaching and increase organic matter, researchers are trying a wide range of techniques for growing vegetables in living mulch. They hope that, in this way, soils can be preserved and improved without taking land out of production. Cornell researchers are screening both legumes and grasses for possible use as living mulches when sown between rows of cash crops such as sweet corn, cabbage, beans and beets. In establishing living mulches in new fields they have found it best to sow the grasses or legumes five weeks after planting the main crop. Vegetation in the paths is then controlled either with herbicides or by mowing.

Tests have shown that frequent mowing early in the growing season provides control equal to the use of herbicides. Once the mulches are established they can either be disced in or left as a permanent cover which is then rototilled with 12 inch wide strips for seeding subsequent crops. Out of the 50 grasses and 50 legumes being screened at Cornell, low-growing white clovers have proven most promising. This, of course, is the same conclusion that Masanobu Fukuoka came to after years of observation and experience—white clovers provide the best living mulch for no or minimum-till grain and vegetable production.

FOR MORE INFORMATION

Holzhauer, Charles. "Minimum Till—Coming on Fast." Oregon Farmer-Stockman 105 (May 20, 1982): 6.

Interview with Dr. Norm Goetze, OSU agronomist, active in development of minimum-till systems. He reviews techniques, equipment, problems encountered, and research needs.

Phillips, S.H., and Young, H.M., Jr. No-Tillage Farming. Milwaukee, WI: Reiman Associates, 1973.

The original text on no-till methods, with emphasis on use of herbicides. Includes a review of minimum till, strip till, and living mulch strategies.

EQUIPMENT

A number of stubble drills are now on the market. Two recommended by Dr. Goetze are the Melroe (Clark Co., Agricultural Division, Bismark, ND 58501), and the Haybuster (Box 1950, Jamestown, ND 59401). John Deere (PO Box 20098, Portland, OR 97220), manufactures a pasture-renovating drill which can be modified for direct stubble sowing. Cole Manufacturing (see below) also produces minimum-tillage equipment.

Cole Manufacturing Company
1318 Central Avenue
Charlotte, NC 28299

Manufacturers of a mulch planting and cultivation system that makes minimum tillage farming possible without herbicides. A specially-adapted subsoiler is used to slice through stubble or cover crop and sow seed in clean soil. Then, as the crop matures, a mulch cultivator spreads composted crop remains around the plants to control weeds and retain moisture.

The Cole system is described in detail by Fred Zahradnik in an article entitled "New Machines Turn Trash to Cash" in the May/June, 1982 issue of The New Farm, pages 25-27.

Weeds

Most of the plants we commonly refer to as "weeds" are not native to the Pacific Northwest. They are immigrants from Eurasia, the "Old World," brought here by good intentions or by accident. These plants are generally fast growing, tough competitors, and will grow in a wide range of soils. Most of them seed prolifically and some have pernicious underground rhizomes. Some of our worst weeds include thistles, tansy ragwort, knapweed, bindweed, quackgrass, puncture vine, wild oats, Kentucky bluegrass (actually native to the Old World), cheatgrass and medusa-head wild rye.

Few people saw the impact that exotic weeds can have on native plant communities as clearly as Aldo Leopold. Below, in an excerpt from A Sand County Almanac, *is an essay Leopold wrote following a visit to Utah and eastern Oregon.*

BIBLIOGRAPHY

Altieri, M.A. Ecology and Management of Weed Populations. Available for $2.50 from Memler, 3287 Dunsmere Road, Glendale, CA 91206.
The role of weeds in agro-ecosystems, with biological control strategies.

Bradley, Joan. Brush Regeneration. 1971. Mosman Parklands and Ashton Park Association, PO Box 164, Spit Junction, New South Wales, Australia 2088.
Practical techniques for the elimination of introduced weeds from native plant communities. Emphasis is on creating conditions which foster spontaneous regeneration of native plants.

Fryer, J.D., and Matsunada, Shooichi. Integrated Control of Weeds. Tokyo: University of Tokyo Press, 1977. (Distributed by International Scholarly Book Service.)
Proceedings of an international workshop on terrestrial and aquatic weed control strategies which integrate cultivation, crop management, and natural enemies.

Gaines, Xerphy M., and Swan, D.G. Weeds of Eastern Washington and Adjacent Areas. 1972. Available from Alice Peterson, Rt 1, Box 59, Edwall, WA 99008. $6.95
Botanical guide to 150 of the most commonly found weeds of the interior Northwest. Color illustrations.

McLeod, Edwin J., and Swezey, S.L. Survey of Weed Problems and Management Technologies in Organic Agriculture. 1979. Organic Agriculture Research Institute, Box 475, Graton, CA 95444.
Review of published literature on biological, cultural, genetic, thermal and mechanical control methods. Includes 113 references.

Muenscher, Walter Conrad, and Hyypio, Peter A. Weeds. Ithaca, NY: Cornell University Press, 1980.
Revised edition of a classic text on weed management. Includes practical details on mechanical control.

Story, Jim. Biological Weed Control in Montana. Bozeman: Montana State University, 1979.
Survey of biological weed management research and experiments.

Cheat Takes Over

ALDO LEOPOLD

Just as there is honor among thieves, so there is solidarity and co-operation among plant and animal pests. Where one pest is stopped by natural barriers, another arrives to breach the same wall by a new approach. In the end every region and every resource get their quota of uninvited ecological guests.

Thus the English sparrow, rendered innocuous by the shrinkage in horses, was succeeded by the starling, who thrives in the wake of tractors. The chestnut blight, which had no passport beyond the west boundary of chestnuts, is being followed by the Dutch elm disease, with every chance of spreading to the west boundary of elms. The white-pine blister rust, stopped in its westward march by the treeless plains, effected a new landing via the back door, and is now romping down the Rockies from Idaho toward California.

Ecological stowaways began to arrive with the earliest settlements. The Swedish botanist, Peter Kalm, found most of the European weeds established in New Jersey and New York as early as 1750. They spread as rapidly as the settler's plow could prepare a suitable seedbed.

Others arrived later, from the West, and found thousands of square miles of ready-made seedbed prepared by the trampling hoofs of range livestock. In such cases the spread was often so rapid as to escape recording; one simply woke up one fine spring to find the range dominated by a new weed. A notable instance was the invasion of the intermountain and northwestern foothills by downy chess or cheat grass *(Bromus tectorum)*.

Lest you gain too optimistic an impression of this new ingredient of the melting pot, let me say that cheat is not a grass in the sense of forming a live sod. It is an annual weed of the grass family, like foxtail or crabgrass, dying each fall and reseeding that fall or the next spring. In Europe its habitat is the decaying straw of thatched roofs. The Latin word for roof is *tectum*, hence the label 'Brome of the roofs.' A plant that can make a living on the roof of a house can also thrive on this rich but arid roof of the continent.

Today the honey-colored hills that flank the northwestern mountains derive their hue not from the rich and useful bunchgrass and wheatgrass which once covered them, but from the inferior cheat which has replaced these native grasses. The motorist who exclaims about the flowing contours that lead his eye upward to far summits is unaware of this substitution. It does not occur to him that hills, too, cover ruined complexions with ecological face powder.

The cause of the substitution is overgrazing. When the too-great herds and flocks chewed and trampled the hide off the foothills, something had to cover the raw eroding earth. Cheat did.

Cheat grows in dense stands, and each stem bears a mass of prickly awns which render the mature plant inedible to stock. To appreciate the predicament of a cow trying to eat mature cheat, try walking through

Weed With Geese

The use of weeder geese is expanding rapidly. Properly managed, they have been used for weeding orchards, nurseries, and field crops such as strawberries, raspberries, beets, potatoes, onions, garlic, carrots, and mint. Geese forage between the rows, eating most weeds, but not the crop.

Weeder geese eat most immature grasses including Bermuda grass *(Cynodon dactylon)* and Johnsongrass *(Sorghum halepense)*. They also eat young sedge and nut grass *(Cyperus* spp.*)*, quackgrass *(Agropyron repens)* and puncture vine *(Tribulus terrestris)*. They will devour clover and chickweed, but hardly touch broadleaf weeds. Geese remove grass and weeds close to plants which cannot be removed by cultivation. They also keep fencerows and irrigation ditches clean and can work fields which are too wet to cultivate.

Chinese geese, both Brown and White, are the geese most often used as weeders. Goslings should be started as weeders at six weeks of age. Introduce weeders to the field while the weeds are still young and succulent. The geese should forage during the day, but must be fed a supplement at night. The amount and composition of the supplement is determined by the amount and species of weeds available to the

geese. If they are feeding on horsetail, for example, a weed which has little nutritional value for geese, a large, relatively rich supplement would be necessary. Shade and drinking water must also be provided, and should be moved regularly for effective ranging.

The number of weeders needed on a field depends on the crop, the weeds and the size of the field. A one-acre field of strawberries, for example, requires at least six geese. Geese can be left on the fields as long as there are weeds to control or until the geese would start damaging mature crops.

BIBLIOGRAPHY

Geiger, Glenn, and Bieller, Harold. Weeding With Geese. Poultry Extension Bldg T-14, Rm 114, University of Missouri, Columbia, MO 65211.
Brief, informative introduction.

Johnson, Clarence. Management of Weeder Geese in Commercial Crops. U of C Extension Service, 128 Madera Avenue, Madera, CA 93637.
Includes an economic analysis of maintaining 120 weeder geese on a 40 acre field of cotton. Result: an annual cost per acre of $13.37, as against estimated returns of $50 per acre.

HATCHERIES

MPI Hatchery, PO Box 7611, Olympia, WA, 98507.
Both brown and white Chinese weeders.

Holderread's Hatchery, 3810 NE Hwy 20, Corvallis, OR 97330.
Brown Chinese weeder geese.

---

it in low shoes. All field workers in cheat country wear high boots. Nylons are here relegated to running boards and concrete sidewalks.

These prickly awns cover the autumn hills with a yellow blanket as inflammable as cotton-wool. It is impossible fully to protect cheat country from fire. As a consequence, the remnants of good browse plants, such as sagebrush and bitterbrush, are being burned back to higher altitudes, where they are less useful as winter forage. The lower fringes of pine timber, needed as winter cover for deer and birds, are likewise being singed back to higher levels.

To a summer tourist, the burning of a few bushes off the foothills may seem a minor loss. He is unaware that, in winter, snow excludes both livestock and game from the higher mountains. Livestock can be fed on valley ranches, but deer and elk must find food in the foothills or starve. The habitable wintering belt is narrow, and the further north one goes, the greater is the disparity between the area of habitable winter range and the area of summer range. Hence these scattering foothill clumps of bitterbrush, sage, and oak, now fast shrinking under the onslaught of cheat fires, are the key to wildlife survival in the whole region. Besides, these scattered bushes often harbor, under their mechanical protection, remnants of native perennial grasses. When the bushes are burned off, these grass remnants succumb to livestock. While the sportsmen and stockmen wrangle over who should move first in easing the burden on the winter range, cheat grass is leaving less and less winter range to wrangle about.

Cheat gives rise to many minor irritations, most of them less important, perhaps, than starving deer or cheat-sores in a cow's mouth, but still worth mentioning. Cheat invades old alfalfa fields and degrades the hay. It blockades newly hatched ducklings from making the vital trek from upland nest to lowland water. It invades the lower fringe of lumber areas, where it chokes out seedling pines and threatens older reproduction with the danger of quick fire.

I experienced a minor irritation myself when I arrived at a 'port of entry' on the northern California border, where my car and baggage were searched by a quarantine officer. He explained politely that California welcomes tourists, but that she must make sure their baggage harbors no plant or animal pests. I asked him what pests. He recited a long list of prospective garden and orchard afflictions, but he did not mention the yellow blanket of cheat, which already extended from his feet to the far hills in every direction.

As is true of the carp, the starling, and the Russian thistle, the cheat-afflicted regions make a virtue of necessity and find the invader useful. Newly sprouted cheat is good forage while it lasts; like as not the lamb chop you ate for lunch was nurtured on cheat during the tender days of spring. Cheat reduces the erosion that would otherwise follow the overgrazing that admitted cheat. (This ecological ring-around-the-rosy merits long thought.)

I listened carefully for clues whether the West has accepted cheat as a necessary evil, to be lived with until kingdom come, or whether it regards cheat as a challenge to rectify its past errors in land-use. I found the hopeless attitude almost universal. There is, as yet, no sense of pride in the husbandry of wild plants and animals, no sense of shame in the proprietorship of a sick landscape. We tilt windmills in behalf of conservation in convention halls and editorial offices, but on the back forty we disclaim even owning a lance.

Wind Shelter

MICHAEL PILARSKI

Shelterbelts and windbreaks are rows of trees and shrubs which break the force of the wind and create protected areas behind them. Hedgerows are in effect low shelterbelts, usually made up only of shrubs. All three are used to limit wind erosion on exposed sites, control drifting snow, fence livestock, and provide fuelwood, poles, posts, wildlife habitat, bee forage, food and fodder. They are also beautiful.

A useful distinction can be drawn between permeable windbreaks and dense shelterbelts. Windbreaks, with 40-60% permeability to wind, provide protection over a greater distance to leeward than a dense shelterbelt because the air filtering through the trees holds the faster moving air aloft. This type of windbreak, usually one or two rows of trees with a shrub understory, is useful for protecting cropland and orchards.

Shelterbelts, on the other hand, are comprised of two or more rows of trees. They provide a greater reduction of windspeed than do windbreaks, but the protected zone is smaller. Because they are solid barriers, they create a zone of calm close to the trees, and a zone of turbulence further downwind. They are used to provide a high degree of shelter in a limited area, such as near barns, homes, livestock pens and equipment yards.

Hedgerows protect crops from damaging winds by blocking and slowing air movement. Close to the leeward side of the hedge, however, air movement can be so slow that frost pockets form. Plants growing there may be damaged by a late spring frost. This can be avoided by thinning the hedge or pruning up the lower branches to allow cold air drainage. Conventional fencing can fill in large gaps as necessary.

DESIGNING WINDBREAKS AND SHELTERBELTS

On relatively level ground windbreaks should be perpendicular to the damaging winds. The windward side of the belt begins with shrubs, followed by medium height and then tall trees. Thus the air flow rises over the belt, rather than confronting it head-on.

Where topography is broken or rolling, windbreaks are placed to take advantage of natural features. For instance, windbreaks planted on ridge crests protect a much larger area than those planted on midslopes or at the base of slopes. Conditions on ridgetops are often too tough for young plants to survive, but planting seedlings just to the leeward side of the ridge allows them to become established before they are exposed to the full force of the wind.

Shelterbelts are planted along the edges of gullies, ravines, streams and permanent roads and along the borders of fields and pastures. The efficiency of

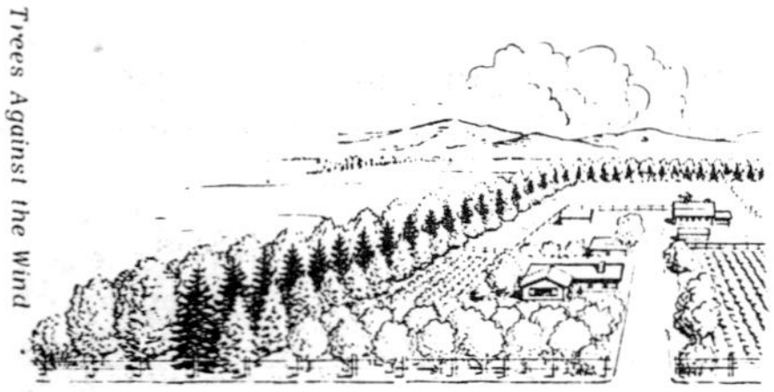

Trees Against the Wind

BIBLIOGRAPHY

Albenskii, A.V. and Nitikin, P.D., eds. Handbook of Afforestation and Soil Melioration. Jerusalem: Israel Program for Scientific Translations, 1967.
Translated from Russian. Design, planting and maintenance of shelterbelts, including a chapter on mixing tree and shrub species. Erosion diagnosis and treatment. Land reclamation by trees. Useful information for the Northwest.

Burlison, Vernon H. Trees Against the Wind. Pacific Northwest Cooperative Extension Bulletin No. 5. January 1975.
Selection of tree and shrub species and windbreak establishment, written specifically for the interior Northwest.

Caborn, J.M. Shelterbelts and Windbreaks. London: Faber & Faber, 1965.
Practical advice on wind shelter design, establishment and maintenance.

Hoag, Donald G. Trees and Shrubs for the Northern Plains. Fargo, ND: North Dakota Institute for Regional Studies, 1965.
Valuable information for landowners in the colder parts of the interior Northwest. Especially good data on windbreak plants.

ITCI. The Windbreak Slide Show. International Tree Crops Institute, Box 666, Winters, CA 95694.
Audio-visual presentation on windbreak design and species selection. Produced as a companion to the California Windbreak Handbook, a detailed manual on windbreak planting, available from the California Association of Resource Conservation Districts, 2103 Stockton Blvd., Sacramento, CA 95817.

shelterbelts can be increased by the addition of hedgerows between the belts, or by planting strips of tall plants such as corn, sunflowers, Jerusalem artichokes or bunchgrasses.

Planting at high densities may be a costly investment but will give protection within a few years. With some species the thinnings can be used for posts, poles and firewood. Planting density can be higher in humid or irrigated locations than in arid areas.

SPECIES FOR WINDBREAKS AND SHELTERBELTS

Native and non-native species of trees and shrubs are used in shelterbelt designs for both diversity and increased productivity. Usually a mixture of deciduous and evergreen species is best.

In some countries, apples, pears, apricots, mountain ash, filberts, *Prunus* species, elderberry and Cornelian cherry dogwood are widely used in shelterbelts. Other hardwoods which might be included in the Northwest are oak, maple, black locust, black walnut, Russian mulberry and ash. Nitrogen-fixing species are also useful in shelterbelts since they can be established on poor sites and prepare the ground for other species. Black locust *(Robinia pseudoacacia)*, Russian olive *(Eleagnus angustifolia)*, and Siberian peashrub *(Caragana arborescens)*, for example, are nitrogen-fixers, tolerate a wide range of soil conditions, withstand drought, and are excellent sources of food for wildlife. The Siberian peashrub is also outstanding bee forage.

DESIGNING HEDGEROWS

Hedgerows are used to fence livestock in many parts of the world. Some hedgerow plants are thorny and stock will leave them alone, but most thorny plants will be cautiously grazed by sheep, horses and goats. If only one side of the hedgerow is exposed to stock, fodder shrubs can be planted on the far side of the hedgerow and periodically trimmed for feed. If planted 3 to 5 feet in back of a fence, the hedgerow will overtake it and be stout enough to resist livestock in 4 to 6 years. Animals also benefit from the hedgerow's protection from hot summer sun and cold winter winds. This protection can be especially valuable during lambing or calving time.

In *The Guide to Self-Sufficiency*, John Seymour provides a list of shrubs and small trees used in traditional stock-proof hedgerows. Included were crabapple, elderberry, heather, holly, honeysuckle, rosa rugosa, quince, medlar, whitebeam (*Sorbus moravica*), and service tree (*Sorbus domestica*). Food bearing plants for humans and wildlife, such as filberts and berries, can also be incorporated in the hedgerow, along with plants which will encourage beneficial insects.

Binda Colebrook includes a brief discussion of hedgerows which provide stock feed in *Winter Gardening in the Maritime Northwest*. Plants she notes for providing winter food for livestock include ash, elm, heather, holly, ivy, and willow.

Logginov, B.I. Principles of Field Protective Forestation. Jerusalem: Israel Program for Scientific Translations, 1961.

Shelterbelt design in the USSR. Stresses multi-story belts with the use of fruit and nut trees. Includes North American trees such as honey locust.

Ministry of Agriculture. "Farm Hedges." In The Farmstead Book, edited by Paul Harmond. Mayne Island, BC: Cloudburst, 1978.

Guide to placement, species selection, establishment, and maintenance of hedges.

Robinette, G.O., ed. Landscape Planning for Energy Conservation. Environmental Design Press, Box 2187, Reston, VA 22090.

Landscape architects become energy planners by using plants for shade, insulation, windbreaks, etc. Environmental Design Press also has an extensive catalog of publications.

Small Farm Energy Project. "Windbreaks Increase Crop Yields." In Small Farm Energy, Box 736, Hartington, NE 68739. May/June, 1982, pp 4-6.

Report on long-range studies on the effect of windbreaks on crop yields conducted by the University of Nebraska's Department of Forestry (101 Plant Industry, University of Nebraska, Lincoln, NE 68583). Fifty plots were established in 1966 to test various tree protection systems to moderate environmental extremes and improve water-use efficiency.

The results have been a dramatic reduction in winter kill in wheat and substantial yield improvements in wheat, soybeans and corn. For example, a $52 per acre increase in income was returned for winter wheat in protected areas over unprotected areas. Soybean yields were 20 to 25 percent higher and dryland corn yields were 450 percent higher on protected areas versus unprotected areas.

Cedar and pine as windbreaks were shown to cause the least moisture competition. Work is also underway on the development of windbreak/woodlot rotations for the coppicing of cottonwood and other hardwoods for cattle feed, bedding and firewood. The harvesting of fruit and fuel tree crops is also seen as a potentially viable alternative to conventional Midwest corn, milo, wheat, and soybean crops.

Thomas, Eric, and White, John T. Hedgerow. New York: Morrow, 1980.

Beautifully illustrated history of hedgerows, with details on their establishment and value.

Grassland Agriculture

MARK MUSICK

Following the ravages of the Dust Bowl and all-out production during World War II, there was great public concern over the need for a sustainable agriculture. In *Grass*, the 1948 USDA Yearbook, Clinton P. Anderson, then Secretary of Agriculture, stressed the need to expand grassland farming, which, he said, is "the foundation of security in agriculture."

Anderson's emphasis on grasses was a recognition of their essential role in building and maintaining soil fertility. "Grassland agriculture" is a phrase that includes all of the grass family and associated legumes—clover, vetch, sainfoin, sweetclover, peas, trefoil, alfalfa, and subterranean clover. All of these plants are well-adapted to various parts of the Northwest. Subclover, for example, has proven to be of special value for non-irrigated pasture lands in southwestern Oregon, where it is planted on over 150,000 acres.

Together, grasses and legumes contribute more organic matter to the soil than do any other crops. Through the action of soil microorganisms, organic matter is transformed into humus, which is essential for soil structure and sustained fertility. Grasslands can be used exclusively for livestock production, they can be integrated in long-term crop rotations, or they can provide the understory in mixed orchards and forests.

The incorporation of grasses and legumes in crop rotations is especially crucial in cultivated fields. As soil is taken out of sod and devoted to cultivated crops, the reserve of fertility is drawn down by each succeeding stage of the rotation. Only a careful recycling of nutrients and the periodic return of the land to grass and legume sod can truly rebuild humus and maintain long-term fertility.

ANIMAL LANDSCAPES

In England temporary grassland incorporated into a rotation is called a ley. Leys are best utilized on diversified farms where there is livestock to take advantage of the forage provided. Newman Turner's *Fertility Pastures* is an eloquent account of the role of what he calls "herbal leys" in dairy farming. Turner's book is a description of a truly biological farm, with grass, clover and cows forming the core of the fertility cycle.

Turner provides detailed information on establishing and maintaining temporary pastures in 3 to 5 year rotations with cultivated fodder crops. He emphasizes the use of crops where they are being grown, either by grazing or by making silage right in the field where it will be eaten. His goal is to maximize in-field nutrient cycling.

BIBLIOGRAPHY

Hafenrichter, A.L., et al. Grasses and Legumes for Soil Conservation in the Pacific Northwest and Great Basin States. US Soil Conservation Service Agricultural Handbook No. 339. Washington DC: USGPO, 1979.
Detailed catalog of grasses and legumes for our region, with specific information on zones of adaptation, uses, and methods of establishment and maintenance. The result of 32 years of research, this is an excellent starting point for identifying which grasses and legumes are best suited to your location.

Heath, M.E., at al. Forages: The Science of Grassland Agriculture. Ames, IA: Iowa State University Press, 1973.
Foremost text on forages and grassland farming. Includes detailed information on grass species and pasture establishment and maintenance.

Lewis, Rebecca. Making Pastures Produce. Emmaus, PA: Rodale Press, 1980.
In 43 pages, this booklet presents a summary of strategies and plant varieties useful for the restoration and maintenance of pastures. Especially useful is a "Calendar of Forage Production," which shows the grass and legume species available for extended forage production. This booklet is available without charge from Rodale's Reader Service Department.

McGuire, W.S., et al. Forage Varieties for Oregon. Circular 617. Corvallis, OR: OSU Agricultural Experiment Station, 1970.
Suggested forage species and varieties for Oregon are based on results of experiments and demonstrations for a multitude of agricultural research projects throughout Oregon. Variety recommendations are arranged by region, making this booklet useful for other parts of the Pacific Northwest.

Robson, John R.K., and Elias, Joel N. The Nutritional Value of Indigenous Wild Plants-An Annotated Bibliography. Troy, NY: Whitston, 1978.
This bibliography contains 796 entries, predominantly of articles and monographs published in scientific journals. It is especially useful for people who wish to increase fodder sources with

Turner uses conventional grasses such as timothy, orchard grass, meadow fescue and perennial ryegrass in his pasture mix. But he recommends adding to these chicory, yarrow, plantain, burnet, and sheep's parsley. He includes these herbs in the ley because they are deep rooted and concentrate a wider range of minerals than do grass and clover. They also provide browse in winter and early spring, the most stressful times of the year for stock.

Special fodder crops and winter gardens can be planted to increase livestock food supply and extend the forage season. Crops useful for this purpose include marrowstem kale, thousand-headed kale, mangle beet, sugar beet, turnip, rutabaga, carrot, forage radish, Swiss chard, Jerusalem artichoke, and comfrey. They can be planted directly in the ley, along hedgerows, or in separate fields with the animals turned in to harvest them when needed.

Animal landscapes can be expanded by establishing hedges around and across pastures. Hedgerows can include livestock forage as well as insectary plants and food and shelter for wildlife. Species which bloom and fruit over an extended period are especially useful.

With careful management, mixed pastures can also be established beneath orchard and forest trees. Honey locust and walnut have fairly open canopies which allow shrubs and pasture to grow below them. Other trees which produce fruit or nuts of good forage value can also be used, including hawthorn, oak, chestnut, filbert, mulberry, persimmon, apple, pear, plum, and cherry. Trees with foliage edible by livestock include alder, willow, hawthorn, chestnut, linden, poplar, and maple. Nitrogen-fixing trees and shrubs, such as red alder and Siberian peashrub, can be scattered in or planted around a pasture to fertilize the grass and adjacent plantings.

Orchardgrass *(Dactylis glomerata)*

Successful management techniques have also been developed for the grazing of livestock in forests. Combined tree farm and pasture systems, known as silvipasture, are catching on in Australia and New Zealand. In these systems sheep and cattle are grazed in plantations of poplar and pine. Reporting in the 1976 edition of *Tree Farmer*, Mary and Geoff Wilson noted that advantages to pasture from planting trees, especially wind protection and higher soil moisture, were evident "even one year after planting." Advantages to farmers and ranchers included quick returns of fenceposts and firewood, with high quality saw-logs available in 15 to 20 years. In the Northwest, experiments have led to the development of successful techniques for grazing sheep and cattle in Douglas-fir, ponderosa pine and juniper forests.

low maintenance trees and shrubs. Some of the studies cited describe plants edible by humans.

Semple, Arthur T. Grassland Improvement. Cleveland, OH: CRC Press, 1970.
Establishment and management of pastures, including information on the use of trees and shrubs for grazing.

Taylor, Tim. "Upgrade Your Pastures By Interseeding Legumes" In The New Farm, September/October, 1979, pp 22-25.
This article describes techniques for grassland renovation by interseeding legumes in an established sod.

Turner, Newman. Fertility Pastures and Cover Crops. 1974. Available from Rateaver, Pauma Valley, CA 92061.
Turner's farming techniques represent the best of English biological agriculture. His book is a valuable guide for establishing and maintaining herbal leys.

USDA. Grass, Yearbook of Agriculture. Washington, DC: USDA, 1948.
One of the best USDA yearbooks. Makes the case for a permanent agriculture based on grassland farming. Essays on grass and legume species for all the major agricultural regions of the United States. Still available.

Wilson, Geoff, and Wilson, Mary, eds. Tree Farmer. Victoria, Australia. Published in three issues–1976, 1977, and 1978. Available from International Tree Crops Institute, Box 666, Winters, CA 95694.
The Wilsons have compiled excellent resource manuals on the integration of livestock with forestry. Grazing has long been practiced in forests, but the conscious planting of trees in pastures and rangelands to produce timber and other tree products while still maintaining a productive livestock operation is a new concept. These manuals describe successful silvipasture systems in Australia and New Zealand, which increase livestock production and yield commercial wood products.

FENCING

The companies listed below offer a full range of coyote- and dog-proof fencing systems.

Action Fence Company, Rt 2, Box 86, Eagle Creek, OR 97022.

Doug Bailey, 14458 168th Ave NE, Woodinville, WA 98072.

Rangelands

Virtually all of the shrub and grasslands of the Interior have been used for livestock grazing. The livestock boom of the open range era from 1860 until 1920 was a period of severe degradation of the rangeland. The resultant soil erosion and invasion of non-native weeds makes it difficult to restore range productivity on many sites. Today, the native bunchgrasses have been nearly eliminated, and almost all rangelands in the Interior are dominated by exotic grasses and weeds.

Appropriate stocking rates combined with the reseeding of rangelands with desirable grasses, forbs, and trees have the potential of restoring the health of grazed ecosystems to a point where livestock production can be increased, along with other resources such as wildlife and timber.

Restoration of our rangelands is one of the greatest challenges facing the Pacific Northwest.

Bedell, T., et. al. Oregon Forage and Livestock References. Special Report 479. Corvallis, OR: OSU Livestock Research and Extension Program, 1977.

This publication lists hundreds of articles, government reports and pamphlets with relevance to rangeland management in Oregon. Many of them will also be applicable to other parts of the Pacific Northwest.

Dayton, W.A., et al. Range Plant Handbook. Washington, DC: USDA, 1937. Reprint PB 168 589. Available from National Tech. Info. Serv., US Dept. of Commerce, Springfield, VA 22161.

Though published in 1937 this book still remains the definitive text on the range plants of the United States. Over 800 range plants are covered. The book contains many excellent line drawings of the plants, along with detailed information on each species' forage value, palatability, occurrence, flowering period, habits, and tolerances. Poisonous range plants are also covered.

Plummer, A.P., et al. Highlights, Results, and Accomplishments of Game Range Restoration Studies. Pub. 70-3. Ogden, UT: Utah Division of Fish and Game, 1970.

The results of many years of research and trials. Though its primary emphasis is on big game, the ecosystem restoration techniques are applicable to livestock management also. The book goes into detail on how to harvest seeds and establish desired plants in the intermountain region.

Rocke, Ben. Range Multiple Use Management. Pullman, WA: Extension Service, 1975.

The philosophies and strategies for multiple use of rangeland are covered by 17 authors. Included are the integration of livestock grazing with silviculture, recreation, wildlife, and soil and water conservation. Plant botany and grasses and legumes for improving rangelands are included. Published by the Washington, Idaho, and Oregon Extension Services.

Trappe, James M. "Mycorrhizae and Productivity of Arid and Semi Arid Rangelands." In Advances in Food Producing Systems for Arid and Semi Arid Lands. New York: Academic Press, 1981: pp 581-599.

Review of the vital role of soil microorganisms in rangeland improvement. Mycorrhiza management can be used to restore eroded, overgrazed, and otherwise degraded lands.

Vallentine, John F. Range Development and Improvements. 2nd ed. Provo, UT: Brigham Young University Press, 1980.

Synthesis of research on rangeland improvement. One of the book's best features is a table including 89 grasses and legumes commonly used for enhancing rangelands. The table details each plant's ease of establishment, drought tolerance, cold hardiness, salinity tolerance, soil tolerances, period of forage usability, grazing tolerance, season of growth, principal cultivars and special considerations and adaptations.

Native shrubs interplanted with grasses on semiarid seeded range can improve the quality of forage consumed by livestock, particularly in fall and winter, and can improve the availability of forage during years of drought or dry snows....

Native shrub species showing greatest promise for revegetation in the Intermountain area have included bitterbrush, cliffrose, winterfat, the mountain mahoganies, selected varieties of big sagebrush, four-wing saltbush, and other saltbush species.

——, ed. United States-Canadian Range Management, 1935-1977: A Selected Bibliography on Ranges, Pastures, Wildlife, Livestock and Ranching. Phoenix, AZ: Oryx Press, 1978.

One of the most extensive bibliographies on range management. Several thousand entries are listed by subject and author index.

Wright, Henry A.; Neuenschwander, Leon F.; and Britton, Carlton M. The Role and Use of Fire in Sagebrush-Grass and Pinyon Juniper Plant Communities: A State-of-the-Art Review. USDA FS Gen Tech Rep INT-58. Ogden, UT: Intermt. For & Range Exp Stn, 1979.

This report gives guidelines for conducting prescribed burns in sagebrush-grass and pinyon-juniper communities. The effects of fire on grasses, forbs, and shrubs are covered in detail. Fire is only one factor in the restoration of range productivity and must be coupled with balanced grazing practices and reseeding. Over 50 references given.

PERIODICALS

Journal of Range Management, and Rangelands
2670 W. Fifth Avenue
Denver, CO 80204

Both of these journals are published by the Society for Range Management. The back issues of the Journal of Range Management contain a wealth of technical detail. Rangelands is its nontechnical counterpart.

Arid Lands

The words Pacific Northwest bring visions of rainy skies and dense rainforests, but it is equally correct to imagine dry, sagebrush landscapes, extensive irrigation systems, dryland farming and juniper savannas. In fact, as the accompanying map illustrates, nearly half of the region is arid, receiving less than 20 inches of annual precipitation.

Desertification is of increasing concern around the world, and its symptoms are becoming evident even in the Northwest. In Desertification of the United States, David Sheridan lists the following symptoms of desertification—declining ground water tables, salinization of topsoil and water, reduction of surface waters, unnaturally high soil erosion, and the destruction of native vegetation. "An arid area suffering from desertification can manifest all five symptoms," he notes, "but the existence of any one can mean that it is undergoing desertification." In essence, desertification means a loss of the land's capacity to support life, and by that criterion large areas of the Northwest are threatened by spreading deserts.

Rising energy costs, severe soil erosion, and increased competition for water from the Columbia and Snake Rivers threaten the future of large-scale irrigation projects. Yet with proper husbandry, the arid lands of the Interior have the potential to support sustainable communities. In his article entitled "Runoff Farming for Dry Lands" (The New Farm, May/June 1981), Gary Nabhan describes some ancient desert farming techniques which may prove useful to dryland farmers today.

As Gary notes, the major water supply in arid and semi-arid areas is storm runoff. Many sophisticated systems have been developed for utilizing this water. Most have three parts: the catchment area, the storage area, and the planting area. A catchment can be a large, natural watershed or a smaller area treated or compacted to increase runoff. Storage areas can range from ponds or tanks to the soil of the area to be planted.

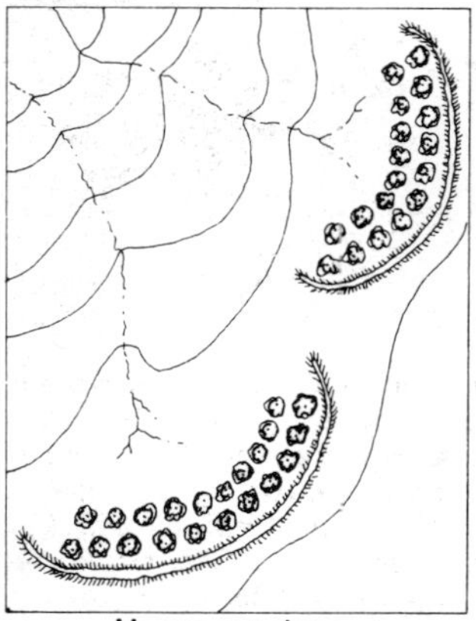

Imagination

Macrocatchment

Macro- and micro-catchments are both practical possibilities for dryland agriculturists. A micro-catchment may be an area compacted uphill of a single plant, such as a shrub. Surrounded by a berm, the catchment funnels water toward a basin around the shrub. A macro-catchment enlarges upon this basic technique (see above).

Gary concludes, "Systems differ in cost, size and efficiency, but...all are within the means of small farmers in the West who have basic tillage implements....Runoff farming, an ancient technique in arid lands, is bound to play an important role again for future desert dwellers."

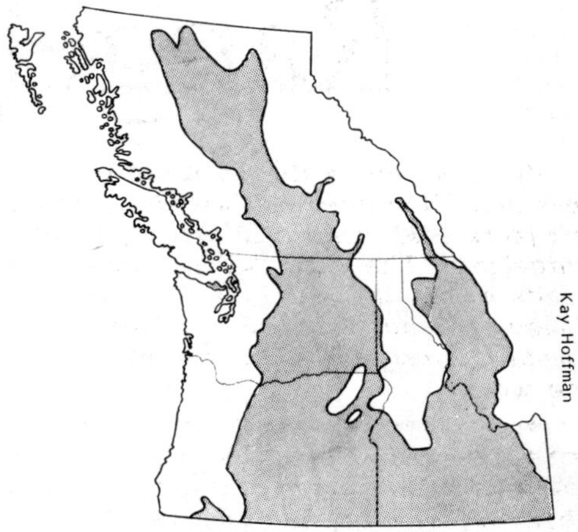

Kay Hoffman

THE ARID NORTHWEST. The map above shows the land area in the region which receives less than 20 inches of annual precipitation.

DESERTIFICATION

Sabadell, J. Eleanora, et al. Desertification in the United States, Status and Issues. Bureau of Land Management. Washington, DC: USGPO, 1982.

Survey of desertification trends in the arid West caused by human activity. Included is a review of current data and a brief discussion of successful rehabilitation projects, including one in Vale, Oregon. Useful appendix to Sheridan's study. 399 references.

Sheridan, David. Desertification in the United States. Washington, DC: USGPO, 1981.

Written for the Council on Environmental Quality, this detailed, readable study analyzes the causes and effects of desertification in the U.S., and suggests solutions. The report is well-referenced, and Sheridan pulls few punches in his vivid portrayal of the degradation of vast sections of our nation's farm and range land. This is a disturbing, important book.

Walls, James. Land, Man, and Sand: Desertification and Its Solution. New York: MacMillan, 1980.

Detailed case studies of desertification around the world and the results of rehabilitation projects. Prepared for the UN Conference on Desertification, the case studies include a report on a large-scale range rehabilitation program in the southeast corner of Oregon.

This is a powerful, rewarding book, providing both an international perspective and practical techniques for combatting desertification in our own backyard.

BIBLIOGRAPHY

Adams, Robert, et al. Drylands: Man and Plants. New York: St Martin's Press, 1979.

Review of techniques being developed around the world to halt desertification, with practical details on arid land irrigation systems and micro-catchments. Procedures are outlined for planning dryland farming systems.

Berkofsky, L.; Faimon, D.; and Gale, J., eds. Settling the Desert. New York: Gordon and Breach, 1981.

A collection of 15 papers summarizing Israel's aproaches to arid land agriculture. Included is a review of ancient terracing, diversion and micro-catchment systems, algae production, solar energy utilization, and ecology as a tool for desert management.

Cannell, Glen H., ed. Proceedings of An International Symposium on Rainfed Agriculture in Semi-Arid Regions. Riverside, CA: University of California, 1979.

Held in Pendleton, Oregon, the symposium covered dryland farming methods in the Northwest and around the world, spotlighting the work of Oregon State University in international research on arid land agriculture.

Dept. of National Development. The Use of Trees and Shrubs in the Dry Country of Australia. Forest and Timber Bureau, 1972. Available from the International Tree Crops Institute, Box 666, Winters, CA 95694.

Information on the use of trees for soil conservation, windbreaks, honey production, livestock fodder, and timber. A good example of multiple-use forestry.

Evanari, M, et al. The Negev, The Challenge of the Desert. Cambridge, MA: Harvard University Press, 1971.

A detailed look at the water-harvesting systems developed in the Negev desert several thousand years ago. Harvesting of dew and runoff waters for maintenance of agriculture in a region receiving less than five inches of annual precipitation.

Goor, A.Y., and Barney, C.W. Forest Tree Planting in Arid Zones. New York: The Ronald Press, 1976.

This book is designed for Third World Countries in the Mediterranean, subtropical zones. Many principles and methods are adaptable to the dry interior Northwest. Good information on establishing nurseries with limited resources.

Perry, Bob. Trees and Shrubs for Dry California Landscapes. 1981. Land Design Publishing, PO Box 857, San Dimas, CA 91773.

This book is an introduction to 360 native and introduced plants which survive in California's dry climates. Many of the species would do well in Pacific Northwest climates as well. A well-produced book with many color photographs.

Riehl, S.; Kinch, Michael; et al., eds. Bibliography of Dryland Agriculture. 3rd edition, 1980. Dryland Agricultural Technical Committee, OSU, Corvallis, OR 97331.

This is one of the most extensive and up-to-date bibliographies on dryland agriculture. Volume One is 485 pages and contains over 4,000 entries with appropriate keywords. Volume Two contains an author and subject index. This useful tool is the result of a five year, global research effort on dryland agriculture which was coordinated from Oregon State University. As a result of the study, the OSU library has the most extensive collection of books and files on dryland agriculture to be found in the Pacific Northwest.

Yeomans, P.A. Water for Every Farm, Using the Keyline Plan. 1981. Second Back Row Press, PO Box 43, Leura, New South Wales, 2781, Australia.

Reprint of two classic texts on water management, Water for Every Farm and The Keyline Plan. Yeomans outlines techniques for designing landscapes which maximize water conservation and soil fertility.

PERIODICAL

Arid Lands Newsletter, Office of Arid Lands Study, University of Arizona, 845 N Park Avenue, Tucson, AZ 85719.

Foremost source of information on deserts and agriculture in North America. The center also publishes Arid Land Abstracts, a monthly review of international literature devoted to arid regions.

Jim Hallas, Bellevue Daily Journal-American

FARMS OF THE FUTURE? The past two decades have witnessed a rapid expansion of irrigation projects in the Northwest. Approximately 8 million acres are currently under irrigation in the region, with nearly 4 million of these acres in Idaho and 2 million each in Oregon and Washington. Another 1.5 million acres are projected for development by the year 2000.

Diane Jones of the Idaho Citizens Coalition recently completed a detailed study of the social and environmental consequences of large scale irrigation developments. Entitled *Water, Energy and Land, Public Resources and Irrigation Development in the Pacific Northwest: Who Benefits and Who Pays?*, the study is available for $5 from the Idaho Citizens Coalition, 216 North 8th Street, Rm 604, Boise, ID 83702. Below are excerpts from the study.

In economic importance in the Pacific Northwest, agriculture ranks second only to the forest products industry; and agriculture depends heavily on irrigation for its survival, with more than 75% of the region's farm produce being grown in the semi-arid regions of the Basin east of the Cascades....

Private large-scale agricultural development in the mid-Columbia region is a relatively recent phenomenon, having begun in the late 1960's. This development took place primarily on land which was considered unsuitable for farming until the dual technologies of high-lift river pumping and center pivot irrigation systems made cultivation possible. The primary areas of private development are the Horse Heaven Hills area and Walla Walla County in Washington and northern Morrow and Umatilla Counties in Oregon.

DIANE JONES

New Crops

Two elements which contribute to the economic survival of small farms are the maintenance of diverse enterprises and the development of markets for specialty crops. Such crops can include fuels, fibers (such as quality flax or wool), oil crops (such as sunflowers or meadowfoam), and food crops new to the region (such as chick peas, amaranth or peanuts).

Of the approximately 80,000 species of edible plants in the world, fewer than twenty now provide 90% of the world's food. Yet most of these edible species were known and eaten by native cultures worldwide until quite recently. J. Sholto Douglas, in *Alternative Foods: A Guide to Lesser-Known Food Plants*, writes that:

"The American Indians, before the coming of the European settlers, lived well on a rich assortment of food plants found in the forests and glades of the western hemisphere. Today, we throw away these plants or label them as 'weeds' and destroy them wantonly. But the time may yet come when we will be very happy to turn back to such neglected and forgotten sources of foodstuff. As the prices of ordinary farm and garden crops rise inexorably, we may be glad to find cheaper and more easily raised alternatives, especially in times of scarcity or need.... The application of horticultural and agricultural expertise to their culture and harvesting could do much to utilize this important but forgotten food resource."

FUEL CROPS

Bernton, Hal; Kovarik, William; and Sklar, Scott. Forbidden Fuel: Power Alcohol in the Twentieth Century. New York: Boyd Griffin, 1982.

The most complete report available on the political and technical background of alcohol fuels. The authors emphasize growing food crops, which bring a higher price than fuel crops, and using the culls or cannery waste for fuel. Specific crops recommended are Jerusalem artichokes and potatoes. As new techniques such as enzyme hydrolysis are perfected, tree crops such as alder, poplar and honey locust will also make good feedstocks.

Tree Crops for Energy Co-Production on Farms. NTIS, U.S. Dept. of Commerce, 5285 Port Royal Road, Springfield, VA 22161. 260 pages, $10.75.

Proceedings of a workshop on the technical, economic, social and ecological feasibility of energy production from tree crops. Includes bibliography and research recommendations.

Oregon Alcohol Fuels Clearinghouse, Eastern Oregon State College, Administration Bldg, Rm 114, 8th and K Streets, La Grande, OR 97850. (503) 963-2171, ext. 370.

Information and assistance for farmers planning alcohol fuel programs, based on experience as well as ideas.

FLAX

Flax has a long history as the fiber plant from which linen is made. Different flax varieties are grown for seed and for fiber, although both belong to the same species, *Linum usitatissimum*. "Cascade" is a flax variety developed in Oregon for commercial fiber production. There is no industrial production of fiber flax in Oregon today, although as recently as 1940 to 1950, some 20,000 acres were grown annually. As interest in natural fiber clothing and goods increases, a new market for flax and linen is developing.

FOR MORE INFORMATION

Calhoun, Wheeler. Fiber Flax Production. Oregon State University, Dept. of Agronomy, Corvallis, OR 97330.

A brief history of and techniques for growing flax. The methods he cites will produce a coarse fiber, such as that used for netting, and need to be modified for clothing-grade fiber.

Zienkiewicz, Henryk, and Frojer, Kare. "Growing and Processing Flax," In The Smallholder, Numbers 22 & 23, 1976 (back issue 80¢). The Smallholder, Argenta, BC V0G 1B0.

Based on 40 years experience, this is the best introduction available for flax growing and processing.

Lee Kirschner, 8005 30th St SE, Everett, WA 98205.

Experienced handspinner/weaver willing to share her knowledge of flax growing, processing, and spinning.

BIBLIOGRAPHY

Douglas, James Sholto. Alternative Foods: A Guide to Lesser-Known Food Plants. Pelham Books, 52 Bedford Square, London, England WC1B 2EF.

Extensive listings of unusual foods, including many tree crops. Over 500 species with extensive plant species index. Good dietary/nutritional information.

Hedrick, U.P., ed. Sturtevant's Edible Plants of the World. New York: Dover, 1974.

First published in 1887 and still one of the best books on the subject. Information on over 3,000 plants used for food.

Masefield, G.B., et al. The Oxford Book of Food Plants. London: Oxford University Press, 1971.

A richly illustrated compendium of information on major and minor food crops of the world. Origin, botany and present distribution.

Tanaka's Cyclopedia of Edible Plants of the World. Tokyo: Keigaku Publishing, 1976.

One of the most comprehensive listings to date, with more than ten thousand food plants given.

Thiesen, Arthur A. Feasibility of Introducing Food Crops Better Adapted to Environmental Stress. NSF/RA-780289. Washington, DC: USGPO, 1978.

The primary objectives of this study were to identify and evaluate 54 lesser-known crops (including elderberry, cattail, and pinyon pine) which have the possiblity of playing a much larger role in US agriculture. This summary gives much information for each crop, including comparison with standard crops, superiority of known cultivars, and sources of additional information.

Uphoff, J.C. Dictionary of Economic Plants. Monticello, NY: Lubrecht and Cramer, 1968.

Among the most complete references to economic plants of the world. Short description of use and native range.

PERIODICAL

Economic Botany
Watson 207-B
New York Botanical Garden
Bronx, NY 10458

A technical journal containing research papers in applied botany and use of food plants, medicinals and pharmaceuticals.

ORGANIZATIONS

Agricultural Development Project
PO Box 323
Arcata, CA 95521
In the late 1970's a team of people in Humboldt County, California conducted an intensive study of the history and potential of agriculture in their area. The results were published in a series of reports covering marketing opportunities, potential crops, processing facilities, and equipment for small farms. Send a SASE for a list of available reports.

Agricultural Marketing Project
2606 Westwood Drive
Nashville, TN 37204
Model for community-based food fairs. List of publications available on many agricultural topics.

Bulk Commodities Exchange
1437 Western Avenue
Seattle, WA 98101
Innovative producer-consumer cooperative established in 1977 to encourage and support local agriculture. Markets farm produce to food-buying clubs and institutions.

Demeter Institute
4214 National Avenue
Burbank, CA 91505
International network promoting the production and marketing of biodynamically grown foods. The Demeter trademark and marketing assistance are available to growers meeting program guidelines.

Direct Marketing Specialist
Oregon State Dept. of Agriculture
635 Capital St. NE
Salem, OR 97310

Farm & Roadside Markets Assoc. (FARM)
c/o Ag. Development Division
Wash. State Dept. of Agriculture
406 Gen. Services Bldg
Olympia, WA 98504
Represents all Farm Map Associations in Washington and Oregon.

Oregon-Washington Farmers Union
10453 4th Plain Blvd NE
Vancouver, WA 98662
Provides technical assistance in starting and managing marketing cooperatives. Instrumental in organizing the Puyallup Valley Raspberry Growers, the Valley Roads Blueberry Growers, the New Farmers Union Hay Marketing Coop, and the Willamette Valley Growers Association.

Pacific Northwest Direct Marketing Conference Association
Box 913
Corvallis, OR 97330
Organizers of an annual conference for direct farm marketers.

Marketing

LOUISE DIX

Producing the crop is only half the work of farming. The other half is marketing what has been grown, often a major problem for farmers. Economic survival requires finding a niche to fill. Diversity of crops and marketing strategies is essential to maintaining a farm's economy. Specialty crops, such as berries or herbs, are often the key to the success of small farms.

Before putting the first seed in the ground, it is essential to develop a marketing plan. Considerations in developing such a plan include: location and accessibility of the farm; proximity to population centers and established marketing outlets; the amount of land available for production; the amount of money and time you want to invest; the demand for the crops you intend to produce; your long range plans for the farm; your own ability to deal with people; your willingness to provide what your customers want; your willingness to market cooperatively with other farmers; and your need for a year round farm income.

Another important factor in a marketing plan is your ability to provide a steady supply of products. Doing this may require succession plantings and/or use of greenhouses or protection devices to extend the growing season. It is a good idea to plant more than you intend to market, as insurance against crop losses. Plan to do some education while you are marketing. For example, one farmer sells his Brussels sprouts on the stalk, and teaches his customers that they actually grow that way.

Depending on your market, proper packaging may also be essential. Check with your warehouse or other major customer to learn about their packaging requirements so that you can prepare well in advance of harvest.

MARKETING OPTIONS

FARMERS' MARKET: A place where a number of growers come together to sell their products by renting a stall or space. Advantages: Low overhead, built-in clientele. Disadvantages: Time away from the farm, transportation costs, most markets open only one day a week.

ROADSIDE STAND: A seasonal stand located on the farm. Advantages: Allows time spent on farm; flexibility in regard to size of operation; can expand to meet consumer demand. Disadvantages: Interruptions of farm work possible at any time during the day; requires diversity and higher volume of crops.

ROADSIDE MARKET: Owner grows a portion of the produce sold, and buys additional products from other farms or wholesalers. Advantages: On-farm location; can provide year round employment and additional income. Disadvantages: Substantial capital requirements; marketing and sales ability and experience required.

U-PICK: The consumer comes to the farm, does the harvesting and pays for produce harvested. Advantages: Numerous sales; low packaging and transportation expenses; good for crops requiring intensive labor for harvesting (e.g. berries). Disadvantages: Inexperienced people on your farm; increased risk of accident and liability; need for parking.

RENT A TREE OR PLOT: Consumer makes contract with grower for the yield of a certain tree or section of a field. Advantages: Crop guaranteed sold; reduced harvest labor. Disadvantages: Possible extra maintenance work; increased management requirements.

GIFT BASKETS/MAIL ORDER: Popular with products of limited perishability such as dried herbs and jams. Advantages: Provides year round income. Disadvantages: Limited to specialty products; requires expensive promotion, processing and packaging labor and expenses.

FOOD SUBSCRIBER NETWORKS: Consumers contract with farmers to provide all of their fruits and vegetables. Advantages: Crop guaranteed sold; some labor costs saved. Disadvantage: Growing to contract specifications and buyers' standards allows little flexibility.

MARKETING COOPERATIVES: Groups of farmers joining together to sell a particular product or to sell a variety of products. Advantages: Provides the volume and variety of produce needed to attract both retail and wholesale customers; shared overhead costs. Disadvantages: Requires efficient management; high capitalization costs.

WHOLESALE OUTLETS: Warehouses, institutions or buying clubs purchasing large quantities at a reduced price. Advantage: Time away from the farm is minimal. Disadvantages: Requires large volumes of crops in steady supply; requires consistently high quality and uniform product; packaging is important; price is lower for the farmer than selling directly; might require hydro-cooled or vacuum-cooled products for maximum shelf life.

SELLING TO A COMMERCIAL PROCESSOR: Advantages: High volume sales; quality need not be consistent. Disadvantages: Low return; fluctuating demand.

ON THE FARM PROCESSING: The farmer processes an item and sells it direct or through wholesale channels. Advantages: Little time spent away from farm; allows use of crops not cosmetically attractive enough for fresh market. Disadvantages: Processing regulations to be followed; may be high equipment or packaging costs.

Provender Alliance
1520 Western Avenue
Seattle, WA 98101
Regional trade association linking food cooperatives, natural food stores, warehouses and producers. Source of information on potential markets for small farmers. Publishes Provender, a bimonthly newsletter featuring current information on the alternative food system.

Washington State Farmers Market Association
508 Security Building
Olympia, WA 98501
Technical assistance, lobbying and promotional activities.

Organic Certification

California Certified Organic Farmers
1920 Marciel Avenue
Santa Cruz, CA 95060
Encouraged enactment of the California Organic Food Bill. Research, education and organic certification program.

Tilth Producers' Cooperative
PO Box 1005
Okanogan, WA 98840
Organic and IPM certification for Washington state. Certification information package available for $5.00.

Willamette Valley Tilth
Rt. 1, Box 308
Sheridan, OR 97378
Oregon certification program. List of grade standards available for $2.00.

BIBLIOGRAPHY

Anderson, Gretchen. A Regulation Handbook for Direct Farm Marketers. Ag Development Div., Dept of Agriculture, 406 General Administration Building, Olympia, WA 98504.
Review of regulations of several agencies which affect direct marketing in Washington state.

Burnell, Diana. U-Pick, U-Profit: Direct Marketing for the Consumer. 1982. Berry Fine Publishing Co, Box 457, Aumsville, OR 97325.
Detailed information on Northwest crops. Useful for growers to learn what consumers are looking for.

Carkner, Richard, et al. Farmer-to-Consumer Marketing. PNW 201-206. Available from WSU, U of O, and U of I Extension Publications Offices.
Series of leaflets on production, marketing and financial management.

Jackson, Bob, and Groder, Roland. Direct Farm Marketing. EC 945. Corvallis, OR: OSU Extension Service, 1978.
Excellent overview of marketing options and management requirements.

Mackay, Susan. Home Storage of Fruits and Vegetables. NRAES 7, 1979. Ag. Engineering Service, Riley Robb Hall, Cornell University, Ithaca, NY 14853.
Diagrams for coolers, storage cellars, etc.

Watts, Gilbert S. Roadside Marketing. London: Orange Judd, 1928.
Although quite old, this book tells everything you would want to know about setting up a roadside market (except current regulations). Available through inter-library loan.

Farm Supplies

By Hand & Foot, Ltd, PO Box 611, Brattleboro, VT 05301.
Importers and manufacturers of "the best in human-powered tools." They publish The Scythe Book and distribute Austrian scythe blades, sharpening equipment, snaths, grain cradles, hayforks, cyclone seeders, flails, a treadle thresher, and a hand-powered cleaner that will process 20 to 30 bushels of grain per hour. They also distribute wood-splitting tools and are developing other tool systems.

Down to Earth Distributors, 947 W Third, Eugene, OR 97402.
Kelp, rock phosphate, Maxicrop, greensand, fish emulsion, and other natural soil amendments "by the bag, ton or boxcar."

Green Earth Organics, 9422 144th Street E, Puyallup, WA 98371.
Full line of rock minerals and organic fertilizers.

Faust Bio-Agricultural Services, PO Box 1150, Twin Falls, ID 83301.
Soil audits, tissue tests, foliar sprays, green manure seeds, and tillage equipment as part of a complete program for biological farming. Catalog $1.

Harmony Farm Service & Supply, Box 451, Graton, CA 95444.
IPM consulting and supplies, hedgerows and insectary plant design, Bio-Dynamic compost, drip irrigation systems, and organic fertilizers.

Harry Sharp & Son, 420 Eighth Avenue N, Seattle, WA 98109.
International distributors of horticultural supplies, including prefab greenhouses, Todd Planter flats, shredders, cyclone seeders, and bedding plants. Catalog $3.

Micro-Ag, Inc., PO Box 125, Steptoe, WA 99174.
Seed inoculants, foliar sprays, soil management assistance.

Multi-Crop Industries, 5307 Pat Bay Highway, Victoria, BC V8Y 1S9.
Seaweed, sulphur, and water soluble fertilizers. Distributes in Canada and the U.S.

NASCO West, PO Box 3837, Modesto, CA 95352.
Extensive catalog of livestock and general farm tools and supplies, including stationary seed cleaners, hand tools, soil testing equipment, and record keeping systems.

Peaceful Valley Farm Supply, 11173 Peaceful Valley Road, Nevada City, CA 95959.
Bio-Dynamic compost, organic transplants, rock minerals, IPM supplies, tools, and much more. Catalog $1.

R & A Plant/Soil, Inc. PO Box 608, Pasco, WA 99301.
Major center for the mass culturing of blue-green algae, which is utilized along with seaweed sprays in a comprehensive fertility program.

Specialty Farm Services, Ltd., 1707 Salton Road, Abbotsford, BC V2S 4N2.
Seaweed and fish fertilizers.

TSC Industries, 14242 "C" Circle Drive, Omaha, NE 68144.
Famous for supplying hard-to-find parts for old farm equipment, plus new tools and supplies.

EXTENSION SERVICES

British Columbia Ministry of Agriculture and Food
Publications Office
Parliament Buildings
Victoria, BC V8W 2Z7

Cooperative Extension
College of Agriculture
University of Idaho
Moscow, ID 83843

Cooperative Extension
Montana State University
Bozeman, MT 59715

Cooperative Extension
Bulletin Service
Industrial Building
Oregon State University
Corvallis, OR 97331

Cooperative Extension
Publications Building
Washington State University
Pullman, WA 99164

Sources of Farm Credit

FARMERS HOME ADMINISTRATION

The FmHA is a U.S. Department of Agriculture agency created to provide supervised credit, at reasonable rates and terms, to farmers unable to obtain loans from commercial lenders. FmHA guaranteed loans are made and serviced by a private lender and backed by the federal government. FmHA insured loans are made and serviced by Farmers Home personnel. Although FmHA is supposed to help beginning and low income farmers, the bureaucratic and lengthy process frequently discourages applicants. Funds are limited, but several new farmers have gotten their start through this program. There are FmHA offices in every area of the Northwest. Check your phone directory for location.

FEDERAL FARM CREDIT SYSTEM

The Federal Farm Credit System consists of three branches: THE FEDERAL LAND BANK loans money for real estate loans; THE PRODUCTION CREDIT ASSOCIATION loans money for operating expenses and capital improvements; and THE BANK OF COOPERATIVES loans for farmer coops. The system operates under a federal charter but is owned by its farmer-members and functions like private banks, except that its interest rates are frequently a point or two lower.

The address of the regional headquarters for the Federal Farm Credit System is W. 705 First Ave, Spokane, WA 99204. There are local offices in many farming communities.

SMALL BUSINESS ADMINISTRATION

Small Business Administration financing is available to farmers. SBA guarantees loans and offers workshops on credit management.

OTHER SOURCES

The Washington Small Farm Network has loans available for low income Washington farmers. The money can be used for capital improvements, operating expenses and equipment. Contact the Washington Small Farm Network, 19 E Poplar, Walla Walla, WA 99136.

Private banks, credit unions and other financial institutions are not adverse to loaning money to small farmers, provided the farmer can convince the lender that he or she has a sound plan, is financially responsible and will be able to pay the money back.

The Garden

Maryanne Caruthers-Akin

MARK MUSICK

The backyard or community garden is one of the primary sources of food for people throughout the world. The garden is as ancient as the village dump heaps (see next page) and as modern as the community gardens now established in many cities and towns.

Work in the garden is on a human scale. Sunshine and human effort are the primary energy sources. Soils are improved by the careful recycling of organic material and, over time, plants can be selected that are especially adapted to local environments. The result of this intensive management has been a continuous supply of food produced from very small spaces by millions of people around the world.

Gardening is always a blend of past, present, and future, as previous years' experiences, today's weather, and the dream that "next year I'm going to...." blend to affect all our garden activities. To reflect this perspective, we look at gardens and gardening from many angles in this chapter. Thus we feature dump heaps, wild foods, and bamboo along with urban agriculture, extending the growing season, and horticulture therapy.

Dump Heaps and the Origin of Agriculture

In **Seeds, Spades, Hearths, and Herds**, *Carl Sauer reconstructs the origins of agriculture. All of our major plants and animals, he notes, were domesticated before the dawn of recorded history. According to Sauer, agriculture did not advance out of deprivation and famine, but rather out of "the leisure for reflection, experimentation, and discussion."*

The first gardeners and farmers, Sauer maintains, were probably people whose lives centered around fishing. Well-supplied with fish and animal protein, they sought out plants as sources of carbohydrates, minerals and vitamins, and to fill a wide variety of domestic needs—medicines, utensils, shelter, and fiber for clothing, fishing lines, and nets. Their first domesticated vegetables and grains were probably taken from close at hand—from wild plants and from the new hybrids that appeared on dump heaps near people's villages and homes.

Edgar Anderson, a student of Sauer's, pursued the study of the origins and evolution of plants. In the process he discovered traditional "dump heap" gardens in Latin America that perfectly fit Sauer's theories on the origins of agriculture. Below is Anderson's description of those gardens from his book, **Plants, Man & Life**.

Sauer, Carl O. *Seeds, Spades, Hearths and Herds, The Domestication of Animals and Foodstuffs*. Cambridge, MA: MIT Press, 2nd ed, 1969.
Classic account of the origins of agriculture.

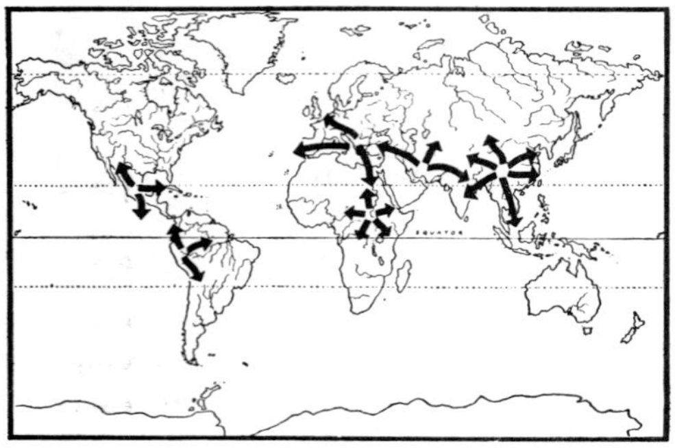

Centers of diversity for cultivated plants, according to Vavilov.

Anderson, Edgar. *Plants, Man and Life*. Berkeley, CA: Univ. of California Press, 1967.
Our landscape has been transformed by plants domesticated in remote corners of the world. Anderson traces their fascinating history and, in the process, provides insight into agriculture in the future.

EDGAR ANDERSON

When I first went to live in San Pedro Tlaquepaque, a small pottery-making town in western Mexico, I was under the mistaken impression that my Mexican neighbors had nothing but dump heaps and a few trees in the yards behind their homes. As I lived there longer and came to know more about the life of the village, I realized that many of these dump heaps were carefully managed gardens and orchards. I saw enough so that several years later in highland Guatemala I played hooky from the cornfields I was supposed to be studying, and spent an afternoon in an Indian village with a Spanish-speaking youth who belonged to one of the two or three Ladino families in that town. Again I came back home with the little more I had learned and thought it over for another two years. By the time of my next visit to Guatemala I was certain that the simple facts concerning these gardens were something worth careful study; and in the few days at my disposal I again visited the village of Santa Lucia on the little height of land between Antigua and Guatemala City and again prevailed upon Señor San Salazar to let one of his sons serve as a guide and interpreter. This time I spent a good part of one day visiting and photographing Indian gardens and finally mapped and measured one of them in detail.

The garden I charted was a small affair about the size of a small city lot in the United States. It was covered with a riotous growth so luxuriant and so apparently planless that any ordinary American or European visitor, accustomed to the puritanical primness of north European gardens, would have supposed (if he even chanced to realize that it was indeed a garden) that it must be a deserted one. Yet when I went through it carefully I could find no plants which were not useful to the owner in one way or another. There were no noxious weeds, the return per man-hour of effort was apparently high, and I came away feeling that as an experienced vegetable gardener (I am one of those strange people who would rather hoe vegetables than play golf or go to the movies) I had gotten more new ideas about growing vegetables than from visiting any other garden anywhere.

Though at first sight there seemed little order, as soon as we started mapping the garden, we realized that it was planted in fairly definite crosswise rows. There were fruit trees, native and European in great variety: annonas, cherimoyas, avocados, peaches, quinces, plums, a fig, and a few coffeebushes. There were giant cacti grown for their fruit. There was a large plant of rosemary, a plant of rue, some poinsettias, and a fine semiclimbing tea rose. There was a whole row of the native domesticated hawthorn, whose fruits like yellow, doll-size apples, make a delicious conserve. There were two varieties of corn, one well past bearing and now serving as a trellis for climbing string beans which were just coming into season, the other, a much taller sort, which was tasseling out. There were specimens of a little banana with smooth wide leaves which are the local substitute for wrapping paper, and are also used instead of cornhusks in cooking the native variant of hot tamales. Over it all clambered the luxuriant vines of the various cucurbits. Chayote, when finally mature, has a large nutritious root weighing several pounds. At one point there was a depression the size of a small bathtub where a chayote root had recently been excavated; this served as a dump heap and compost for the waste from the house. At one end of the garden was a small beehive made from boxes and tin cans. In terms of our American and European equivalents the garden was a vegetable garden, an orchard, a medicinal garden, a dump heap, a compost heap, and a beeyard. There was no problem of erosion though it was at the top of a steep slope; the soil surface was practically all covered and apparently would be during most of the year. Humidity would be kept up during the dry season and plants of the same sort were so isolated from one another by intervening vegetation that pests and diseases could not readily spread from plant to plant. The fertility was being conserved; in addition to the waste from the house, mature plants were being buried in between the rows when their usefulness was over.

I was so impressed by the apparent efficiency of the garden that I have since tried out several of its basic principles on my own vegetable plots with considerable success....From these experiences I suspect that if one were to make a careful time study of such an Indian garden, one would find it more productive than ours in terms of pounds of vegetables and fruit per man-hour per square foot of ground. Far from saying that time means nothing to an Indian, I would suggest that it means so much more to him that he does not wish to waste it in profitless effort as we do.

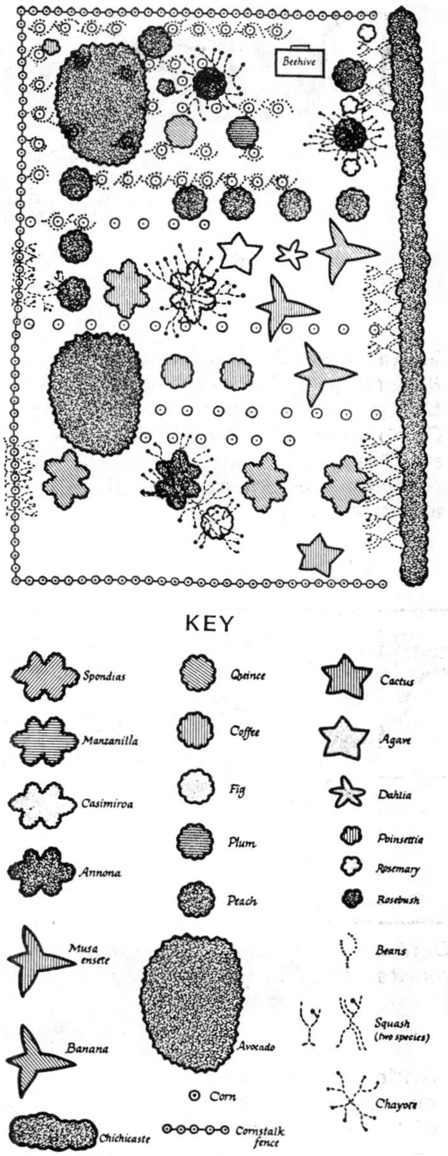

Diagrammatic map of an orchard-garden in the Indian village of Santa Lucia, Guatemala. The glyphs listed above not only identify the plants as shown in the plan, they indicate by their shapes in what general category the plants belong. Circular glyphs indicate fruit trees (such as plum and peach) of European origin; rounded irregular glyphs indicate fruit trees (such as the manzanilla) which are of American origin. Similarly, dotted lines are for climbing vegetables, small circles for subshrubs, large stars for succulents, and an irregular wedge-shaped figure for plants in the banana family. The long irregular mass at the right-hand side represents a hedge of "chichicaste," a shrub used by the Mayas.

Plants, Man & Life

Community Agriculture

Coates, Gary J. *Resettling America, Energy, Ecology and Community.* Brick House, 34 Essex St, Andover, MA 01810. 555 pp, $14.95.

It took a hefty volume to do it, but *Resettling America* unites in one volume 21 major essays on the movement toward community self-reliance. The book covers both urban and rural alternatives emerging across the country. Included is a lengthy analysis of Eugene's Whiteaker Neighborhood and original essays by Amory Lovins, William and Helga Olkowski, Peter Van Dresser, Murray Bookchin, and New Alchemist Earle Barnhart.

Resettling America is a textbook for students of the transformation now sweeping every aspect of our culture.

Most people in the Northwest live in towns, cities and suburbs. We live in an industrial society cut off from the land and from the practice and traditions of agriculture. The widely touted figure of three per cent of the population on farms feeding the urbanized 97 per cent can be seen as either a mark of success or as a symptom of a culture out of balance. But attitudes are changing. In communities throughout the region there is a growing sense that neighborhoods can once again become exciting, productive places. Rather than clinging to a romantic notion that everyone can or should go "back to the land," many people share a new commitment to digging in right where they live, making their community a better place to live.

A simple inventory of any town, city or suburb would reveal acre upon acre of vacant lots, school lands, yards, roadsides, rooftops, and parklands, all with potential for productive community agriculture. Adding greenhouses, backyard chicken, rabbit, fish and bee culture, woodlots, fruit and nut forests, and food processing and storage facilities

Nancy Walz

Finding a piece of ground for a garden can be hard in any urban area, but it was especially difficult in Seattle's International District, a diverse neighborhood caught between downtown skyscrapers, the railroad depot, and the city's industrial area. Most neighborhood residents are from China, Japan, Southeast Asia, and the Philippines. The people come from many lands and speak many languages, but one thing that unites the community is a love of gardening.

After a lengthy search, in 1976 the International District Improvement Association discovered that the only garden space available was a rubble-strewn, clay hillside overlooking the neighborhood. Using hundreds of donated railroad ties, nearly 200 volunteers helped build a terraced garden for the elderly residents of the neighborhood. Within weeks after completion the site was filled with raised vegetable beds, and dwarf fruit trees were soon added to the expanding garden.

The intensive, raised garden beds provide an abundance of fresh food year-round. Many of the gardeners use traditional techniques handed down for generations, often interplanting a half dozen or more different vegetables in each bed. Today the International District Garden is a unique demonstration of the art of Asian food production.

BIBLIOGRAPHY

Chan, Peter, and Spencer, Gill. *Better Vegetable Gardens the Chinese Way*. Portland, OR: Graphic Arts Center, 1977.

Excellent how-to introduction to the Chinese raised bed system of gardening. Beautiful photos.

will greatly increase the ability of communities to sustain themselves.

Bringing food production into the backyard and the neighborhood has many advantages. First, gardens produce a wide variety of fresh foods right at home, with a minimum of outside energy inputs. And second, by re-integrating food production and distribution in our communities, we have the opportunity to achieve the closed nutrient cycles essential for sustained productivity. Wastes will be seen as a resource to be treasured rather than as a burden to be disposed of. As the beginning of this process, many community gardens have already established on-site composting systems, recycling all weed and crop residues for return to the soil. Several cities are also getting into the act through municipal composting programs.

In part because of the value of their manure, small livestock are becoming increasingly important components of backyard and neighborhood "farmsteads." Not only do chickens, quail, ducks and rabbits convert foods inedible by humans into high quality protein, but their manures add an additional source of fertility to the compost pile. Animal manure stimulates biological activity in the soil and contributes phosphorus and other essential nutrients.

As we overcome our cultural abhorrence of human wastes, a much greater integration of nutrient cycles will be possible. The resource most readily available is our own urine, which can become a valued source of fertility for the garden. In *The Integral Urban House* members of the Farallones Institute describe methods they have devised for utilizing urine safely and effectively. According to the Farallones book, each year "the average adult produces enough nitrogen in his or her urine to fertilize approximately fifteen hundred square feet of soil (a garden about forty feet by forty feet)." This is equivalent to the application of 300 pounds of nitrogen to an acre of soil. And, while the proper use of urine will enrich compost and feed soils, increased use of composting toilets will make it possible to recycle nutrient-rich human manure as fertilizer for trees and shrubs.

The percentage of a community's food that can be produced within its boundaries will vary greatly, but home and neighborhood production will become increasingly significant, especially for fresh vegetables, honey, fish and small livestock. At the same time, encouragement of public plantings of fruits, nuts and berries will greatly widen the sources of food available in every community.

USE IT AGAIN SAM

The Seattle City Council has established a goal of assisting residents throughout the city in composting *all* yard wastes. Since publication and adoption of the city's *Composting Strategy* in 1980, efforts to accomplish this goal have included the promotion of composting on radio and TV and the establishment of a special Compost Hotline to give advice and answer questions. Using recycled lumber, the city has also established 11 community garden composting centers and has initiated 5 neighborhood-scale demonstration projects.

For more information on the program contact the Seattle Solid Waste Utility, Rm 606, Municipal Building, Seattle, WA 98104, (206) 625-2089.

Kari Berger

Skyscrapers and sunflowers grace Seattle's skyline. Here Gabriel Baudista takes time to water a neighbor's eggplants in the International District Garden.

BIBLIOGRAPHY

Farallones Institute. *The Integral Urban House*. San Francisco, CA: Sierra Club, 1979. 494pp, $12.95.

Detailed manual for the design of self-reliant houses and communities. Covers energy, urban gardening, livestock and fish production, and pest management. This book is filled with practical, down-to-earth information unavailable anywhere else. Well-researched and illustrated, *The Integral Urban House* is the best introduction to urban agriculture available.

Horticulture Therapy

the art of healing with plants

Becky Burns is a horticulture therapist, one of a new breed of gardeners who understands the power of plants to heal. In the Fall, 1981 issue of *Tilth* (Vol 7, No 3), Becky wrote that horticulture therapy is "the rehabilitation of people through their contact with nature. Today," she continued, "horticulture therapy is used successfully with patients in mental hospitals, nursing homes, hospitals, reformatories, and drug rehabilitation centers."

Becky works with twenty people in a horticulture therapy program near Everett, Washington. She wrote that "the average age of the gardening clients I work with is 40. Many of them have been institutionalized for many years. They have been dependent on the system and on others all their lives. Here in the garden the dependency roles are reversed, however. Many who cannot care for themselves take pride in the fact that a plant needs them. A lot of the clients wish they could have had children. Here," she notes, "they are finally able to care for and nurture a plant as if it were a child. They experience an emotionally meaningful relationship with living plants and their life cycles."

An important aspect of Becky's project is that it is a self-supporting market garden, with clients marketing produce to several restaurants in the city. "The clients are paid a salary for their work in the garden," she notes, "and for many, this is the first time they've ever had a paying job."

"If there is a universal quality of disability," Becky wrote, "it is that its victims suffer from a lack of self-esteem. Above and beyond the fruits of the harvest are the changes that are observed in gardening participants....As they become involved in the work, many of them for the first time begin to feel good about what they are doing."

Through the growing season participants in the program gain new skills and begin to share in the vitality of the plants in their care. According to Becky, "horticulture therapy is an ideal profession for someone who likes working with plants and also enjoys the challenges and rewards of working with disabled people."

FOR MORE INFORMATION

NW Chapter, National Council for Therapy and Rehabilitation through Horticulture
Crestview Conference Center
16200 42nd Avenue South
Seattle, WA 98188

Source for information on horticulture therapy programs in the region, educational opportunities, and the registration of Horticulture Therapists. They have a speakers bureau and a newsletter. Membership is $5.00 for students and $10.00 for non-students.

BIBLIOGRAPHY

Burns, Becky. "The Healing Power Within the Plant Kingdom," In Tilth, Vol 7, No 3 (1981):24-25.

Barber, Andrew. "The Nature of Change in Horticulture Therapy." Tilth, Vol 5, No 3-4 (1980):18-19.
Review of programs at the Menninger Foundation, with a discussion of why horticulture therapy is effective.

University of British Columbia. Gardening as Therapy, Technical Bulletins No 5 and No 6. $4.50 each from the Office of the Botanical Garden, 6501 NW Marine Dr, Univ of BC, Vancouver, BC V6T 1W5.
Resource manuals for the development of horticulture therapy programs.

Community Gardening

For humans, the next step of evolution may necessarily involve entire groups of people at once elevating themselves and others around them by changing perspectives, by paying more attention to nature and less to human happenstance. One way toward this evolutionary leap might be found through the very gates of opportunity opened by community gardens.

Jamie Jobb, The Complete Book of Community Gardening

BIBLIOGRAPHY

Boston Urban Gardeners. A Guide to Community Gardening. New York: Charles Scribner's Sons, 1982.
Detailed information on every aspect of starting and maintaining a community garden.

Community Gardening: A Handbook. Brooklyn Botanic Garden, 1000 Washington Ave, Brooklyn, NY 11225, 1979.
Guide for starting and maintaining a community garden, based on the experience of projects throughout the US.

Gardens for All. Community Garden Organizer's Kit. 1982. Gardens for All, Dept P120, 180 Flynn Avenue, Burlington, VT 05401. $12.00.
Eight-piece kit covering all aspects of community garden organizing, including details on setting up and managing a successful program.

Jobb, Jamie. The Complete Book of Community Gardening. New York: William Morrow, 1979.
Lively introduction to community gardening around the country, with an emphasis on the people and plants that make them work.

The Urban-Rural Connection

Along with closed nutrient cycles within their neighborhoods, many communities are establishing direct links with the countryside. One of the most ambitious examples of an urban-rural cooperative community comes from Japan. In the mid-1970's a Tokyo food buying club made up of 300 families rented a small farm to provide for its own needs. Known as Tamago-no-kai, or "A Group for Eggs," the coop was first formed to purchase eggs wholesale. With their own farm the group was soon raising 3,000 chickens for eggs and meat, 3 cows and 40 pigs, plus goats, rabbits and bees. In addition, 1½ acres of vegetables and 2 acres of rice and other grains were grown. Five families resided on the farm to manage its production and coordinate volunteer labor from the city. Kitchen wastes and human manures were returned from the city for composting. Within three years the farm was producing 100 per cent of the members' eggs and 80 per cent of their vegetables, plus meat and grains.

An example of urban-rural cooperation closer to home is the Community Alternatives Society in British Columbia. Over 50 people live within the community, which includes a three-story cooperative house built by its members and a 10-acre farm in the Fraser Valley. Their urban center is landscaped with intensive gardens and food-producing trees and shrubs. Meanwhile the farm is being developed to supply produce for a new restaurant the community is planning in Vancouver. As more people seek to secure their own sustenance, and as land becomes increasingly expensive, similar experiments will take hold throughout the region.

Eden in the City

REGINA HUGO

No account of sustainable agriculture in the Northwest would be complete without a visit with Cahil in one of his Seattle gardens. His gardens do not have the conventional beauty of rational order, with neat rows of plants and naked brown soil. His gardens, and those inspired by them throughout the city, have the beauty of natural order. Each embodies a complex ecology, established through conscious yet intuitive understanding.

"Gardening is not like a chess game," he says. "There's no sense of what's coming, or what's happened before. Everything is right in front of you. When the gardener does act, it must be with 100% concentration." This sounds more like the advice of a yoga master or spiritual teacher than a gardener, but there is no contradiction. Cahil comes to gardening from a yoga/spiritual background.

The appearance of Cahil's garden is that of a small forest. The plants are so unusual that one cannot be sure offhand what most of them are. Some are simply perennialized versions of common annual vegetables, such as three-year-old broccolis and kales. Others are wild plants like nettle and mallow which Cahil has collected on his foraging trips through the urban wilderness.

"Most of what I eat from plants are flowers, or buds that would be flowers if I left them, or seed pods. In the stalk and leaf-making season, I eat the outer leaves." Cahil states that the main work in his gardens is "browsing," harvesting the edible parts. In this way, he keeps his radishes producing succulent pods, his kales providing spicy edible flowers.

When it comes to tillage, Cahil's ideas are just as unconventional, but sensible. Digging potatoes is one example of what he feels is an appropriate form of tillage. Weeding is another example. "Sometimes," he says, "I do tillage by pulling out morning glory roots. That tills the soil beautifully for planting peas in early spring."

In addition, Cahil uses mulch whenever possible. Over time, the end product is like a forest floor—loose and rich, diggable with the hands. His "urban food forest" gives plants a chance to become adapted to their immediate environment. In Cahil's term, they've mongrelized. "I want plants to become mongrels, I want them to hybridize, to cross-breed right away, instead of keeping the lines pure...I like plants that can take care of themselves, so that the garden could conceivably be there forever, taking care of itself without me."

Cahil is a teacher. He teaches how to live consciously, caringly, holistically. His garden is a teaching metaphor, but it is also an ongoing experiment on how to raise food without creating soul- and soil-depleting environments. "Any other kind of work in my book is meddling with nature."

What is Cahil's ultimate vision for the city garden? "I'm doing this in order to feed people, birds, insects, rodents, worms, beetles and spiders. My idea is to provide a healthy environment for everyone. There can be plenty of food and you don't have to worry about predators. A jungle-like environment provides forage for lots of different creatures. We keep each other's balance, with a little help and guidance from nature. I literally want to create the Garden of Eden on the physical plane.

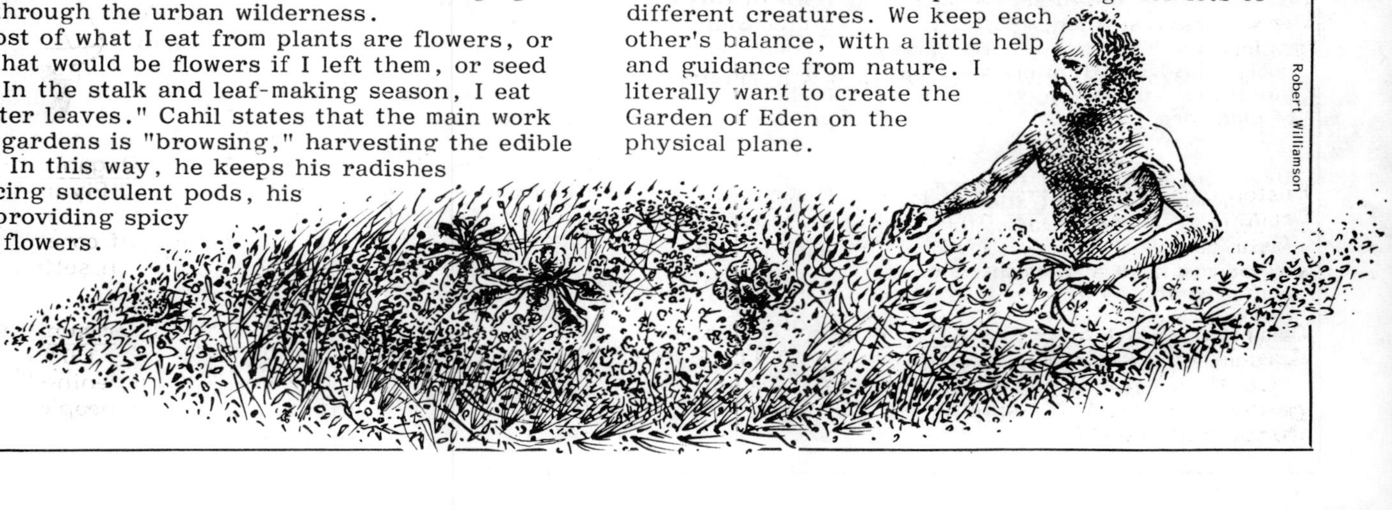

Robert Williamson

The Garden Bibliography

Antonelli, Art, el al. Organic Gardening. Extension Bulletin 0648. Pullman, WA: WSU, 1982.
Introduction to organic soil amendments and non-chemical disease and pest control. Emphasis is on selection of disease-resistant crops. Discusses cages which proved successful against root maggots in tests at the Puyallup Research Center.

Beckett, Kenneth A. Growing Hardy Perennials. London: Croom Helm, 1981.
An authoritative dictionary of hardy herbaceous and evergreen perennials. Included is a chapter on hardy perennials in their native habitat, plus a listing of cultural conditions and propagation methods for all the main genera, species, and cultivars of hardy perennials. Emphasis is on ornamentals.

Bloom, Alan. Perennials for Your Garden. Chicago: Floraprint USA, 1981.
North American edition of a classic British text on the design and establishment of perennial flower gardens. Over 500 species of hardy perennials are described and, while most are ornamentals, a few (such as daylilies and yucca) are edible. The stunning color photos offer a delightful invitation to a new world of gardening with one of Britain's contemporary masters.

Browse, Philip McMillan. Plant Propagation. New York: Simon & Schuster, 1979.
One of a series produced in cooperation with the Royal Horticultural Society. This is an excellent guide to propagation techniques for herbaceous and woody plants. Thorough and authoritative.

Bubel, Nancy. The Seed-Starter's Handbook. Emmaus, PA: Rodale Press, 1978.
A full, practical manual, based on the author's own experience. Information on growing almost every garden vegetable from seed, including soil mixes, sprouting and germination, and transplanting techniques. Much advice on saving seed, also.

Brown, George E. The Pruning of Trees, Shrubs and Conifers. London: Faber & Faber, 1972.
This book is quite comprehensive, covering 450 different plant genera. The characteristic habits of growth and flowering for each genera are introduced along with its most common species and their variations in pruning needs.

Clark, D.E., ed. New Western Garden Book. Menlo Park, CA: Lane, 1979.
Perhaps the most complete and useful guide to annual and perennial gardening in the West. Detailed cultural information and zone maps, plus landscaping techniques and lists of plants for special effects. Excellent reference tool.

Evans, G.E., and Murphy, J.L. A Woody Ornamental Plant Inventory of Selected Montana Communities to Evaluate Length of Freeze-Free Season as an Indicator of Plant Hardiness. Research Report 96. Bozeman, MT: Montana Agricultural Experiment Station, Montana State University, 1976.
During the summer of 1968 a survey of woody plants growing in 17 Montana cities was conducted. The results of the survey allowed the researchers to create a plant hardiness zone map which is more detailed than any previously published. Unusual plants encountered were especially noted and mentioned in the report. Included are several climate maps, 24 references, and appendices containing 250 plant species, their distribution in Montana, and their hardiness rating.

Gessert, Kate Rogers. The Beautiful Food Garden. New York: Van Nostrand Reinhold, forthcoming (1983).
Design manual for edible landscaping with vegetables, herbs, flowers, fruits, and nuts. Includes details on specific plant varieties based on research in cooperation with University of Oregon, Eugene.

Hakala, Sue. Drought Gardening. Charlotte, VT: Garden Way, 1981.
One of Garden Way's excellent Country Wisdom Bulletins, this is a concise introduction to gardening during the dry time of the year.

Edible Flowers

Many plants have edible flowers which will brighten any meal. Some, like Daylilies and Squash Blossoms, are cooked as vegetables, while others can be used in confections or used to top off a summer salad. A sampling of edible flowers is provided below. *Bon Appetit!*

Borage *(Borago officinalis)*
Calendula *(Calendula officinalis)*
Chamomile *(Matricaria recutita* or *M. chamomilla)*
Chives *(Allium schoenprasum)*
Chrysanthemum *(Chrysanthemum* spp.*)*
Comfrey *(Symphytum officinale)*
Dandelion *(Taraxacum officinalis)*
Daylily *(Hemerocallis* spp.*)*
English Daisy *(Bellis perennis)*
Gladiolas *(Gladiolus* spp.*)*
Hops *(Humulus lupulus)*
Lavender *(Lavandula officinalis)*
Lilac *(Syringa vulgaris)*
Mallow *(Malva neglecta)*
Marigold *(Tagetes* spp.*)*
Mariposa Lily *(Calochortus macrocarpus)*
Mustard *(Brassica* spp.*)*
Nasturtium *(Tropaeolum majus)*
Pansy *(Viola* spp.*)*
Primrose *(Primula vulgaris)*
Red Clover *(Trifolium pratense)*
Rose *(Rosa* spp.*)*
Squash Blossom *(Cucurbita* spp.*)*
Viola *(Viola* spp.*)*
Yucca *(Yucca* spp.*)*
Zinnia *(Zinnia* spp.*)*

Nursery Management

Bailey, L.H. The Nursery Manual. New York: MacMillan, 1967.
First published in 1896, this fine book is loaded with information and illustrations of how to do all types of plant propagation.

Davidson, Harold. Nursery Management: Administration and Culture. Englewood Cliffs, NJ: Prentice-Hall, 1981.
The best modern text on starting and managing a nursery. Emphasizes commercial operations, and goes into a wide range of practical business and plant propagation details. Features latest information on containerization. Well illustrated. Good bibliographies.

USDA. Seeds of Woody Plants in the United States. USDA Handbook #450. Washington, DC: USGPO, 1974.
The nurseryperson's bible. Detailed information on collection, processing, storage, germination tests, stratification requirements and nursery practices for all major species of woody plants in the US.

Williams, R.D., and Hanks, S.H. Hardwood Nurseryman's Guide. USDA Handbook #473. Washington, DC: USGPO, 1976.
The most comprehensive text available on growing nursery stock of native North American hardwoods.

Halpin, Anne Moyer, ed. Gourmet Gardening. Emmaus, PA: Rodale Press, 1981.
For market gardeners seeking new specialty crops, or home gardeners wishing to experiment with unusual vegetables, this book includes well-written introductions to over 40 possibilities.

Hartmann, H.T., and Kester, Dale E. Plant Propagation: Principles and Practices. Englewood Cliffs, NJ: Prentice-Hall, 1975.
A guidebook for professionals. Standard text covering all methods of propagation.

Jeavons, John. How to Grow More Vegetables Than You Ever Thought Possible on Less Land Than You Can Imagine. Berkeley: Ten Speed Press, 1979.
Best introduction to intensive wide bed plantings. Plant spacings must be modified to adapt to microclimates.

Morse, Harriet K. Gardening in the Shade. Beaverton, OR: Timber Press, 1982.
Design manual for growing trees, shrubs, vines, ground covers, and herbs in light to heavy shade. Extensive species index.

Raymond, Dick. Down-to-Earth Vegetable Gardening Know-How. Charlotte, VT: Garden Way, 1975.
Practical, step-by-step guide to intensive food gardening. Wide-row technique is ideal for gardening with a rototiller. The section on green manuring is detailed and useful.

Seymour, John. The Self-Sufficient Gardener. Garden City, NY: Dolphin Books, 1979.
Detailed and beautifully illustrated book on the intensive production of vegetables, herbs and fruits in the home garden.

Salter, P.J.; Bleasdale, J.K.A.; et al. *Know & Grow Vegetables.* New York: Oxford University Press, 1979.

Market and home gardeners seeking maximum yields from limited space will appreciate the precise detail provided in Know & Grow Vegetables. Based on studies at the National Vegetable Research Station in England, the authors present results from the latest scientific studies on establishing and maintaining garden vegetables. Two noteworthy sections cover plant spacing and watering, with charts provided on the most efficient spacings for many vegetables and on watering techniques for optimum yields. Also included is a detailed explanation of a do-it-yourself fluid seed sowing system which ensures rapid germination.

While their point of view is not organic, the authors do suggest non-chemical methods for avoiding or controlling insects and diseases before recommending poison sprays. One significant example is their use of foam carpet padding around the base of cabbage family crops as a barrier against the root fly maggot. They found the foam padding to be a great advance over tarpaper collars, providing control equal to insecticides while serving as a moisture-conserving mulch.

Efficient watering: limiting the quantity of water given to the crop.

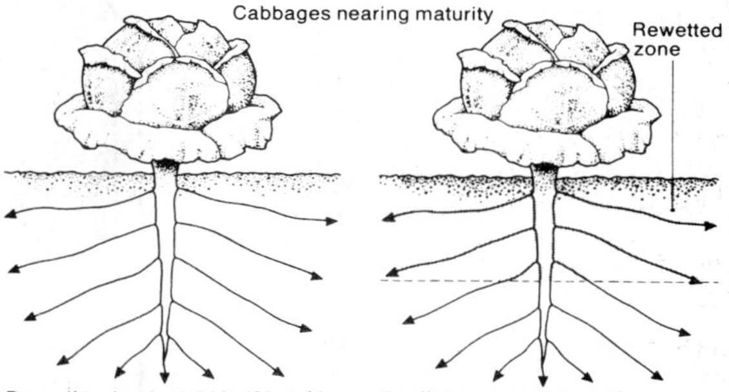

"Fluid sowing of pre-germinated seeds has now been tested experimentally on a wide range of vegetables and for all of them it gives earlier emergence, particularly with those crops such as carrots and parsnips which are normally slow to germinate, or those such as tomatoes which germinate only slowly at low temperatures. In fact, with tomatoes emergence and growth are advanced so much that it is possible to produce ripe crops by sowing directly into the garden soil even where the season has been regarded as being too short to get a satisfactory crop, for example in the United Kingdom."

"For maincrop potatoes water should not normally be given until the tubers start to form. This stage coincides approximately with that of flowering and for most varieties a single, heavy watering of 4-5 gallons per square yard when the tubers are at the marble stage can be very beneficial."

Harrison, John B. Good Food Naturally (How to Grow It, Cook It and Keep It). Vancouver: J.J. Douglas Ltd, 1977.
A good introduction on how to make the most of a garden or small farm by an experienced British Columbia organic farmer.

Hills, Lawrence D. Grow Your Own Fruit and Vegetables. Salem, NH: Merrimack, 1973.
Organic techniques from Britain, especially valuable for Maritime gardeners.

——. Comfrey: Fodder, Food and Remedy. New York: Universe Books, 1976.
Thorough overview of the many virtues and uses of comfrey. Hills provides an appreciation for the value of comfrey as food, fodder and nutrient-rich mulch.

Organic Gardening Staff. The Encyclopedia of Organic Gardening. Emmaus, PA: Rodale Press, 1978.
Updated, this volume continues to be a valued garden companion.

Pellegrini, Angelo. The Food Lover's Garden. Seattle, WA: Madrona Publishers, 1970.
A delightfully written guide for growing your own food. Cultural information on obscure as well as common vegetables written by a connoisseur of good food and the good life.

Philbrick, J. and Philbrick, H. Gardening for Health and Nutrition. Blauvelt, NY: Steiner Publications, 1971.
A straightforward introduction to Bio-dynamic agriculture.

Shewell-Cooper, W.E. The Complete Vegetable Grower. London: Faber & Faber, 1975.
A British book covering many gardening methods with instructions on the specific culture of each vegetable. Especially valuable for Maritime gardeners, by an author who has written more than 40 gardening books.

Solomon, Steve. Growing Organic Vegetables West of the Cascades. Seattle, WA: Pacific Search, 1981.
A practical and concise guide by the owner of Territorial Seed Company.

Stout, Ruth. How to Have a Green Thumb Without an Aching Back: A New Method of Mulch Gardening. New York: Simon & Schuster, 1973.
Wise words on mulch gardening from a delightfully eccentric exponent of no-till gardening.

ORGANIZATIONS

The Chadwick Society
Green Gulch Farm
Star Route #1
Sausalito, CA 94965
Former students and friends of Alan Chadwick have formed The Chadwick Society to carry on his work. They publish a quarterly newsletter to maintain communication with gardening projects around the country. The newsletter also serves as a continuing source of ideas and inspiration from the courses Alan Chadwick taught in North America.

City Farmer
Suite 801, 318 Homer Street
Vancouver, BC V6B 2V3
A lively center for information and inspiration on urban agriculture in British Columbia. Projects have included developing a demonstration food garden, producing curriculum materials, and putting on a series of urban agriculture courses.

Ecology Action of the Midpeninsula
2225 El Camino Real
Palo Alto, CA 94306
Center for research and education on biodynamic/French intensive gardening. They sponsor workshops and publish reports on intensive backyard food production.

Gardens for All
180 Flynn Avenue
Burlington, VT 05401
National voice for home and community gardening. Membership includes subscription to a monthly news magazine filled with interesting articles and useful gardening techniques and tools. This non-profit organization's current projects include the promotion of gardens in schools, hospitals, prisons and mental institutions, and around industrial and office areas.

Hardy Plant Society
11907 Nevers Road
Snohomish, WA 98290
Pacific Northwest chapter of an English organization devoted to the promotion of hardy perennials. Membership includes a yearly bulletin, infrequent newsletters, a seed exchange, lectures, and garden tours. Emphasis is on ornamentals.

Northwest Ornamental Horticulture Society
Univ. of Washington Arboretum XD 10
Seattle, WA 98195
The Society's activities include educational programs, an annual plant sale, a seed exchange, and horticultural tours. Their journal includes detailed information on native and exotic plants.

Let Your Worms Do the Digging

Maryanne Caruthers-Akin

Maryanne Caruthers-Akin, a Portland artist, has developed a highly productive gardening system in which earthworms do the digging. The mulched beds have also eliminated the problems of frequent weeding and struggling with resistant soil. Maryanne believes her method is ideal for the city gardener who wants to transform his or her backyard into a food factory.

Garden beds (shown above) are four feet wide, in a shape or length that fits the site. After outlining the area with stakes or string, she evenly applies to every 40 square feet one pound or more of dry molasses to attract the earthworms, and one-half pound or less of kelp powder to provide nutrients to the soil. She then spreads newspapers, about five sheets thick, over the surface, using a thicker layer if the lawn is poor and plagued by weeds.

After wetting down the newspaper to keep it in place, Maryanne covers the bed evenly with damp straw, which tends to be less expensive and contain fewer weed seeds than hay. For the initial bed, one bale of straw adequately covers 40 square feet. After that, she replenishes the straw as needed.

Instead of planting seeds directly into the beds, Maryanne uses Todd Speedling trays, in which each seedling develops an individual, pyramid-shaped root ball. To set the transplants in the garden, she pulls back the straw, exposing the soil. Then she adds two to four cups of compost, one-quarter cup of bone meal and a pinch of dried alfalfa leaves to the exposed soil. She sets plants in deep, because until the earthworms become actively involved in the tilling, the plants are, in effect, growing in straw flower pots, and can easily dry out.

To encourage growth, Maryanne uses a garden tea made of one tablespoon fish emulsion, a tiny dribble of seaweed emulsion, and one tablespoon of liquid molasses, which she mixes in one gallon of water. Vegetables are fed with this mixture every two weeks during the growing season.

Maryanne finds that once this system is set up, the rest is easy. She has found that, with the exception of turning the compost tumbler and processing kitchen scaps, she and her two housemates work only one afternoon a week from mid-February to mid-November to maintain the garden. Maryanne is pleased that the garden produces food for her household all year round, and as an artist, she also appreciates the fact that the design of the garden holds up all year.

For more details on Maryanne's garden, see her article "No-Till Gardening Easy and Productive," in Tilth, Vol. 7, No. 4.

PERIODICALS

Gardens for all Newsmagazine
180 Flynn Avenue
Burlington, VT 05401
Lively tabloid with news of gardening activities across the continent. Regular articles on gardening, edible landscaping, church and community gardens, and horticultural therapy. Subscription included with Gardens for All membership.

Horticulture
300 Massachusetts Avenue
Boston, MA 02115
Practical and well-illustrated articles on all aspects of gardening, berry and fruit culture, and tree management. An exceptionally beautiful magazine.

HortScience
American Society for Horticultural Science
Alexandria, VA 22314
A technical journal of interest to the advanced horticulturist.

Organic Gardening
Rodale Press
33 E Minor Street
Emmaus, PA 18049
The pioneer of the present-day organic movement, reaching millions of North Americans. Always filled with useful ideas and information for the home gardener.

Pacific Horticulture
PO Box 22609
San Francisco, CA 94122
Journal of the Pacific Horticultural Foundation, this is a beautifully illustrated, authoritative quarterly review of ornamental horticulture and Western native plants.

The Sproutletter
PO Box 10985
Eugene, OR 97440
Fresh sprouts and indoor gardens have brought food production inside for many people. The Sproutletter is a lively source of information on sprouts, raw foods and nutrition, including details on gathering seeds and growing wild plants indoors.

Sunset Magazine
Menlo Park, CA 94025
Excellent articles on western gardening and landscaping. A special Northwest edition provides detailed information for gardeners throughout the region.

The Herb Garden

Cleves, Bernard, ed. Herb Growing: A Visual Guide. Berkshire: Sampson Low, 1978.
From Britain, this book covers how to identify, propagate, harvest and use a wide range of herbs.

Coon, Nelson. Using Plants for Healing. Emmaus, PA: Rodale Press, 1979.
Revised edition of a standard text on herbal medicines. It includes brief history of the use of plants in healing and an extensive survey of medicinal herbs.

Gibbons, Euell. Stalking the Healthful Herbs. New York: David McKay, 1966.
Delightful introduction to the properties and value of wild herbs. The book covers a wide range of plants, introducing each one with Euell Gibbons' wealth of knowledge, experience, and enthusiasm.

Gilbertie, Sal, and Sheehan, Larry. Herb Growing At Its Best. New York: Atheneum, 1980.
Written by a commercial herb grower, this book organizes herbs by propagation method, and gives clear instructions for growing both common and uncommon herbs. Especially inspiring are the many herb garden plans.

Grieve, M. A Modern Herbal. 2 Vols. New York: Dover, 1971.
Reprint of 1931 edition. Although some of the information is now archaic, this is a classic reference for the medicinal qualities of herbal plants from around the world.

Jacobs, Betty E.M. Growing and Using Herbs Successfully. Charlotte, VT: Garden Way, 1981.
Useful information on the culture and uses of herbs. Nearly half the book is devoted to small-scale commercial herb production, processing and marketing. The author has developed a successful herb farm on Vancouver Island. Her book is personal in tone and is filled with practical experience.

Millspaugh, Charles F. American Medicinal Plants. New York: Dover, 1974.
The subtitle for this classic work, first published in 1892, is "An illustrated and descriptive guide to plants indigenous to and naturalized in the United States which are used in medicine." 180 species are described in great detail with description, origin, habitat, flowering time, parts used, preparation and physiological action.

PERIODICALS

The Herb Quarterly
Newfane, VT 05345
Elegant and informative journal covering all aspects of herb culture and use.

The Herbarist
The Herb Society of America
300 Massachusetts Avenue
Boston, MA 02115
Annual anthology of information on herb gardening and uses. The Herb Society promotes the cultivation and study of herbs and the establishment of herb gardens.

MEDICINAL GARDENS

Garden of Medicinal Plants
School of Pharmacy
University of Washington
Seattle, WA 98195
Two-acre garden of medicinal herbs.

The Physick Garden
University of British Columbia
Vancouver, BC V6T 1W5
Established in 1978, the Physick Garden was modeled on a sixteenth century European apothecary's garden. It features plants that have been used in medicine for centuries.

FOR MORE INFORMATION

Otto Richter & Sons
Goodwood, Ontario L0C 1A0
In addition to seeds for herbs, gourmet vegetables and wildflowers, Richters also carries an extensive listing of books on medicinal plants. Included are Uses of Plants for the Past 500 Years, by Charlotte Ericksen-Brown, Herbs, Their Culture and Uses, by Rosetta Clarkson, Ginseng and Other Medicinal Plants, by A.R. Harding, and Indian Herbology of North America, by Alma Hutchens. The Richters catalog, including an annotated listing of titles, is available for $1.

Economic Botany Laboratory
US Dept of Agriculture
Washington, DC 20250
Emphasis is on medicinal plants, with computer files of over 4000 species of plants used in folk medicine, some of which are also used for food, fiber, dye, or other products. Most plants listed are tropical.

Extending the Season

CHANGING THE PACE

Year-round gardening strengthens body, mind, soul and checkbook. Beyond avoiding higher food prices, it inspires a seasonal diet, innovation in the kitchen, and a year full of celebrations. It increases our self-reliance, conserves energy, strengthens our regional economy, and is just plain fun and interesting.

For people who already garden from April to October, the first step is to expand the garden's productive time into the other six months of the year with new varieties and planting dates. This step requires a change in perspective that must be similar to the mental shift required of a runner who switches from sprints to marathons. The gardener/cook must set a pace that is possible to sustain in the long term.

Karol Stevens, in "Year-Round Gardening Makes Good $ense." Tilth, Volume 7, No. 4.

BIBLIOGRAPHY

Aquatias, A. Intensive Culture of Vegetables on the French System. London: L. Upcott Gill, 1913. Reprinted in 1978 by Solar Survival Press, Harrisville, NH 03450.
Reprint of classic text on use of cloches and other techniques for intensive vegetable production. Solar Survival also sells solar cones and passive solar frames.

Colebrook, Binda. Winter Gardening in the Maritime Northwest. Arlington, WA: Tilth, 1977.
Well-researched guide to growing over 60 different vegetables and herbs through the winter in the Maritime. Out of print but available in many libraries. Extensively revised and expanded edition to be published in early 1983 by Madrona Publishers. Will also include information on winter gardening in harsher climates.

Doscher, Paul; Fischer, Timothy; and Kolb, Kathleen. Intensive Gardening Round the Year. Brattleboro, VT: The Stephen Greene Press, 1981.
Tools and techniques for year-round gardening in cold climates, including a good chapter on solar devices. Well suited to the Interior Northwest. Includes a section on no-till gardening.

Vegetables and herbs fresh from the garden make a wonderful winter or early spring salad. Some are relatively easy to grow. Siberian kale is as hearty as it is rugged. Corn salad makes a perfect complement and, depending on your climate, can be grown with little or no protection.

Extending the growing season is a challenge that is amply rewarded, and does not necessarily require fancy or expensive protection devices. Many salad greens besides those mentioned above can be kept growing throughout the winter and spring with simple cold frames or portable coverings called cloches.

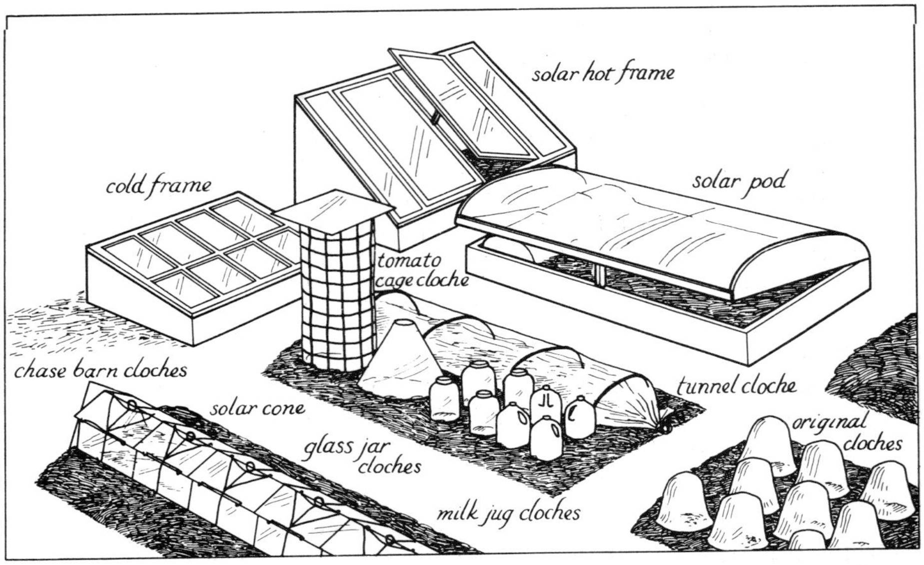

Intensive Gardening Round the Year

"Cloche" is a French word for bell. Cloches, or "bell glasses," were an essential part of French intensive gardening. First introduced in the 17th century, bell-shaped jars of blown glass were set over plants in the garden to extend the growing season in the fall or to force lettuce in the early spring. Use of the term cloche has since expanded to cover a wide variety of temporary structures that can be put over plants in the garden, opened to allow for ventilation, and then removed when no longer needed.

Unlike a greenhouse, in which you create an artificial environment, the idea of a cloche is to work with nature, creating a warm micro-climate right around the plants in the garden. The difference between a cloche and a cold frame is that the cold frame is usually a permanent structure while cloches are temporary and portable.

A simple modern equivalent of the traditional French bell-jar is a one-gallon clear plastic or glass jug with the top or bottom cut out. These jugs can serve as hot caps over tender young plants. Air circulation is essential. To allow for proper circulation, either prop open or remove the cloches on sunny days and then replace them at night.

In May 1980, the Seattle Tilth chapter received a $5,000 grant for a coldframe and cloche demonstration project. In his article "Cold Frames and the Year-Round Gardener" (*Tilth*, Vol. 7, No. 3), project coordinator Carl Woestendiek described some varieties of lettuce and Oriental greens which were most successful in the structures he built. His article makes the point usually stressed in discussions of extending the season—that the proper choice of varieties can mean the difference between success and failure. Excerpts from Carl's article follow.

> *"Of the Oriental greens, the Green Wave, Mizuna, Florida Broad Leaf and Chinese Broad Leaf gave the most satisfactory growth. Tendergreen mustard tended to bolt readily in the lengthening days after the winter. Still later in the season Mizuna mustard showed some tendency to bolt, but was valued highly until then for its mild flavor and attractive appearance. The edible chrysanthemum, though not as fast growing as the mustards, did grow at a reasonable pace and was valued for its unusual flavor. Chinese Broad Leaf and Florida Broad Leaf mustards were the most spectacular in their rate of growth, and the Green Wave mustard was the most resistant to bolting.*
>
> *"Winter Density was the only romaine-type among the lettuces. All-Year-Round, Winter Marvel and Arctic King are all Bibb types, and of these, All-Year-Round was both the fastest growing and least cold hardy.*
>
> *"The most destructive pest in the frame and cloche plantings was the slug. Warmer winter outside temperatures coupled with the warmth of frame and cloche served to increase their numbers. The broad leaf mustards with some hair on their leaves (Florida Broad Leaf, Chinese Broad Leaf and Green Wave) showed the least slug damage."*

Varieties mentioned are available from Abundant Life Seed Foundation. Some other sources of seed for winter crops are Alexander & Brown, Johnny's Selected Seeds, Sutton Seeds, Territorial Seed Company, Thompson & Morgan, and William Dam. See **Seed and Nursery List** for addresses.

Hill, Lewis. Successful Cold-Climate Gardening. Brattleboro, VT: Stephen Greene Press, 1981.
Concentrates on techniques for gardeners in short-season areas, but most of the information would be useful to every gardener who wants to extend the growing season. The book does address some problems specific to Northern areas, such as snow loads and late summer frosts. Especially useful are its lists of fruits, vegetable and ornamental varieties suited to Northern climates, with a list of seed sources included in the Appendix.

Moody, Tere; Morton, Kenneth; Shields, Richard; and Steggell, Dorene. Cold Frames and Cloches: Extending Your Growing Season. Willamette Community Design Center, PO Box 10273, Eugene, OR 97440. 1981. $3 plus $.85 postage.
Introduction to design principles for cloches and cold frames, with detailed plans for four designs. Emphasis is on design and not on cultural techniques.

Organic Gardening Magazine, eds. Getting the Most from Your Garden. Emmaus, PA: Rodale Press, 1980.
A chapter on "Coldframes and Other Season-Extending Devices" has good information on coldframe design, particularly solar frames. Also some good general cultural practices for coldframe growing.

Ruttle, Jack. "Grow Your Own Midwinter Sun Salads." Organic Gardening Magazine, November 1980.

Shewell-Cooper, W.E. Basic Book of Cloche and Frame Gardening. London: Barrie & Jenkins Ltd., 1977.
Much good information on using cloches year round, for maximum efficiency. A dazzling array of cropping possiblities.

Wolfe, Ray. "Garden All Winter with Solar Grow Frames." Organic Gardening Magazine, August, 1979.

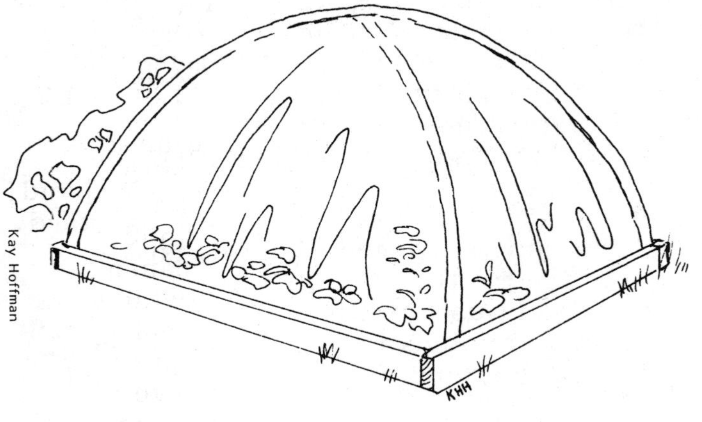

Kay Hoffman

Of the several cloche and cold frame designs Carl tested, the tunnel cloche shown at left was his favorite. It is lightweight, yet sturdy, encloses a good amount of growing space (about 4 feet by 6 feet), and is easy for one person to move. Plans, directions and a materials list are available from **Seattle Tilth, 4649 Sunnyside N, Seattle, WA 98103.** Enclose a SASE with your request.

The Wildfood Trailguide

The Wild Garden

Nettles

A delicious, nutritious and under-utilized food native to the Pacific Northwest is the stinging nettle *(Urtica dioica)*. A wild perennial, nettles are often found at the edges of woods or near people and their livestock, in rich, moist, shady places. Nettles improve the soil while providing some of the best spring greens around. They are high in protein, vitamins and minerals, especially iron. Picked carefully (wearing gloves) and steamed or boiled briefly to remove the sting, fresh nettles are a rich gourmet treat in early spring.

Dried nettles make an excellent tea, a winter soup vegetable and, in quantity, a mineral-rich animal feed. Nettle roots build humus in the soil, the fresh plants are a good addition to compost, and dried plants are used as a nitrogen-rich mulch. Nettles were an important source of fiber for the native people of the Northwest coast, and in Europe nettle fibers were used to make cloth, rope and paper. If the area was well-drained, pioneers often chose places where nettles were growing as sites for orchards and gardens.

"In some parts of the world you can sleep between nettle sheets, eat off of a nettle tablecloth, dine on nettle-enriched steaks and eggs ordered from a nettle-paper menu, in an emergency fish with a nettle line, and in the springtime especially revel with delectable nettle dishes washed down with nettle beer."

Bradford Angier, *Field Guide to Edible Wild Plants*. Harrisburg, PA: Stackpole, 1974.

Wild foods are our link with the roots of agriculture. They are a reminder that we need not be totally dependent on garden crops and supermarkets, and that we can learn to trust in the bounty of the natural world. Wild plants offer a nutritional supplemental food source, especially in early spring when so little is growing in the garden.

Exploring for wild fruits, berries, herbs and vegetables can provide a surprising amount of food, even in urban areas. The best way to learn about wild foods (and identify ones that may be poisonous) is to go on walks with foraging friends, or to read one or more of the many books on harvesting wild foods. You will soon discover new foods beneath your feet and tucked into the margins of every field, stream, roadside, and vacant lot.

With a little help, many garden vegetables can also be grown as wild plants. Masanobu Fukuoka, a Japanese farmer, has demonstrated that many food plants can be grown like wild herbs. Fukuoka's approach to growing vegetables in what he calls a semi-wild fashion is described in his book, *The One-Straw Revolution*.

Sensing intuitively that it must be possible to grow food in the wild, Fukuoka experimented with scattering seeds of herbs and vegetables throughout his hillside orchard in early spring and late summer. His technique was to broadcast seeds of plants such as peas, beans, radishes, garlic, onions, leeks, cabbage, turnips, carrots, squash, and mustard, and then cut back the orchard's clover ground cover using his kama, or Japanese sickle. He mulched the seeds with the clover cuttings, and cut back competing vegetation as needed until the seedlings became established.

Fukuoka's objective was to allow herbs and vegetables to select their own spots in the orchard and revert to the hardy growing habits of their wild ancestors. His objective was reached, as many of his plants did reseed themselves, and now come up on their own year after year, with only a minimum of effort.

In *The Complete Book of Community Gardening*, Jamie Jobb describes another approach to encouraging wild foods throughout the landscape, called "guerrilla gardening." The guerrilla gardener makes vacant lots and marginal lands productive again by broadcasting the seed of edible weeds and hearty domestic vegetables onto lightly cultivated ground. Plants suggested for guerrilla gardening include fava beans, clover, dandelion, marigold, sheep sorrel, vetch, alfalfa, borage, sunflowers, mustard, nasturtiums, flax, plantain, shepherd's purse, lamb's quarters, nettles, and amaranth.

Learning to identify and use wild foods, and encouraging herbs and vegetables to grow in a semi-wild fashion, can change your perspective on gardening and food. Wild and semi-wild foods offer a glimpse of the abundance we can create with conscious effort and vision, and help us make the entire landscape our garden.

"Lamb's-quarters contains over three times more calcium than spinach and at least five or six times more calcium than most of the other vegetables.... Amaranth, dandelion greens, purslane, and watercress are also very high in calcium. Lamb's-quarters also has the highest phosphorus content. Amaranth and purslane contain more iron than either of the well-known high-iron vegetables, spinach and chard. In potassium content, chicory is exceeded only by chard and spinach. Dandelion greens, curled dock, and lamb's-quarters all contain far more vitamin A than spinach or any other garden vegetable, and curled dock, watercress, lamb's-quarters, and amaranth are all rich in vitamin C, exceeded in concentration only by green peppers....

"Utilized to their fullest, weeds can increase the potential of your garden and can certainly make it more interesting. If you do not have a garden, try volunteering to weed those of your friends and relatives—they will thank you for it and you can gather your supper at the same time!"

Szczawinski, Adam F, and Turner, Nancy J. ***Edible Garden Weeds of Canada***. National Museum of Natural Sciences, National Museum of Canada, Ottawa, Canada K1A 0M8.

BIBLIOGRAPHY

Benoliel, Doug. Northwest Foraging. Edmonds, WA: Signpost Books, 1974.
Guide to over 50 of the most common and tastiest wild edibles in the Northwest, with information on "look alike" poisonous plants.

Domico, Terry. Wild Harvest, Edible Plants of the Pacific Northwest. Saanichton, BC and Seattle, WA: Hancock House, 1979.
Thirty-nine edible plants of the Northwest, each identified with a beautiful color photo. Information on habitat, gathering and preparation is provided.

Elliot, Douglas B. Roots: An Underground Botany and Forager's Guide. Old Greenwich, CT: Chatham Press, 1976.
Sixty useful wild roots, tubers, and rhizomes of North America discussed in detail.

Hardin, James W., and Arena, Jay M. Human Poisoning from Native and Cultivated Plants. Durham, NC: Duke University Press, 1974.
Plants can sicken or kill as well as feed and heal. This book provides an excellent background for understanding and responding to problems that can arise.

Hall, Alan. The Wildfood Trailguide. NY: Holt, Rinehart and Winston, 1973.
Popular introduction to 70 wild foods, with details on habitat, plant identification, and uses. Excellent drawings of each plant by Katie Matthysse (including the stinging nettle at the top of the left page).

Kirk, Donald. Wild Edible Plants of Western North America. Happy Camp, CA: Naturegraph Publishers, 1975.
Thorough and useful guide to wild edibles. Especially good coverage of plants in the Interior.

Manning, David; Jason, Nancy; and Jason, Dan. Some Useful Wild Plants for Nourishment and Healing. Vancouver, BC: Talon Books, 1972.
Practical, folksy, with fine illustrations by Robert Inwood.

Turner, Nancy J., and Szczawinski, Adam F. Edible Wild Plants of Canada. 4 Vols. 1978-1980. National Museum of Natural Sciences, National Museums of Canada, Ottawa, Canada K1A 0M8.
Nancy Turner's deep knowledge of Indian uses for native plants has been combined with Adam Szczawinski's knowledge of economic botany to produce an authoritative and elegant series intended to encourage greater understanding and use of wild plants. The four volumes include Edible Garden Weeds, Wild Coffee and Tea Substitutes, Edible Wild Fruits and Nuts, and Wild Green Vegetables.

PERIODICAL

Coltsfoot, Rt 1, PO Box 529, Elkins, WV 26241.
Bimonthly exchange of information on edible and useful wild foods.

Camas Lilies

The camas lily *(Camassia spp.)* was a staple food for the native people of the Northwest. Early pioneers who ate cooked camas bulbs described them as having the flavor of baked pears, quince preserves, figs, molasses, or licorice.

Luther Burbank, the noted plant breeder, took special interest in camas lilies. Beginning with wild plants found on Vancouver Island, at the turn of the century he was conducting breeding trials with over 20,000 bulbs.

Lilies of B.C.

Great Camas
(Camassia leichtlinii)

After 20 years of selection, Burbank concluded that, "In the end, the camassia will prove to be an ornamental of distinctive value, highly prized for its flowers. But it will be prized also for its bulb...which has undoubted food value, surpassing the potato even, both as to nutriment and flavor."

FOR MORE INFORMATION

Burbank, Luther. "The Camassia—Will it Supplant the Potato?" In Whitson, John, et al, eds. Luther Burbank, His Methods and Discoveries, Vol 7. New York: Luther Burbank Press, 1914.

Smith, Harriet Louise. Camas: The Plant That Caused Wars. 1978. Smith, Smith and Smith Publishing Co, 17515 SW Blue Heron Rd, Lake Oswego, OR 97034.
Descriptions of camas harvest and use, plus the tragic story of the destruction of camas fields. Annotated bibliography.

THE RUSSELL GRAHAM catalog lists seven varieties of camas lilies. (See the Seed and Nursery List)

THE MARIPOSA FOUNDATION is seeking to re-establish camas and mariposa lilies throughout their native range. The Foundation is also seeking seed of wild camas and is willing to trade bulbs for seed. For details on seed collection methods and additional information, send a SASE to The Mariposa Foundation, 415 Residence St, Moscow, ID 83843.

Seeds of the Future

FOREST SHOMER

Most plant species in temperate regions reproduce and evolve through their seeds. Wherever frosts occur, plants which bear seeds predominate over those which reproduce vegetatively.

During the 19th century, seed production was a local or regional activity. Some of the more skilled growers formed their own small companies and encouraged neighbors to grow seed. Shaker communities in upstate New York and elsewhere saved seeds and made them available to each other, and commercial efforts prospered in Connecticut, Pennsylvania and Michigan. Eventually, plant diseases drove seed production to specialized areas such as parts of Oregon, Washington, Idaho, Arizona and California, where favorable climate and soil conditions make it possible to grow vast quantities of seed for worldwide distribution. In order to maintain hardy, locally adapted varieties, however, it is necessary to decentralize seed raising and to re-establish it in the home garden, where seed selection began centuries ago.

For the sake of a healthy and diverse agricultural future in our own region, we need seed foundations, seed exchanges and local seed banks. Every small valley, island and regional watershed contains valuable native plants and adapted cultivars. Individuals and groups would do well to inventory, collect and catalog the flora of their neighborhoods for the benefit of future generations.

Seed collectors, selectors and plant breeders develop plant species from plants most like the ones wanted for cultivation in following seasons. Commercial food-plant breeders tend to select for yield and uniformity, home gardeners for earliness, hardiness, yield, taste and size, and flower growers for height, color, texture and duration of bloom. In other situations concern is with drought-hardiness, nutrient content, and insect- and disease-resistance.

Almost all of our primary food plants, cultivated herbs and flowers in the Pacific Northwest originated somewhere else. Because ill-adapted plants are difficult to maintain, finding or developing varieties of non-native plants which will naturalize can ease a gardener's labor. For this reason, gardeners will find their seed selection and saving efforts amply rewarded.

Some cultivated plants will persist for generations in the wild if their seeds fall on disturbed ground such as unused marginal land, broken sod or a roadside embankment. In a home garden many plants will be able to reproduce by themselves, or self-sow, if allowed to go to seed. In one or two plant generations this naturalizing process becomes apparent as volunteers begin to compete more vigorously with the weeds than their immigrant parents did.

VARIETAL TESTING

Selection and preservation of locally adapted varieties is key to the development of an indigenous agriculture. Each area's microclimate gives it unique advantages and disadvantages in the crops that can be grown. Selection of winter-hardy or drought-resistant vegetable varieties are two examples of the potential people have to choose food plants specifically adapted to their soils and climates. In addition, by selecting for insect and disease resistance, storage qualities, flavor and nutrition, important steps will be taken toward local and regional self-reliance.

A good example of people working together to develop locally-adapted vegetable varieties is in the Rogue Valley of southern Oregon. A group of farmers within Rogue Valley Tilth has initiated a testing program to evaluate which varieties of melon, corn, cucumber, onion, potatoes, peppers, carrots and tomatoes are best adapted to the area. Notes on the merits of various varieties and production techniques are swapped each week when participating farmers bring their produce into the local farmers' market.

FOR MORE INFORMATION

Wheeler, Mark. "Rogue Tilth Sends Down Roots." Tilth, Vol 6, No 2 (1980): 4-5.
Overview of the Rogue Tilth varietal testing project, with a sample of their testing calendar and program guidelines.

SAVING SEEDS

Seed collecting is best done during dry weather. Seed contained in a fruit may either be dried or fermented for a few days in the pulp to slough off the seed coats. Seed borne atop a flower stalk may be shaken when fully ripe into a container such as a breadpan or tight-woven basket, or the tops may be gathered and the seed rubbed out. The seed should then be winnowed by pouring between two containers in a light breeze. The wind will carry off the chaff and dust. For winnowing it is helpful to have a screen mounted on a wooden frame (#16 mesh is easiest to obtain).

Thorough cleaning is good protection against seed-borne pests and diseases. Stored seed must be very dry to discourage mold. Paper and cloth are adequate storage materials if humidity remains low. Glass helps protect against humidity and rodents.

Seeds metabolize in their resting state, gradually using up their carbohydrate reserves. With relatively constant environmental conditions, however, seeds will remain viable for a surprisingly long time. In some cases very old seeds which can still germinate produce superior plants.

The portion of the plant's life cycle spent in seed form might be considered a kind of sleep, in which the experiences of the preceding lifetime, including pollination and fertilization, are integrated in preparation for the next unfoldment. This time might also be considered meditation, where the life-force withdraws from the leaves and flowers to concentrate in the seed, or essence. The seed germinates and returns to plant form, with a slightly evolved consciousness that enables the species to survive in a changing world.

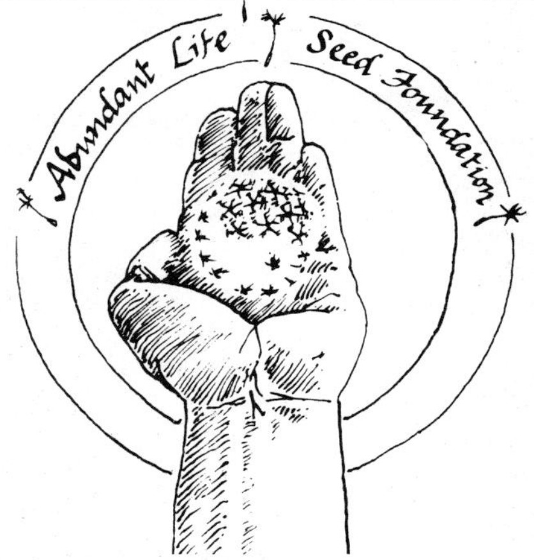

The Abundant Life Seed Foundation was founded in 1975. Its purposes are:

1) to acquire, propagate and preserve the plants and seeds of the native and naturalized flora of the North Pacific Rim, with particular emphasis on those species not commercially available, including rare and endangered species;

2) to provide information on plant and seed propagation;

3) to aid in the preservation of native and naturalized plants through cultivation.

The Foundation is a non-profit, federally tax-exempt educational and scientific institution. An annual seed catalog and quarterly newsletter are available for $2/yr U.S. and $3/yr in Canada.

Abundant LIfe Seed Foundation, Box 772, Port Townsend, WA 98368.

SEED PEOPLE

Seed Savers Exchange
c/o Kent Whealy
Rural Route 2
Princeton, MO 64673

A network of farmers and gardeners dedicated to saving and sharing rare vegetable varieties. The Exchange is a grass-roots effort to preserve the genetic heritage of plants handed down from generation to generation. Directories are published each year listing hundreds of members offering and seeking special vegetable varieties. Information on seed preservation and updates on the growing movement to save endangered varieties is provided in each issue. Membership is $6/year US and $8/year Canada.

Seed Banks Serving People. 1982. Meals for Millions, PO Box 42622, Tucson, AZ 85733. 75 pp, $3.50.

Proceedings of the first conference of North American seed savers. The gathering was a direct outgrowth of efforts (thus far unsuccessful) to limit seed patenting legislation. As a result people are seeking alternative solutions to preserving endangered varieties. Included in the proceedings is an overview of the world's seed crisis by Cary Fowler of the National Sharecropper's Fund, summaries of workshop presentations, and a proposal for a non-profit vegetable seed storage laboratory. The report serves as an introduction to seed saving and to the many groups struggling to preserve genetic diversity in wild and domestic plants.

Fowler, Cary. Graham Center Seed Directory. Frank Porter Graham Center, Rt 3, Box 95, Wadesboro, NC 28170. 1979. 17 pp, $1.00.

Guide to sources of traditional vegetables, herbs, fruits and nuts. Good overview of the crisis looming from the loss of genetic plant diversity.

Snyder, Barbara. "Seed Supplement." Tilth, Vol 5, No 3-4 (1980). $2.00

Collection of articles on the loss of genetic diversity, the implications of plant patenting laws, and growing your own seed, with references to literature and seed sources.

SEED SAVING BIBLIOGRAPHY

Hawthorn, Leslie R., and Pollard, Leonard H. Vegetable and Flower Seed Production. New York: Blakiston, 1954.

A detailed textbook for commercial seed growers. Includes information on harvesting and milling equipment.

Johnston, Rob. Growing Garden Seeds. Johnny's Selected Seeds, Albion, ME 04910. 1976. 32 pp, $2.30.

Thorough guide to growing, harvesting and cleaning vegetable seeds. For each species information is provided on spacing, culture, flowering, pollination, harvest, cleaning, and preservation. This is the best single introduction to seed growing and saving, of value both to home gardeners and small farmers.

Rogers, Marc. Growing and Saving Vegetable Seed. Charlotte, VT: Garden Way, 1978.

Step-by-step introduction to raising vegetables for seed, with instructions for over 50 vegetables. Includes details on hand pollination and germination testing.

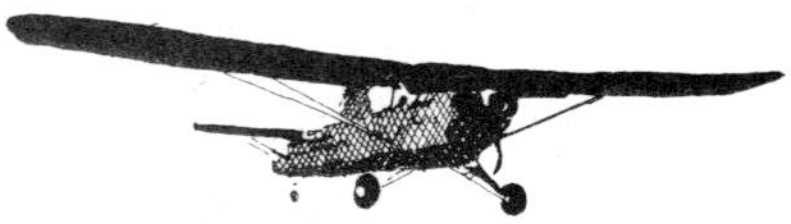

Woven bamboo airplane, Philippines, 1952.

Bamboo 竹

JIM ORJALA

Human association with bamboo goes back to the beginning of Asian culture. For 45 centuries, the hardy, rapidly-spreading bamboo has proven ideally suited to stitching together exposed, erosion-prone soils devastated by slash and burn agriculture or clearcutting.

But bamboo's present global omnipresence principally has been due to its usefulness to people. In the face of other resource shortages, cultures of Asia, Africa and Latin America have found bamboo a reliable provider of food, agricultural tools, and materials for manufacture and housing. It is estimated that one-third of humanity still makes use of bamboo.

GROWTH

More than 1200 species of bamboo grow around the world, to latitudes 50° north and south of the equator, from sea level to snow line. They range in form from shrubs several inches high to 120' giants more than a foot in diameter. Probably more than 50 bamboo species are adapted to the Pacific Northwest.

All species consist of an underground network of "rhizomes" which absorb, store and transport nutrients and fluids throughout a community of aerial stems, or "culms." In tropical regions "sympodial" types of bamboo have a short, horn-like rhizome, and culms emerge from their tips to form tight clumps.

Bamboo suspension bridge across the Min River, Szechuen, China.

Temperate, "monopodial" species have long, slender rhizomes. From late summer through fall they extend horizontally through the soil for up to 20 feet. Buds at the joints swell through winter, emerging as shoots in the spring. The culm then elongates rapidly (up to 2" per hour) to achieve full height by mid-summer. The culm grows no more, but lives for about a decade, changing its leaves each spring. The rhizome, however, continues to branch and produce new shoots, until the grove flowers. After flowering, the entire grove often dies.

On August 6, 1945, the Nuclear Age exploded over Hiroshima, leaving 200,000 dead and one lone survivor: a clump of bamboo scorched on one side, near the epicenter of the blast, on the site of the present Memorial to Peace.

HABITATS

In mountainous regions of China, natural stands of temperate bamboo form a middle story in deciduous and coniferous forests. The shallow rhizome competes little with the deep roots of mature trees. But bamboo is most common in relatively pure stands between forest and farmland, on steep banks of streams and rivers, where the rhizomes bind erosion-prone soil and the culms are convenient for use in adjacent villages and farms.

Most species appreciate warmth, humidity and partial shade. Unprotected south to west slopes should be avoided. Bamboos will grow in a wide variety of soils, but prefer slightly acid, well-drained sandy loam.

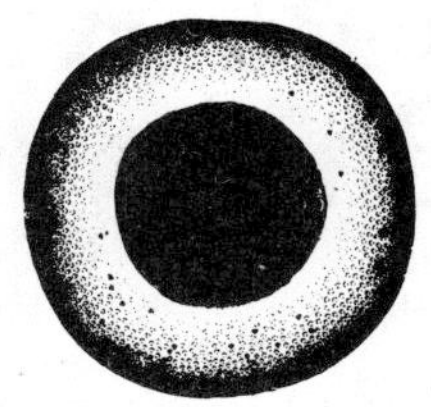

Bamboo culm cross section reveals optimal structural configuration; strongest fibers are found in increasing density towards the periphery, where stress is greatest.

PROPAGATION AND MANAGEMENT

Bamboo is generally propagated by rhizome cuttings. Rhizomes radiate from the base of a culm in the top two feet of soil. They are best dug while dormant, in February, cleanly severed and protected from dehydration. They should be young (yellowish in color), with good roots, at least ten good buds, and preferably with several one or two year old culms attached. After planting them to their original depth in a bed of compost mixed with soil, they should be well mulched, and watered regularly.

New shoots are left alone for three years. Then, selective thinning will hasten the development of full-size culms. This process can take five to ten years, as the rhizome grows and is able to support larger shoots and culms.

Many bamboos are rapid spreaders and can become invasive. Proper site selection is important. Bamboo generally spreads downhill. Water, marshes and rock provide natural containment. Continual harvest of shoots or tilling discourages spread. In small plantings, rhizomes can be checked by burying fiberglass, concrete, or galvanized metal sheeting two or more feet deep, extending about a foot above ground.

HARVEST AND USE

Bamboo fits well as an agro-forestry crop for small farms. A grove can be quickly established on land unsuited to other economic crops, and provides a high yielding, continual harvest with minimal care.

A bamboo culm matures at four years of age. It is best cut in mid-winter and left standing upright in the grove, with leaves intact, for two weeks. Culms are often submerged in water for about a month to leach out starches, then air-dried under cover for several months.

Shoots of all *Phyllostachus* species are edible, but "Moso" is the most popular. In Japan and China, Moso is intensively cultivated for shoots on humid, clayey soils, and yields three to ten tons per acre. Raw shoots are toxic, however, and should be boiled at least twenty minutes. They compare to apples in texture, artichoke in taste and onions in nutritive value.

Culms of some species have excellent technical properties, exceeding Douglas-fir in compressive strength while approaching steel in tensile strength. Lightweight, easily harvested and worked, bamboo is useful for stakes, orchard poles, fences, irrigation tubing, flutes, cups, furniture, and virtually every component of shelter.

Easily split when green, bamboo is twisted into rope, woven into baskets, or used as lath for reinforcement in plaster or concrete. It can be bent by subjecting it to heat or steam. When cool, it will retain its shape.

But most of all, bamboo exudes a calm vitality. Seeking enlightenment, Buddha found final shelter, and died, in a bamboo grove.

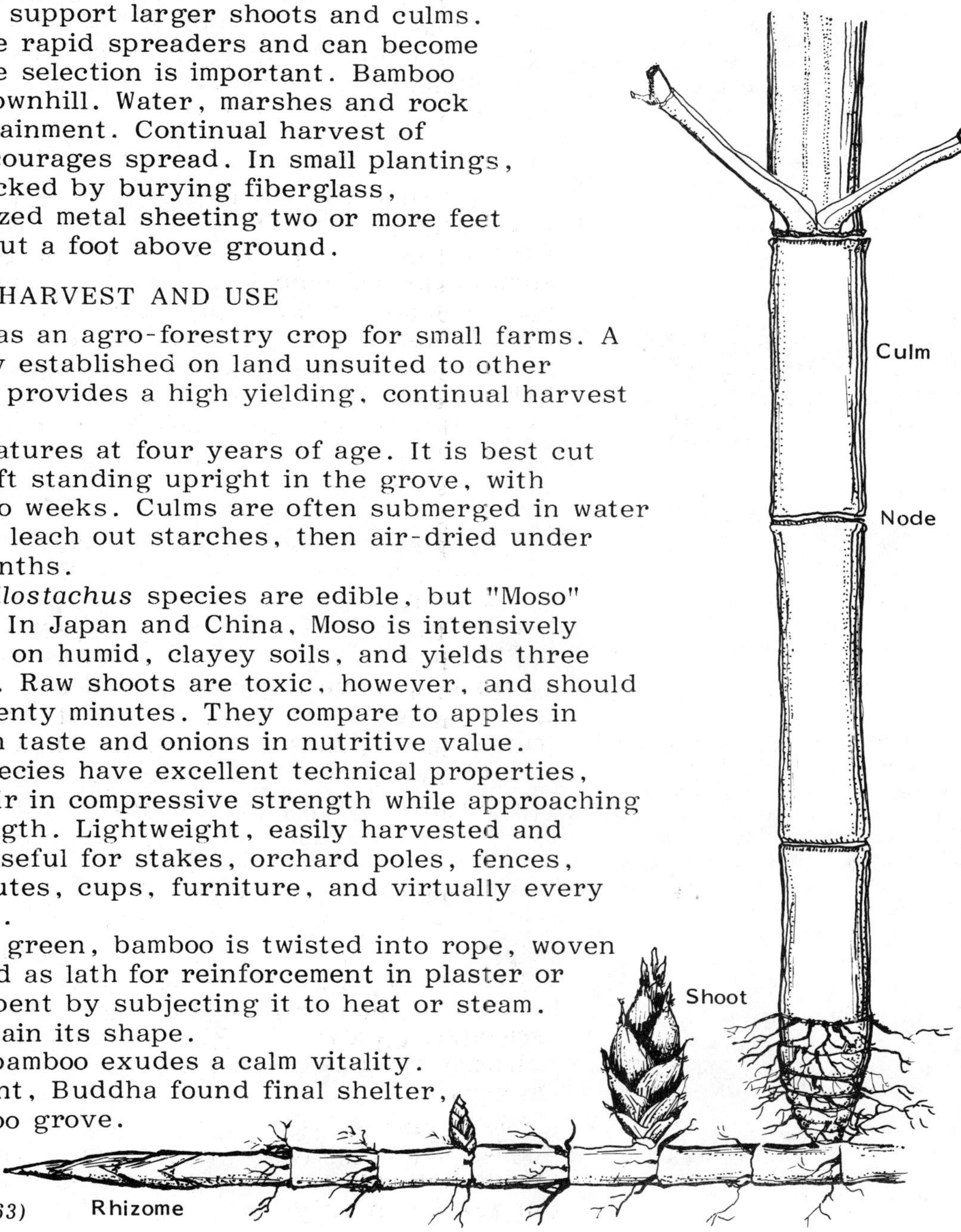

(See also the Bamboo Species Index, pp 162-163)

The Aprovecho Bamboo Project

Jim Orjala, author of the article on the preceding pages, is working with Aprovecho, an international appropriate technology group, to explore potential uses of bamboo in our region.

Begun in 1979 with the acquisition of a large grove of timber bamboo, the goal of Aprovecho's Bamboo Project is to explore the value of bamboo as a source of material for cottage industries, as a structural material for building, and as a food crop. Aprovecho is now evaluating species best adapted to our region, their cultural requirements, and the best techniques for establishment and maintenance.

Bamboo is one of the central elements of Aprovecho's new Rural Project being established on forty acres of land near Cottage Grove, Oregon. The Rural Project will include experiments in agriculture, forestry, and appropriate technology. A number of bamboo species are being planted on a variety of sites to test their compatibility as an understory in both Douglas-fir and alder forests, and to test their soil retaining characteristics in erosion-prone gullies. The new center will also feature experiments with bamboo as a building material.

Aprovecho is serving as a resource center for research and information on bamboo in the Northwest. They have begun a series of workshops on bamboo cultivation and usage, and will be distributing starts of selected species. For more information, send a self-addressed, stamped envelope to the Aprovecho Bamboo Project, 80574 Hazelton Rd, Cottage Grove, OR 97424.

FOR MORE INFORMATION

American Bamboo Society
1101 San Leon Ct.
Solana Beach, CA 92075

The liveliest, most authoritative source of information on bamboo in North America. The American Bamboo Society promotes the full utilization of bamboo through research and the dissemination of information. Membership, which includes a quarterly journal and occasional newsletters, is $12 per year.

"Bamboo Species Sources" is a combined species index and source list. It includes 65 species of bamboo and 15 sources of plants. For a free copy of the list send a self-addressed stamped envelope to the American Bamboo Society at the address above.

See also the Bamboo Species Index, pp. 162-163.

BIBLIOGRAPHY

Haubrich, Richard. "A Selected, Annotated Bibliography on Bamboo." In The Journal of the American Bamboo Society, Vol 2, No 3, August, 1981, pp 51-60. $5.00.

Detailed review of 26 books and articles which provide information on how bamboo is grown and used around the world.

Marden, Luis. "Bamboo, the Giant Grass." National Geographic. 158: 503-529 (October, 1980).

Beautiful introduction to the many forms and uses of bamboo, with stunning photos and drawings.

Uncommon Mushrooms

MIKE MAKI

The common button mushroom *(Agaricus bisporus)* is about the only species now grown commercially in North America. In Asia and Europe, other species are also grown on a large scale, including the oyster mushroom *(Pleurotus ostreatus)*, shiitake *(Lentinus edodes)*, and "velvet ears" *(Flammulina velutipes)*. Other choice edible mushrooms, such as morel and chanterelle, have not been successfully domesticated because their associations require that the host tree be living in the soil.

Harvest and culture of mushrooms for purposes other than food are also becoming popular in the Pacific Northwest. The well-known psychoactive properties of some mushrooms, especially members of the genus *Psilocybe*, have led to research and development of cultural techniques suitable for home production. The literature on mushroom culture, until recently, has been devoted almost exclusively to *Agaricus bisporus*, although informative and practical guides on culturing other mushrooms are now available.

MUSHROOM BIBLIOGRAPHY

Chang, S.T., and Hays, W.A., eds. The Biology and Cultivation of Edible Mushrooms. New York: Academic Press, 1978.

One of the most comprehensive books on mushroom culture. Includes shiitake, oyster mushrooms, and other species with commercial potential.

McKenny, Margaret, and Stuntz, Daniel E. The Savory Wild Mushroom. Seattle: U of W Press, 1971.

The most popular guide to wild mushrooms in the Pacific Northwest.

Stamets, Paul, and Chilton, Jeff. The New Age Mushroom Cultivator. Seattle: Homestead Book Co, in press.

An up-to-date guide to mushroom growing, with valuable (and much original) cultural information.

Stamets, Paul. Psilocybe Mushrooms and Their Allies. 1979. Homestead Book Co, PO Box 31608, Seattle, WA 98103.

An excellent study of several related psychoactive species found in the Pacific Northwest. Cultural information is suitable for growing other wild species. Classic in its field.

MUSHROOM SOURCES

THE KINOKO COMPANY
PO Box 5425
Oakland, CA 94603

Kits for producing five species of edible mushrooms, plus other mushroom products.

KURTZMAN'S MUSHROOM SPECIALTIES
815 S Harbour Way, #12
Richmond, CA 94804

Extensive catalog of mushroom spawn, equipment and literature.

MUSHROOMPEOPLE
PO Box 158
Inverness, CA 94937

Source for books, supplies and spawn of uncommon mushrooms, including *P. cubensis*.

Fruits of Sustainable Agriculture

MICHAEL PILARSKI

Had Johnny Appleseed a compatriot named Jimmy Cornseed, Jimmy's memory would have survived through only one, maybe two, harvests, for short is the life cycle of the corn plant. Johnny, on the other hand, sowed the seeds of his legend by planting trees—long-lived, dependable, fruit-bearing trees. So people remember him, through decades of apple pies, apple cider, and apples to-keep-the-doctor away, harvest after harvest from those first trees.

Establishing an orchard, vineyard, or berry patch requires a substantial investment of care, capital and patience, for the rewards are not quick to come. But, like all wise investments, the return over time is substantial, and at a yearly cost of only a fraction of annual crops.

In the pages to follow, you will learn about varieties of nuts, berries and fruits that may not be familiar to you. While we have not neglected entirely the more common domestic varieties, we have chosen to emphasize the lesser-known wild cultivars. You'll find that varieties native to your region within the Pacific Northwest are well-adapted to modest but dependable food production, with fewer cultural demands than domestic varieties. And you'll discover new foods, new flavors: salal jam on pine-nut bread, huckleberry pie and rose-hips tea for dessert.

We begin with berries, wild and domestic.

Native Berries

Furlong, Marjorie, and Phil, Virginia. Wild Edible Fruits and Berries. Healdsburg, CA: Naturegraph, 1974.
An introduction to a world of delicious wild edibles.

There are native berries adapted to almost every Northwest landscape, from timberline to ocean shore, from dense forest to open plains. In many forest ecosystems, berry plants play a pioneer role by quickly and densely colonizing disturbed land. After a fire, clearcut, or natural landslide, there is often a profusion of berries until the forest trees grow through and shade them out. In some cases this will take only five or ten years—in others it takes much longer.

US Forest Service

There are over 100,000 acres of wild huckleberry fields in the mountains of Oregon and Washington. Thousands of people gather the fruit each summer. Research is now underway on techniques for maintaining and improving the wild huckleberries.

Some berry species are intolerant of shade and die out quickly, but others, such as salal *(Gaultheria shallon)*, some huckleberries *(Vaccinium* spp.) and thimbleberries *(Rubus parviflorus)* persist for many years in the gradually deepening shade. Thimbleberry grows in dense thickets under a full canopy of alder, birch or aspen.

Berries are an excellent food, containing many vitamins and easily available sugars. The edible berries of the Northwest were well known to the native peoples who dried them into cakes or used them in pemmican, a high calorie, easily preserved food made from berries mixed with dried meat and fat, and pounded into cakes. Almost all wild berries are good processed into jams, jellies, concentrates, juices and syrup.

Edible berries are found in a wide variety of genera. They vary in their growth habits from cane plants to small shrubs to large woody shrubs and even small trees. Berry plants can be propagated by root cuttings, suckers or tip layering. Plants which are noted for the quantity, size, or tastiness of their berries, or for their resistance to pests or adverse conditions, can be used for breeding improved varieties.

Following are descriptions by genus of some of the most important or useful native berries.

RIBES

There are over 200 *Ribes* species in the world, of which over 30 are native to the Northwest. Gooseberries and currants are of this genus. Native species are found in almost every type of habitat in the Northwest. The berries are small, and either acid or insipid. The berries contain pectin and are good additions to other berries when making jelly or jam.

Unfortunately, the *Ribes* are host plants for white pine blister rust, which has infected white pines throughout its range. The Forest Service conducted a campaign in northern Idaho in the 1930's to eliminate the *Ribes* from areas where western white pine is an important timber tree. The campaign was not successful, and the disease remains a serious problem. Check with local officials to see if there is a quarantine in your area before introducing any *Ribes* plants.

COMMON WILD GOOSEBERRY *(R. divaricatum)* is armed with spines at all joints of the stems. Gooseberries are found in coastal forests and along roadways from British Columbia to California. The berries may be used green or ripe for jam, jelly, pies and wine.

FLOWERING CURRANT, RED FLOWERED CURRANT *(R. sanguineum)* is a thornless bush two to ten feet high. It is best known as a flowering shrub growing in low elevations along the coast. The berries are better cooked than fresh, and are most often dried.

VACCINIUM

The *Vaccinium* genus includes the huckleberries, blueberries, cranberries, bilberries and whortleberries. There are about 130 species of *Vaccinium* in the northern hemisphere, of which 16 are native to the Pacific Northwest. These species range from eight-inch shrubs to six-foot bushes. They flower in the spring, and the berries ripen from July to October.

Huckleberries are the most widely used wild berry in the region. The US Forest Service estimates that huckleberry fields occupy over 100,000 acres in Washington and Oregon, with several hundred tons of berries harvested annually in the Puget Sound region alone. In the Maritime there are species found at every elevation, and as a general rule, the higher the elevation the tastier the huckleberries. There are fewer species of huckleberries in the Interior, and they are usually found at elevations above 3,000 feet as an understory in moist forests. Some *Vaccinium* species of note are:

BIG HUCKLEBERRY *(V. membranaceum)* is related to the commercial blueberry. It was the most widely used *Vaccinium* by the Native Americans of the Northwest. They kept the huckleberry stands open and productive by periodically burning them. Yields of Big Huckleberry can reach 100 gallons an acre. At the 1981 market price of $10 to $20 per gallon, picking huckleberries can be a rewarding seasonal job.

BLUELEAF HUCKLEBERRY, ALPINE BLUEBERRY or RAINIER BLUEBERRY *(V. deliciosum)* is a common berry found near the timberline in the subalpine forest zone of the Cascades. It is almost always found above 5,500 feet. The clear mountain air and cool climate help make these berries especially sweet and flavorful. The shrubs form a dense, scrubby ground cover 12 inches high so picking is done on hand and knees.

Minore, Don. The Wild Huckle-Berries of Oregon and Washington: A Dwindling Resource. USDA FS Research Paper 143. Portland: PNW Forest & Range Exp. Stn, 1972.

In 1972, an estimated 160,000 acres supported huckleberries in Oregon and Washington, but the area has been dwindling. This pamphlet gives descriptions and an identification key of the 12 Northwestern *Vaccinium* species. The author also describes management techniques and habits of the lowbush and highbush blueberries from eastern North America. Besides offering possibilities of production of these eastern species in the Pacific Northwest, their management techniques may have some application to our western species. There are over 50 references covering Western and Eastern *Vaccinium*.

Minore, Don; Smart, Alan W.; and Dubrasich, Michael E. Huckleberry Ecology and Management in the Pacific Northwest. USDA FS Gen Tech Rep PNW 93. Portland: PNW Forest & Range Exp Stn, 1979.

The production of big huckleberry *(Vaccinium membranaceum)* is declining in many Northwestern huckleberry fields as they are invaded by subalpine trees. Seeking ways to halt this invasion and increase berry production, the authors studied huckleberries in the Cascade Range of Oregon and Washington from 1972 through 1977. This illustrated report includes descriptions of the experiments performed, results, conclusions, and management recommendations. This report also lists the titles of other recent pamphlets on huckleberry research.

The two publications above offer some basic groundwork for a strategy to increase *Vaccinium* species in the Northwest. There are dozens of huckleberry, blueberry, and cranberry species that are or could be grown in the Northwest. At this time extensive research has only been done on the eastern highbush blueberry and the eastern cranberry. Increased research offers the potential to bring other *Vaccinium* species under cultivation. There is also a need for increased experimentation into intensively managing wild *Vaccinium* stands and the establishment of new patches in recent clear-cuts.

"Huckleberry Heaven...the Cascades." In Sunset Magazine. August, 1981, pp 26-27.

A detailed guide for when and where to harvest the native huckleberries in the wild.

AMELANCHIER

There are many local variations, but at least 25 different species of this genus are recognized, with two native to the Pacific Northwest, *Amelanchier alnifolia* and *A. utahensis*. *A. alnifolia* is found throughout the Northwest, while *A. utahensis* is found only in the dry, southern part of the region. They are both commonly called serviceberry, but some say "sarviceberry," and in western Canada they are called Saskatoon berries.

The importance of the serviceberry among the native fruits of the Interior is demonstrated by the following quote by Nancy Turner from Food Plants of British Columbia Indians:

> "Of all the berries and fruits eaten by the Interior Indians of British Columbia, this was the most popular and most widely used, especially in the southern and central Interior. Among the northern groups, such as the Beaver and Slave, Saskatoons were often the only type of vegetable food available in any quantity. So important were these berries that the various interior Salish groups have developed a classification for different varieties of Saskatoon which is far more detailed and complex than that of professional taxonomists. On the basis of habitat, blooming and ripening time, growth form, and size, colour, seediness, and taste of the berries, they distinguish many varieties, each with its own particular advantages and disadvantages as a food."

The serviceberry commonly grows 10 to 15 feet tall, but on good sites it can reach 25 feet. It is remarkably hardy, being found in open woodland and in the driest of landscapes, as long as there is a rock fracture or slide into which it can sink its tenacious roots. This is a plant which could be of great value in rehabilitating the arid and semiarid landscapes of the Interior.

SAMBUCUS

BLUE ELDERBERRY *(Sambucus caerulea)* is the blue elder of the Interior Pacific Northwest, native from British Columbia to California, and as far east as Montana and Utah. It is considered the best of the six edible species found around the world. Blue elderberry grows to thirty feet in height on good sites. It sometimes takes the form of a tree, but usually grows as a many-stemmed shrub. The valleys on the eastern slopes of the Washington Cascades contain the region's densest stands. A good way to harvest elderberry clusters is to pull them down with long, hooked poles or with long-handled pole loppers such as those used in orchard pruning.

RED ELDERBERRY *(Sambucus racemosa)* was used by the Northwest native peoples. The berries were usually mixed with other berries to make them palatable. The sap of this plant is poisonous.

VIBURNUM

There are about 120 species of *Viburnum* scattered across the Northern Hemisphere, of which three are native to the Pacific Northwest. The *Viburnums* are considered among the most valuable genera of ornamental shrubs in the world. Besides showy flowers and decorative fruit, *Viburnums* have handsome foliage which turns to bright colors in autumn. A number of North American species, including two in the Pacific Northwest, yield edible berries.

SQUASHBERRY *(V. edule)* is native throughout the forests of British Columbia, Washington, Idaho, Montana and northern Oregon. It forms a shrub from 1 1/2 inches to 8 feet tall in moist woods, along streambanks and swamps, and at the edges of bogs. The bright red berries are tart, with a cranberry-like flavor. They were picked by the native peoples while hard and sour and stored for winter use, or allowed to ripen on the bush (usually after frost) and then eaten at feasts.

HIGHBUSH CRANBERRY *(V. opulus)*, also known as wild guelder-rose or snowball, is found primarily in British Columbia. It prefers moist woods, and grows in thickets along streams bordering low woods, or by fences and walls. It has tasty, acid berries.

RUBUS

There are over 400 species in the *Rubus* genus worldwide. The genus includes domesticated blackberries, raspberries, loganberries and boysenberries, as well as native dewberries, salmonberries and thimbleberries. At least 12 species are native to the Pacific Northwest. They are an important source of wild food for animals and birds.

SALMONBERRY *(R. spectabilis)* is common on wet sites throughout much of the Maritime. It is especially widespread near coastal areas and in the Coast ranges. Salmonberries form dense thickets under red alder in shady swamps and in moist sites along roadsides. The yellow or orange berries are generally tasty. The Native Americans ate them fresh but considered them too watery for drying.

THIMBLEBERRY *(R. parviflorus)* has large, maple-shaped leaves which make the 3 to 5 foot thornless bushes easily identifiable throughout the season. The tasty red berries were eaten fresh and dried by the Native Americans. Thimbleberries grow in clearings, along roadsides and shorelines, and in open woods. They are among the most shade tolerant of the native berries.

DEWBERRY or TRAILING BLACKBERRY *(R. ursinus)* is an indigenous blackberry that forms trailing canes up to 20 feet long. It grows in clearings throughout the Maritime and in the moist forests of Idaho. Several select cultivars have been developed, including thornless varieties.

HIMALAYA BLACKBERRY *(R. procerus)* and EVERGREEN BLACKBERRY *(R. lacianatus)* are non-native berries which have become prominent members of the landscape throughout the Maritime lowlands. They are a widely available free food during their season. The Himalaya blackberries (developed by Luther Burbank) are sweeter and softer then the Evergreen. Both are excellent eaten out of hand or processed. Jam makers buy them from local pickers, providing many people with an important source of seasonal income. There are tasty blackberry varieties available from nurseries which are small in form and not as invasive as the wild plants.

BLACKCAP RASPBERRY *(R. leucodermis)* is a raspberry-like plant whose berries turn purple-black when ripe. The fruits are drier than red raspberries but have a good flavor. The plant prefers sunny locations, and can take a drier soil than any of the other Rubus species.

ROSA

Wild roses are widespread throughout the north temperate regions of the world, with six species native to the Northwest. Though few people consider them berry plants, the fruits, or rosehips, are well worth harvesting for their Vitamin C content. They can be dried whole and crushed to use for tea, or the seeds can be scraped out and the pulp preserved with honey and added to a small amount of brandy for cough syrup. Rosehips also make a fine wine. Strings of red "hip-beads" are often hung on Christmas trees and walls for winter decoration or used by children to make necklaces.

NOOTKA ROSE *(Rosa nutkana)* yields the largest and tastiest native rosehips. It is common along roadsides, shorelines, and in meadows of moist, open woods, where it often forms dense thickets. Nootka rose also has the largest flowers of the native Western roses.

WOODS ROSE *(Rosa woodsii)* is found on the edges of the sagebrush/bunchgrass areas in the Interior. Their summer green and autumn red help brighten up the arid landscape.

OTHER BERRIES

SALAL *(Gaultheria shallon)* is common throughout the Maritime, where it forms dense evergreen thickets of oval, glossy-leaved bushes in disturbed sites. It also grows in thin stands in humid forests up to 2,500 feet in elevation, and in the moist forests of the Interior. The berries are sweet and juicy and can be processed into jams and jellies. They are sometimes combined with Oregon grape, gooseberries or other more acid fruits. Salal berries were dried in large quantities by natives for winter use. They do not stay on the bushes very long once they are ripe, so they must be picked promptly.

OREGON GRAPE *(Berberis spp.)* is very drought tolerant and has been widely used as an ornamental both here and abroad because of its glossy, holly-like leaves and attractive yellow flowers. The dark blue berries are generally ignored, except by the few who like the fine flavor beneath their puckeriness. They contain pectin and are good for jellies, either by themselves or with other fruits or berries. They also make one of the finest native berry wines. The roots of the Oregon grape are bright yellow and are used to make dye of the same color.

BUNCHBERRY *(Cornus canadensis)* is a diminutive relative of the Pacific Dogwood. It is a perennial ground cover plant which seldom reaches 12 inches in height. In moist woods it spreads by underground rhizomes, often forming dense mats. The bright red, pulpy berries are sweet and tasty and the flowers are exquisite. There is a demand for this plant as an ornamental ground cover for shady locations.

BIBLIOGRAPHY

Adams, Blair. Small Fruits for Home Gardens. Extension Bulletin 0708. Pullman, WA: Washington State University, 1979.

Overview of selection and culture of strawberries, raspberries, blackberries, blueberries, currants and gooseberries for home gardens. Useful introduction, with lists of varieties. Of value throughout the region.

Blueberry Propagation. Extension Bulletin 124. Pullman, WA: WSU, 1966.

Because of the difficulties associated with propagating blueberries from root cuttings, almost all blueberries are propagated from hardwood and softwood cuttings. This pamphlet explains how.

Childers, Norman F; Eck, Paul; et al. Blueberry Culture. New Brunswick, NJ: Rutgers Univ. Press, 1966.

Containing a wealth of detail, this is the most extensive textbook available on blueberry production.

Clark, J. Harold. Growing Berries and Grapes at Home. New York: Dover, 1981.

Thorough guide to the culture of berries and grapes. Excellent coverage of all major berries.

Darrow, G.M.; Draper, A.D.; and Scott, D.H. Commercial Blueberry Growing. USDA #2254. Washington, DC: USGPO, 1978.

The basic facts to get your berries in the ground and growing.

Darrow, George M., The Strawberry: History, Breeding and Physiology. New York: Holt, Rinehart and Winston, 1966.

The most comprehensive textbook on strawberries in print.

Dodge, John C., and Doughty, Charles C. Cranberry Production in Washington. Extension Bulletin 645. Pullman, WA: WSU, 1972.

"Cranberries are grown near the Washington Coast in the marsh or bog areas not adapted to other crops. Approximately 1,000 acres produce an average of 5,000 tons of berries annually. Washington's cranberry crop amounts to about 10% of the total United States production. The first cranberry bog in the state was planted in 1882."

Doughty, Charles C., and Robbins, Jo Ann. Gardening with Blueberries Extension Bulletin EM 3885. Pullman WA: WSU.

Quick but comprehensive; a good start.

Hall, J.W. Thornless Blackberries for the Home Garden. USDA #207. Washington, DC: USGPO, 1975.

"For years home gardeners hesitated to grow blackberries in the backyard because of the annoyance of the thorns. Then two thornless varieties, Smoothstem and Thornfree, were developed by the USDA. Popular with home gardeners and proprietors of U-Pick enterprises, these productive blackberries are easy to handle and require little summer pruning. The fruits are firm and highly flavored."

Hendrickson, Robert. The Berry Book. Garden City, NY: Doubleday, 1981.

Well-researched, witty, with a basketful of obscure tidbits in addition to the expected cultural information. One chapter, "Alternate Lifestyles for Strawberries," shows you how to grow strawberries where you didn't think you had the room.

Logsdon, Gene. Successful Berry Growing. Emmaus, PA: Rodale, 1974.

All the major berries are covered, along with grapes and some lesser-known fruits.

Scheer, P.A., ed. Strawberry Production Guide. Extension Bulletin 3905. Pullman, WA: WSU, 1978.

"The cultivated strawberry belongs to the family *Rosaceas*, the rose family and to the genus *Fragaria*. This genus has approximately twelve native species throughout the world. The history of the strawberry goes back as far as the ancient Romans and possibly the Greeks. The modern strawberry, however, did not materialize through long cultivation of select wild types in Europe. Rather, it originated as something entirely new from the genetic mixture arising from the crosses between two American species."

Scheer, William, and Garren, Ralph. Commercial Red Raspberry Production. Extension Bulletin 126. Pullman, WA: WSU, 1981.

One cup of red raspberries provides 50% of the recommended daily allowance of Vitamin C, one-tenth the iron, and useful amounts of other vitamins and minerals, but only 70 calories. This publication is an outstanding compilation of all available information related to red raspberry production in the Pacific Northwest.

Fruit of the Vine

Kiwi

The kiwi fruit we see in markets, *Actinidia chinensis*, is native to China. Formerly called the Chinese gooseberry, it was transported to New Zealand, where it was renamed kiwi fruit. The United States still imports kiwi fruit from New Zealand, but increasing quantities of this fruit are being grown in California. The kiwi is high in Vitamin C and contains beneficial enzymes similar to those found in papaya. Kiwi fruits can be stored for months in a cold cellar and ripened a few at a time at room temperatures.

Actinidia chinensis is only one of the 36 named species of *Actinidia* which are native across Asia, from Java and the Himalayas through much of China, Japan, Korea, Siberia and the Kurile Islands. Tanaka's Encyclopedia of Edible Plants lists sixteen species as bearing edible fruit. The three species acknowledged as bearing the tastiest fruit are *A. chinensis*, *A. kolomikta* and *A. arguta*.

The Sunset Western Garden Book lists *A. chinensis* as hardy in Zone 4, which is borne out by fruiting vines in a number of locations around Puget Sound, including Seattle, Bainbridge Island, Tacoma, Issaquah and Snohomish. A vine in Snohomish has survived 0°F two winters with minimal die-back. *A. kolomikta* and *A. arguta* are hardier than *A. chinensis* and have survived the cold winters at Lawyer's Nursery on the Clark Fork River in western Montana. *A. kolomikta* is grown commercially in central Asia.

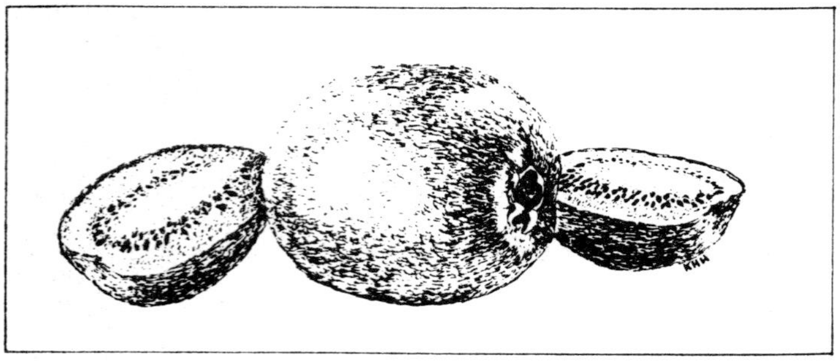

Actinidias are handsome climbing vines which can grow to great size and length. They prefer a somewhat moist and rich soil. They do not do well in shallow or chalky soils, or in soils with low humus content. *A. chinensis* needs full sun to do well in the Northwest. *A. kolomikta* and *A. arguta* can be grown in partial shade in the Maritime, but will need full sun in the Interior.

The *Actinidia*'s climbing vines are good for covering trellises, arbors and the walls of buildings. If the vines are too tender to survive the winter they can be gently taken down and covered. Temperatures which will kill the vines often leave the roots unharmed, and they will resprout in the spring. The *Actinidias* are easily propagated by winter cuttings. They can be grown from seed planted in flats and started in the greenhouse.

Commercial sources of plants include Forestfarm Nursery, Burnt Ridge Nursery, Bountiful Ridge Nurseries, Lakeland Nursery, Wayside Gardens, Calchico Kiwi Company, and Mellingers. See the Seeds and Nurseries List for addresses.

Fletcher, W.A. Growing Chinese Gooseberries. Bulletin 349. New Zealand Ministry of Agriculture and Fisheries, Wellington, New Zealand, 1976.

Planting Guide Supplement for Chestnuts, Filberts, Citrus, Kiwifruit, & Roses. Dave Wilson Nursery, 4306 Santa Fe Avenue, Hughson, CA 95326. $2.

This pamphlet gives much information on planting, irrigating, fertilizing, pests and diseases, trellising, training and pruning of the kiwi vine.

Kiwi Growers of California, Rt. 1, Box 445, Chico, CA 95926.

They sponsor a kiwi conference in early May each year during kiwi bloom time, as well as provide information.

Fruit of the Vine

Grapes

The extensive planting of vineyards in the lower Columbia Basin has led Washington to become second only to California in grape production. Small commercial vineyards are now successfully expanding into the Willamette Valley and western Washington.

Clore, Walter J. and Tukey, Ronald B. Grape Varieties for Eastern Washington. Extension Bulletin #394. Pullman, WA: WSU.

A brief circular describing several grape varieties for warm, arid climates.

———. Grapes: Their Characteristics and Suitability for Production in Washington. Extension Bulletin #FB 635. Pullman, WA: WSU, 1973.

Excellent specifics relating grape variety to site conditions throughout both the maritime and arid Northwest.

Ordish, George. Vineyards in England and Wales. London: Faber & Faber, 1977.

Pearkes, Gillian. Growing Grapes in Britain. Amateur Winemaker Publishers, South Street, Andover, Hamps, UK, 1972.

Britain's cool, wet climate, similar to that of the maritime Northwest, long discouraged serious grape growing, until a recent surge of interest and experimentation established the young British grape industry. The two books above detail many of the methods and varieties that British grape growers are using to overcome their climatic handicaps.

Winkler, A.J. General Viticulture. Berkeley: University of California, 1974.

The most comprehensive text on grape production. Oriented toward commercial growers, but of value to the home grower as well. It covers propagation, pruning, irrigation, pests, and diseases of table, wine and raisin grapes.

Galet, Pierre. A Practical Ampelography. Ithaca, NY: Comstock Publishing Associates, 1979.

Ampelography is the art of identifying the thousands of different varieties of grapes. Pierre Galet is the world's leading ampelographer and this book is the most recent revision of what has been the standard identification guide for grapes. Hundreds of varieties of European and American grape cultivars and rootstocks are described, along with the qualities of each for wine and their resistance/susceptibility to pests and diseases.

Norton, R.A. Growing Grapes for Wine and Table in the Puget Sound Region. Extension Bulletin #EB 0775. Pullman, WA: WSU, 1979.

This publication discusses some of the unique features of growing grapes in the Puget Sound region, with special emphasis on suitable cultivars to make the most out of four different climate zones within the region.

Northwest Tree Fruits

In the early years of immigration and settlement, fruit varieties proliferated as seeds and seedlings brought from Europe and other parts of the world were propagated in the Northwest. Amateur and professional plant breeders contributed to the creation of new varieties and the naturalization of others.

During the past half-century, plantings of fruit trees, especially on a commercial scale, have become less diversified, causing the number of varieties available through nurseries to decline sharply. However, renewed interest in homesteaders and backyard gardeners are showing in antique and locally adapted varieties has recently stimulated nurseries to again expand their listings.

Following is a brief list of fruit trees and their preferred habitats.

APPLES prefer a sunny location in meadows, on hillsides, along roads or on the edges of forests. Soil type is not as important as depth and drainage. Standard size trees are fairly drought tolerant.

CRABAPPLES are the most cold-hardy domesticated fruits, and are sometimes used as the rootstock for grafted apples. Crabapple fruit is suitable for cooking, jams, pickles, and ciders. Breeding experiments with the native Oregon crabapple suggests the promise of better-producing varieties for the Maritime.

APRICOTS are surprisingly tough for trees which produce such a luscious, soft fruit. Apricots are very drought resistant and can survive in rocky soils. They prefer deep soil for their extensive root system, which is the secret of their drought tolerance. Although able to withstand harsh winters, apricots tend to bloom early and are apt to lose their crop to spring frosts if not sited correctly. Hot dry weather is best for disease-free ripening. Apricots are thus better suited to the arid Interior than to wet Maritime zones.

PEARS are almost as cold-hardy as apples and can tolerate heavy soils better than other domesticated fruits. They are also not as demanding of full sun as other fruits. The damp

"Crabapples were an important food for all Coastal Indian groups of the Province. They were generally harvested in late summer, when full-sized but still green, and allowed to ripen during storage.

"The Straits Salish, Holkomelem, Squamish, and Nootka often hung harvested crabapples in cat-tail bags until they were ripe, then ate them raw or cooked and sometimes mashing them and mixing them with other fruit such as salal. ...In Nootka territory, crabapples were most plentiful and of greatest importance at the head of the Alberni Canal, although they grew sporadically along the outer Coast.

"According to Sproat (1868), when the white settlers started to encroach on Indian land, the Nootka, bitterly resentful, cut down all the wild crabapple trees around the white settlements so that their last crop of fruit would at least be easy to harvest."

Nancy Turner, *Food Plants of the British Columbia Indians, Part 1, Coastal Peoples*. British Columbia Provincial Museum, Victoria, 1975, pp 202-203.

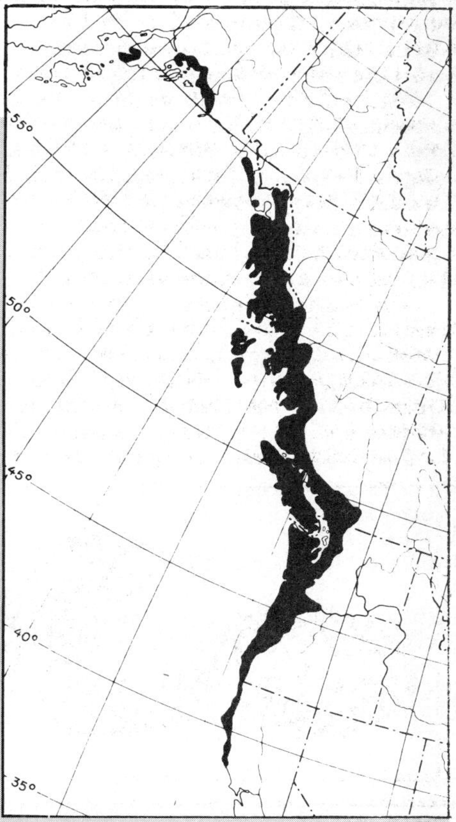

E.L. Little, *Atlas of US Trees*, Vol 3

Wild Crabapple

HABITAT—Moist woods, streambanks, swamps and bogs, often in dense thickets.

Fruit & Nut Inventory

As part of a long range effort to preserve rare germ plasm, the USDA's Beltsville Fruit Laboratory is compiling a computerized inventory of all known fruit and nut varieties being grown in Europe and North America. Nearly 40,000 varieties have been identified to date.

Sources of uncommon fruit and nut varieties are provided to both professional and amateur orchardists and to plant breeders. For more information write to the Fruit Laboratory, Beltsville Agricultural Research Center, Beltsville, MD 20705.

climate of the Maritime, however, predisposes pears to leaf diseases unless they are in full sun.

PLUMS of Asian, American and European varieties all tolerate a wide range of sites and conditions. There are plum varieties suitable for almost every Northwest site. Some plums sucker readily and are thus useful in hedgerows.

PEACHES require hot, dry climates. The trees are at their best in deep, sandy, well-drained soils. Peaches need plenty of warm weather and sunlight to ripen properly. Nectarines are similar in requirements and habits.

SWEET CHERRIES are one of the largest standard-sized fruit trees. A well drained soil is essential. Cherries need full sunlight and adequate moisture to ripen well. Mulberries are often planted adjacent to them to divert birds from devouring ripening cherries.

SOUR CHERRIES are hardier than sweet cherries. They can withstand drought and cold. Sour cherries can be kept low by pruning and are therefore useful in mixed plantings or as middle story trees.

PROFILES ON LESSER KNOWN FRUITS

There are a number of temperate and subtropical fruits which are not well known or widely grown but which are adaptable to the Pacific Northwest. A few of these plants will require some pampering to set fruit and ripen, but may be worth consideration by the backyard gardener with a yen for the rare and succulent in fruits.

FIGS *(Caricus* spp.*)* have been grown and ripened in the Puget Sound region for many years. The brown turkey fig is most common and breeders in Oregon offer several superior varieties. They have also been grown in the Interior, but need special care in order to produce and ripen fruit.

PERSIMMONS *(Diospyros* spp.*)* are very popular in the Orient where hundreds of varieties have been developed over the centuries. Oriental persimmons are producing well in Ashland, Oregon, and in parts of the Puget Sound region. The hardier, American persimmon ranges as far north as Michigan and Wisconsin and is worthy of trial in the lower Columbia Basin. Persimmon is in the Ebony family and its wood is highly valued for veneer and specialty items.

MEDLAR *(Mespilus germanica)* is a hardy pome fruit, the size of a small apple, which requires a long season to ripen. It may be gathered green after frost and stored until it ripens. Some varieties are spiny and thus make an excellent hedge.

QUINCE *(Cydonia oblondata)* is a suckering plant which will form a shrub unless pruned into the shape of a tree. Its fruit is hard and astringent to taste until fully ripened. The best varieties are Mediterranean cultivars which can produce a fruit weighing more than four pounds. Quince is generally used for jams, jellies or other preserves, often mixed with apples or pears. It is hardy in any apple climate. The Japanese quince *(Chaenomeles lagenaria)* bears a slightly smaller and more aromatic fruit than the common quince and is widely used as an ornamental.

Karen Hayden

Connie Yenne lives on the same island in Puget Sound where she was born. The many plants around her home bring back memories of family and friends.

Connie's home is surrounded by apples, walnuts, grapes, plums, pears, and figs. The figs, planted by her husband, Arlis, twenty-five years ago, are among her favorite trees. The trees were originally brought to the island as cuttings from Eastern Europe. Protected with a deep sawdust mulch the first few winters, the figs have thrived, producing delicious fruit every year.

Figs

George Pinyuh, King County Extension Horticulturist, has written a bulletin on the selection, siting, pruning, and management of fig trees west of the Cascades. He also includes notes on how to safeguard fig trees in colder climates. Figs, Bulletin #25, is available from the King County Extension Service, 312 Smith Tower, Seattle, WA 98104.

Persimmons, Anyone?

Griffith, E, and Griffith, M. Persimmons for Everyone. 1982. NAFEX, c/o D. Nichols, Rt 2, Box 13, Arcola, MO 65603. $6.75.

The taste of a tree-ripened persimmon is the best advertisement for this book. The persimmon will grow in the milder climates of the Maritime and Interior Northwest and this book, based upon decades of experience by the North American Fruit Explorers, is the best introduction available.

SAND PEAR *(Pyrus pyrifolia* var. *culta)* the Japanese apple-pear, or nashi, is shaped like an apple but tastes like a pear. It is a temperate fruit, with a range similar to apples and pears. The fruit is harvested in the fall and will store for long periods.

MULBERRIES *(Morus* spp.*)* can be grown in all but the coldest parts of the Northwest. The white mulberry (a native of China) and the red mulberry (a native of the continental US) have fruit of variable quality. The black mulberry is less hardy, restricted to the milder parts of the Northwest, but it yields the best fruit. The Russian mulberry, a drought and wind resistant tree, is particularly suitable for shelterbelts. It is valuable as a decoy for birds, who will desert other fruits, such as cherries, in favor of the mulberries.

STRAWBERRY TREE *(Arbutus unedo)* is a relative of the madrone. Native to Chile, this evergreen ornamental shrub has fruit the size of a cherry. It has an agreeable taste with a consistency and interior similar to the kiwi. It grows well in Seattle.

WESTERN CHOKECHERRY *(Prunus demissa)* is a native shrubby tree found throughout the Northwest. The astringent cherry-like fruits make jellies, jams and wine.

Tree Fruit Organizations

British Columbia Fruit Growers' Association, PO Box 160, Kelowna, BC V1Y 7N6.

Home Orchard Society, c/o Winnifred Fisher, 24640 Wallen, Estacada, OR 97023.
A society of backyard orchardists who grow fruit and nuts in the Maritime Northwest. They sponsor a spring scionwood exchange and a fall fruit show. They have a newsletter, and are a membership organization.

International Dwarf Fruit Tree Association, Department of Horticulture, Michigan State University, East Lansing, MI 48824.
Disseminates information on dwarf fruit trees by the world's top researchers. Publishes an annual journal, Compact Fruit Trees, plus a newsletter.

North American Fruit Explorers, 1848 Jennings Drive, Madisonville, KY 42431.
NAFEX is an organization of more than 2,100 backyard fruit growers, testing, developing and discovering better fruit varieties for various climatic conditions. They publicize the resulting information in their quarterly newsletter, Pomona, one of the best sources of information on rare and native fruits and nuts. NAFEX maintains a mail order horticultural library, a scionwood and seed exchange, and consulting staff for specific fruit and nut crops.

Northern California Fruit Growers Society, PO Box 27073, West Portal Station, San Francisco, CA 94116.
An organization bringing growers together to share information.

Phoenix Orchards Association, Fulford Harbor, BC, Canada.
This local association locates and caretakes abandoned and neglected orchards on Saltspring Island. Many of the trees are owned by older people unable to maintain them. Phoenix prunes the trees, picks the fruit and turns much of it into juice which is then shared with the owner.

Western Cascade Tree Fruit Association, 420 92nd Street, Everett, WA 98204.
"The goal of the Western Cascade Tree Fruit Association is to bring together all persons interested in growing tree fruits west of the Cascade Mountains for the purpose of assembling and disseminating information for the benefit of all."

PEST MANAGEMENT

Bio Lure, 64550 Research Road, Bend, OR 97701.
Controlled-release pheromone traps for over 30 pests.

Brunner, Jay F., and Howitt, Angus J. Tree Fruit Insects. MSU Extension Service, Box 231, East Lansing, MI 48824.
Survey of harmful and beneficial insects and mites. While written for Michigan and eastern conditions, the color photos and detailed descriptions make this a valuable tool for all orchardists seeking a better understanding of insect ecology.

——., Hoyt, S.C., and Wright, M.A. Codling Moth Control, A New Tool for Timing Sprays (EB 1072). Pullman, WA: Cooperative Extension Service, 1981.
Orchardist's guide to implementing a new computer model for control of codling moth in apples and pears in Washington state. Based on weather data, the model is used to predict each life stage of the codling moth (egg, larva, pupa and adult) as a basis for precision control. Bulletin includes an introduction to the modeling concept, degree day tables, and codling moth development charts. Essential for orchardists seeking to reduce sprays for control of the codling moth.

Omnidata International, PO Box 3489, Logan, UT 84321.
Microcomputers which record degree days, chill units, rainfall, and solar radiation, providing data for predictive models. Tested under orchard conditions and found very useful in IPM programs.

Page, Steve. Spray Saver Orchard Almanac. Spray Saver Almanac, Box 355, Camden ME 04843.
Thoughtful guide to orchard biology with specifics on how to control apple pests using a minimum of chemical sprays. Lists most selective and least toxic pesticides, guidelines for protecting beneficial insects, information on resistant apple varieties, and additional references. A practical and pragmatic sourcebook.

Tree Fruit Insect and Disease Fact Sheets. IPM Programs, New York Agricultural Experiment Station, Geneva, NY 14456.
A series of low-cost one page fact sheets which outline the biology, monitoring and control of tree fruit insects and diseases, with color photographs to aid in identification. Included are apple scab, pear psylla, codling moth, and apple maggot.

Tree Fruit Bibliography

Bartham, Richard D., and Snyder, John C. Grafting Fruit Trees. Extension Bulletin #0062. Pullman, WA: WSU.
The details of six different methods of grafting, with helpful photographic illustrations for them all.

Blodgett, Earle C. Why Cherry Trees Die. Extension Bulletin #668. Pullman, WA: WSU.
A comprehensive discussion of the many maladies which can befall cherry trees throughout the Northwest, with ample photographic documentation.

Childers, Norman F., ed. Mineral Nutrition of Fruit Crops. New Brunswick, NJ: Rutgers University, 1959.
Prepared for the small fruit grower who does not have easy access to information. What fruit trees need for best quality production.

——. Modern Fruit Growing. Somerville, NJ: Somerset Press, 1973.
One of the most comprehensive and useful textbooks available. Written specifically for commercial orchardists, but useful for the home grower also.

Fogle, H.W., and Winters, H.F. Fruit and Tree Nut Germplasm Resources Inventory. USDA Publication #1406. Washington, DC: USGPO, 1981.
This publication contains an up-to-date listing of fruit and nut cultivars and evaluations of them.

Fruit Tree Cultivars in British Columbia. Publication #1609. Ottawa: Canadian Department of Agriculture, 1977.
An exhaustive list of several hundred cultivars, as useful south of the border as north. Everything you need to know to choose which varieties to plant.

Garner, Robert J. The Grafter's Handbook. New York: Oxford University Press, 1979.
This book has been revised several times by one of the present-day English masters. Well-illustrated. Sufficiently technical, yet readable.

Hill, Lewis. Fruits and Berries for the Home Garden. New York: Alfred A. Knopf, 1977.
A recent book on home fruit growing, especially useful for beginners. Clear and thorough, with excellent illustrations. Lists named varieties and their characteristics.

Home Orchard Society. A Home Orchard Guide for Growing Fruits and Nuts. Handbook No I, 1980. Home Orchard Society, 24640 Wallen, Estacada, OR 97023.
A valuable guide for Maritime growers, listing the best rootstocks, varieties and cultivars. Emphasis is on small urban plots.

In 1847, Henderson Lewelling and William Meek crossed the plains and mountains by ox team from Iowa to Oregon with some 700 grafted sprouts of trees, shrubs, and vines in prepared soil. In their boxes were varieties of apples, pears, plums, and cherries. They also carried assorted seeds.

In Milwaukie, Clackamas County, just south of Portland, they established the first nursery business in the Northwest. Ralph G Geer, another 1847 pioneer immigrant and fruit grower, later wrote, "That load of living trees brought more wealth to Oregon than any ship that ever entered the Columbia River."

These men, together with Seth Lewelling, who joined his brother in 1850, produced some of Oregon's finest varieties of fruit and helped in the State's renowned horticultural development.

Charles E. Randall and Henry Clepper, *Famous and Historic Trees*. American Forestry Association, 1976. p 65.

Larson, Fenton E., and Stebbins, Robert L. Rootstock for Apples in the Pacific Northwest. Extension Bulletin #208. Pullman, WA: WSU, 1981.
Evaluates the suitability of seven rootstocks for apple cultivation in the Pacific Northwest.

Logsdon, Gene. Organic Orcharding, A Grove of Trees to Live In. Emmaus, PA: Rodale Press, 1981.
Good introduction to the design, establishment and maintenance of fruit, nut and hardwood trees. Written from an ecological perspective.

Norton, Robert A. Tree Fruit Cultivars for Western Washington Homes and Orchards. Extension Bulletin #0937. Pullman, WA: Cooperative Extension Service, August, 1981.
The recommendations in this bulletin are largely based on research done over the past 18 years at Washington State University's Northwestern Washington Research and Extension Unit in Mt. Vernon. The objective of the research has been to find varieties adapted to the unique climate of western Washington.
Nineteen apple cultivars are discussed, 12 pear cultivars, 12 plum cultivars, 12 cherry cultivars, 10 peach cultivars, and 3 apricot cultivars.

Shewell-Cooper, W.E. The Compost Fruit Grower. London: Pelham, 1975.
Shewell-Cooper is one of England's most prominent horticulturists. This book emphasizes biological growing methods suitable for maritime climates.

Stebbins, Robert L. Pruning the Home Orchard. Extension Bulletin #650. Pullman, WA WSU, 1974.
Includes information about pruning berries.

——. Training and Pruning Apple and Pear Trees. Extension Bulletin #156. Pullman, WA: WSU, 1976.
Well illustrated, succinctly stated.

Storey, William. The Fig. 1977. Jurupa Mountains Cultural Center, 7621 Highway 60, Riverside, CA 92509.
Detailed cultural information. The most comprehensive source available on figs.

Tukey, H.B. Dwarfed Fruit Trees. Ithaca, NY: Cornell University Press, 1964.
The definitive work on dwarfed fruit trees.

Walheim, Lance, and Stebbins, Robert L. Western Fruit, Berries and Nuts: How to Select, Grow and Enjoy. Tucson, AZ: H.P. Books, PO Box 5367, Tucson, AZ 85703.
Extensive lists of the most commonly available fruit varieties are given, along with information on each variety's hardiness, origin, season of ripening, fruit description, and pollenizers. Many color photos of varieties are shown. Besides all the common fruits, nuts and berries, other lesser known fruits such as persimmon, quince, kiwifruit and jujubes are also covered. It includes step-by-step guides to planting, pruning, propagation, and pest control. Highly recommended for all western North America fruit regions.

Westwood, M.N. Temperate Zone Pomology. San Francisco: W.H. Freeman, 1978.
A current, worldwide overview of orchard practices for pome fruits. An in-depth textbook.

Northwest Nut Crops

Nuts can be an important element in a balanced diet. They are rich in protein, oils and carbohydrates, and can serve as a substitute for animal protein. There are nut trees adapted to almost every part of the Pacific Northwest, including walnuts, filberts, hazelnuts, chestnuts, beechnuts, pine-nuts, almonds, pecans, sweet-pit apricots, and gingko nuts.

Nut trees are long-term investments, and success depends on selecting the best species and cultivars for each location. Unfortunately, with the exception of walnut and filbert, there has been little research on nut culture in the Northwest, so choosing nut varieties is somewhat more difficult than choosing fruit varieties, for example. If you are thinking of growing nuts, is would be wise to consult experienced neighbors, regional nut tree nurseries, and your county Extension office.

Beverly Reed

Leo Givens is a tree farmer who is helping revive the walnut industry in the Northwest. A Texan transplanted to Raymond, Washington, Leo is establishing a nut tree nursery and testing station.

According to Leo, his breeding program "is aimed at development of Persian walnut cultivars suited to Pacific Northwest conditions by combining late-leafing tendencies of the Spurgeon variety with the heavy-bearing Manregian and the superior flavor of the Carpathian. Also, use is being made of stock from the Schaefer tree in Clarkston, Washington, which survived the '57 freeze and is precocious—beginning to bear in the third year—as well as Holton, a hybrid that combines the hardiness of Broadview and the nut size and flavor of Hanson."

Leo is testing dozens of walnut cultivars at his Willapa Hills Nursery, but because this is a new program, neither scion wood nor grafted trees will be available until after the 1983 season. For more information, write to Leo Givens, Willapa Hills Nut Tree Nursery, Rt 1, Box 530, Raymond, WA 98577.

NUT CULTURE

Nut trees lend themselves to a multi-level, diversified agriculture. The taller-growing nut trees can form the upper story, while others, such as filberts, can be grown as middle story elements. Walnuts, butternuts, chestnuts and hickories do not leaf out until late in the spring, and thus can be used in situations where winter and early spring sun, combined with summer shade, is desired. Numerous shrubs and herbs are adapted to the partial shade of nut trees.

Some types of nut trees come fairly true to seed, while others show a great degree of variance. For this reason, many of the trees sold through nurseries have been grafted to select nut cultivars. Some nurseries, however, also sell lower-priced, non-grafted nut seedlings. Commercial orchardists and backyard growers with limited space will want to plant grafted trees.

Most nut trees develop deep, strong taproots in their first year of growth. This makes them difficult to transplant successfully, especially if grown for two years in the nursery bed before transplanting. To avoid transplant shock, nuts may be planted in the spot you wish them to grow. If fall-planted, a deposit of fresh cow manure will discourage rodent pilferage plus act as fertilizer and mulch. Mark each seed with a stake so it can be located easily. After one or two years of growth the seedling may be grafted. Allowing seedlings to grow ungrafted increases the genetic base of the species and allows new cultivars to come into existence.

Rodents frequently cause heavy damage to young nut trees and some method of winter screening is advisable. Trees can be mulched with stones to good advantage since water is conserved and competing vegetation and mice are discouraged. Water requirements of different nut trees vary. Most prefer moist, rich soil, but some oaks, pines, almonds and sweet-pit apricots are drought tolerant.

Filberts & Hazelnuts

There are fifteen *Corylus* species in the north temperate regions, two of which are native to North America, *C. americana* and *C. cornuta*. Both species are referred to as wild hazelnuts. *C. americana* is native in the eastern forests while *C. cornuta* is found across much of Canada and throughout the Maritime Northwest. Its habitat includes clearings and forest understories. The nuts are tasty but small, and a human gatherer has to be quick to compete with the wide variety of wildlife which feed on them.

The cultivated filberts are the result of hybridizing between the native European species, primarily *C. maxima* and *C. avellena*. Most *Corylus* species are multi-stemmed shrubs. The Turkish hazel or columnar hazel *(C. colurna)* and the Chinese hazel *(C. chinensis)*, however, form columnar trees up to 100 feet tall. They are being crossed with domesticated filberts to create a hardier, tree form of filbert.

The long, moderate growing season of the Maritime is ideal for filbert production. In fact, the Willamette and the lower Columbia valleys produce 98% of the US filbert crop. Filberts also grow well in the Interior wherever the climate is suitable for sweet cherries. Filberts grown in the hotter parts of the Columbia Basin will benefit from an overstory of larger trees for shade and wind protection.

The Barcelona has been the leading filbert cultivar for many years, but the recently introduced Ennis cultivar (developed by researchers at Oregon State University in Corvallis) will most likely outrank it in the next decade, as Ennis gives significantly higher yields and a higher percentage of filled shells. Butler and Hall's Giant are good pollinators for Ennis.

"Hazelnuts have been used for centuries by North American Indians, who used to search for the nut caches of squirrels to save themselves the trouble of harvesting and cleaning the nuts. Usually a handful of corn or other food would be left in place of the nuts so that the squirrels would not go hungry. Hazelnuts were a common Indian trade item. They were often pounded with berries, bulbs, or animal fat to make cakes, or were boiled in water to extract their oil, which was then used to flavour soups and dress vegetables. The straight, young shoots of hazel wood were used to make arrows, and the buds and roots were used in some areas to make a blue dye....

"In Britain and Europe, hazels are planted in hedgerows, and the slender shoots are used for making hurdles, crates, and thatch-fastening rods, as well as baskets and wicker fences. Hazel charcoal is used by artists, and hazelnut oil is used as a base for fragrant oils and perfumes."

Turner, Nancy J. and Szczwinski, Adam F. Edible Wild Fruits and Nuts of Canada, page 52. 1979. National Museum of Natural Sciences, National Museums of Canada, Ottawa, Canada K1A 0M8.

The Eastern filbert blight, a fungus disease, was introduced into southwest Washington 10 to 15 years ago and is just beginning to spread into Oregon. The blight is especially injurious to young trees and causes tip die-back on older trees. It is partially controlled by proper pruning and spraying with fungicides.

Hickories & Pecans

There are 22 *Carya* species, most of them native to eastern North America. Its most famous member, the pecan, is a southern tree with little possibility of adaptation to the Northwest outside of the most favored locations in the lower Columbia Basin. The shagbark and shellbark hickories are hardier, but still require a long, hot growing season for their nuts to ripen. Superior cultivars which have thin shells and large nutmeats are available and deserve further trial in the warmer zones of the Interior Northwest. Hickories need a deep, rich soil for best development.

Chinquapins

The chinquapins are closely related to the chestnuts. There are only three species, the largest of which is the golden chinquapin *(Castanopsis chrysophylla)*, which is native to the Pacific Northwest. The golden chinquapin grows up to 200 feet in height in the Klamath mountains, but gradually reduces in size as it nears the top of its range, becoming a small shrub in the Puget Sound area. Its nuts are small and sweet, and it is a fine ornamental.

Almonds

Almond trees require long, hot summers to ripen their nuts, and are best suited to areas where peaches can be grown successfully. A favorable location with a deep, light soil is a necessity. Recent introductions of dwarf trees (such as the Zaiger dwarf almonds offered by the Dave Wilson Nursery) can be grown as container plants.

Sweet-Pit Apricots

Sweet-pit apricots are relatively unknown in North America, but have the potential of becoming an important crop in regions too cold for almonds but suitable for apricots. The most famous sweet-pit apricots are the Hunza-cots. The people of the remote, mountain country of Hunza have made sweet-pit apricot kernels a main part of their diet for centuries. Most apricot kernels contain a form of cyanide poison which accounts for their bitterness. Sweet-pit apricots, which taste like almonds, do not contain this substance and are safe to eat.

Sweet-pit apricots are multiple-use trees yielding fruit, nuts, carving wood, and a source of spring bee food. Two sources at present are Bear Creek Nursery and Grimo Nut Nursery.

Northwest Chestnut Project

Maryanne Caruthers-Akin

European chestnut (Castanea sativa) near Vancouver, Washington. The champion American chestnut (C. dentata) is in Oregon City, Oregon. It is 82 feet high, with a spread of 93 feet.

Only a century ago, the American Chestnut *(C. dentata)* was one of the most prized of the eastern hardwoods. Because its wood was durable and rot resistant, it was used for home siding and shingles, furniture, telephone poles, ships' masts, railroad ties and farm fencing. And, of course, it bore chestuts, sweet nutmeats of fine flavor and important nutritional value.

Here in the Northwest, Maryanne Caruthers-Akin has created the Northwest Chestnut Project to collect, organize and publicize information about chestnuts. For an information packet, send $1 and a SASE to The Northwest Chestnut Project, 2519 NE 14th Avenue, Portland, OR 97212.

Maryanne has also written an article entitled "The Chestnut Connection," in *Stepping Out Northwest*, #12 (1421 NE Highland, Portland, OR 97211). The following is a quote from her article.

"What appeals to me about the edible chestnut is that it is an excellent 'survival' food. A hardwood tree, the chestnut is stately and powerful in shape. Chestnut trees can occupy land inappropriate for grain or vegetable cultivation. Chestnut trees 'suit' city parks, arboretums, and private gardens, and they can also be used to reclaim marginal sites or to reforest mountainous areas all over the Pacific Northwest. Plus, the chestnut does not depend on an annual seed supply. Once the trees start to bear fruit, they produce and produce and produce."

Maryanne Caruthers-Akin

Chestnuts

Along with a rich supply of carbohydrates, chestnuts contain about 5% oil and 5% protein, making them the "potato" of the nut family. Because of their relatively low oil content, chestnuts can be used as a diet staple, dried and ground into flour.

The American chestnut *(Castanea dentata)* was once the dominant tree of eastern North America. Unfortunately, the chestnut blight *(Endothia parasitica*, introduced in 1904) has virtually eliminated the American chestnut tree from its original range, and is now attacking the European chestnut *(C. sativa)* in western Europe. However, the Chinese chestnut *(C. mollisima)* and the Japanese chestnut *(C. crenata)* are fairly blight-resistant. Some progress has been made in developing blight resistant American chestnuts, and breeding between species is underway with the hope that this forest giant can be restored to its former range.

All four of the chestnut species mentioned above are being grown in the Pacific Northwest. The best yields of sweet, well-filled nuts are obtained in areas where peaches do well. The trees will grow in cooler areas but are less productive. Breeding and selection are underway to produce satisfactory cultivars for a wider variety of climates. Chestnuts grow best in full sun on deep, well-drained, loamy soil. They begin bearing within several years of planting and rarely self-pollinate, so two or more trees should be planted close together.

A diminutive relative of the chestnut is the Allegheny chinquapin *(C. pumila)*. This small tree or shrub produces a sweet nut which was used by the Indians to produce a chocolate-like confection. It is adapted to a variety of habitats, including dry, rocky hillsides and streambanks.

Of particular interest are hybrids between the American and Chinese chestnut. These trees are blight resistant, have larger nuts than the American, stronger growth than the Chinese, and produce valuable straight trunks in addition to the nuts. Several valuable varieties have been reported from China, including *C. sequinii*, a dwarf Chinese chestnut of three to eight feet, and *C. vilmoriniana* a large, American type with superior taste.

Pine-Nuts

Pine-nuts are quite tasty and nutritious. They have been a staple food for many people around the world. Some species which produce seeds large enough to be worthwhile harvesting include the Italian stone-pine *(Pinus pinea)*, pinyon pine *(P. cembroidies)*, Korean stone pine *(P. koreansis)*, Swiss stone pine *(P. cembra)*, Siberian stone pine *(P. siberica)*, sugar pine *(P. lambertiana)*, digger pine *(P. sabiniana)*, limber pine *(P. flexilis)* and whitebark pine *(P. albicaulis)*. Sugar pine and whitebark pine are native to the Northwest. Pinyon pine and digger pine stop just south of its boundaries. Swiss stone pine and Siberian stone pine are extremely hardy and slow growing, and are used as urban ornamentals. The Korean stone pine offers the most promise in the Northwest because of its hardiness, fast growth and large nuts.

Walnuts

There are fifteen species of walnuts, the most famous of which is the Persian or English walnut *(Juglans regia)*. Since earliest recorded history, people have been selecting and planting the best of these trees, producing thousands of named cultivars in the process. Walnut breeders are now going back to wild genetic stock from Europe and Asia to breed in hardiness, disease resistance and other desirable characteristics. The Carpathian walnut (a hardy strain of *J. regia)* was discovered by a plant explorer in the Carpathian mountain region between Greece and Yugoslavia.

The Manchurian walnut *(J. manchurica)* produces a small, hard-shelled nut which has little value for commercial nut production but which is proving valuable for rootstocks and for breeding work with Persian walnuts.

The Japanese walnut *(J. sieboldiana)* has been bred over centuries to produce a rich, sweet nut called the Japanese heartnut. Though it has one of the thickest shells of the walnuts, it is easy to crack. The heartnut is one of the most ornamental walnut trees, with lush foliage reminiscent of tropical plants and large, pendulous strings of nuts. The heartnut is similiar in hardiness to the Persian walnut.

Superior cultivars of the black walnut *(J. nigra)* and butternut *(J. cinera)* are available which are thin-shelled and have large nutmeats. Through the efforts of Carl Weschke and other nut breeders, cultivars of exceptional hardiness have been developed.

The high value of walnut wood is well known. The quality of the lumber of any walnut, however, is dependent on the speed of growth. The highest grades of walnut lumber are obtained from slow-growing trees which produce a very tight, dense grain. Walnuts grown in the Maritime Northwest, or other locations where they make high annual increments of growth, do not make the best grades of lumber.

All walnuts need deep, well-drained soil and an adequate supply of moisture. Their deep root systems enable them to draw on subsurface water, thus minimizing irrigation requirements.

Wild Fruit & Nut Forests of Central Asia

Fruit and nut trees were not always grown in pampered orchards. In fact, large areas still exist where these trees grow naturally. Further study into how these trees survive in the wild, what conditions they prefer and what plant associations they form can be helpful in designing multiple-crop systems based on tree crops. Nature has been experimenting for millennia and has many lessons for those who observe her results.

One of nature's most extensive experiments in growing wild fruit and nut forests is in the Caucasus mountains and neighboring parts of Central Asia. In the Soviet Union there are approximately one million acres of forests where wild fruit and nut trees are dominant. These include walnut, chestnut, pistachio, almond, apple, pear, fig, pomegranate, persimmon and apricot. Most of these species are found as an admixture in forests, but some form pure stands over extensive areas. To illustrate a few of the lessons that nature has to teach us about natural tree crop ecosystems, here is a quote from The Forests of the USSR by V.P. Tseplyaev. It is interesting to note that many of the understory species mentioned are also native to, or have close relatives in, the Pacific Northwest.

"Walnut forests may be of pure stands, or with some admixture of apple and maple. The underbrush is usually sparse and consists of cherry-plum *(Prunus cerasifera*, var. *divaricata)*, barberry, Altai hawthorn *(Crateagus altaica)*, buckthorn, honeysuckle, dog-rose and others.

"An important biological peculiarity of walnut is its highly ramified root system, which may reach a radius of 20 m [66 feet] and a depth of 3 m [10 feet]. This enables the Persian walnut to make use of an enormous volume of soil, at various depths at different times, and thus survive the unfavorable periods. In mountain regions which are exposed to water erosion, landslide, mud, streams, and other agencies of slope destruction, the Persian walnut reinforces the soil by means of its ramified root-system.

"Usually only half of the walnut [trees] in a stand are fruit-bearing. In thinned stands, the yield is higher although there are fewer trees.

"The low productivity of the walnut forests is due to poor husbandry. If tending is timely, if reforestation is practiced, if better strains are selected, and if the collection of nuts is better organized, the actual productivity may be raised by a factor of two or three."

According to the Soviet Academy of Sciences, the productivity of walnut stands in the USSR averages 110 to 165 lbs. per acre, although yields reach as high as several thousand pounds in favorable years.

Tseplyaev, V.P. The Forests of the USSR. Moscow, 1961. Translated from Russian into English by A. Gourevitch. Jerusalem: Israel Program for Scientific Translations, 1965.

Zarubin, A. Reclamation and Development of Walnut and Fruit Forests in Southern Kirghizia. Jerusalem: Israel Program for Scientific Translation, 1968.
Translated from Russian. Thorough discussion of natural fruit and nut forests in a climate similar to the Interior Northwest. Hard-to-find information on coppice regeneration of old walnut stands, collecting seeds from wild nut trees, direct seeding, and handling pest and disease problems.

Oaks

There are over 300 oak species in the world. Their acorns furnish an important wildlife food and at one time they were a food source for many native peoples. The oaks can be separated into two main groups, the white oaks and the black oaks (which includes red oaks). Acorns from the black oak group contain high amounts of tannin and are seldom used for food. The acorns of the white oak group contain smaller amounts of tannin, which is removed by leaching. One of the oaks most frequently suggested as a source of edible acorns is the burr oak *(Quercus macrocarpa)*, a native of eastern North America. The acorns of the Garry or Oregon white oak *(Q. garryana)* were widely used by the Northwest natives. There are oaks which can be grown in various Northwest climates to produce fodder for livestock and quality timber.

The cork oak *(Q. suber)* can also be grown in the Northwest, as is demonstrated by specimens in Seattle and Portland. Its bark is a source of cork and it provides an edible acorn as well.

Beechnuts

Beech trees are native to both Europe and eastern North America. The small nuts offer little promise for human food, though they have been used for producing an edible oil. They do have value as food for hogs, domestic fowl, and other livestock and wildlife. They are adapted to shallow, wet soils.

Nut Crops Bibliography

Baron, Lloyd C., and Stebbins, Robert. Growing Filberts in Oregon. Extension Bulletin #628. Corvallis, OR: OSU, 1981.
Current information on varietal selection, orchard design, management and harvesting techniques. The best source of information on the Northwest's most important nut crop.

Flint, M.L., and Kobbe, B. Integrated Pest Management for Walnuts. California IPM Project Pub. 3270. Division of Agricultural Sciences, University of California, Berkeley, CA 94720.
Life histories and management guidelines for integrated control of over 50 diseases, weeds, and insect pests of English walnuts. Written for California orchards, but much of the information, including photos of insects and disease symptoms, are of value to all walnut growers.

Jaynes, Richard, ed. Nut Tree Culture in North America. New York: Charles Scribner's Sons, 1981.
Up-to-date information on all the edible nut trees grown in North America. Twenty-five essays by leading specialists cover techniques for cultivating nut trees and developing nut varieties. Covers use of nuts for food, landscaping, wildlife and lumber.

Lanner, Ronald M. The Pinyon Pine: A Natural and Cultural History. Reno, NV: University of Nevada Press, 1981.
This recent book brings together much useful information on the pinyon pine in a readable form.

Lassoie, James P., et al. Acorns from Natural Oak Woodlands and their Utilization. 1980. Tree Crops Project, Cornell University, Roberts Hall, Ithaca, NY 14853.
Review of the potential for increasing acorn yields for use as food and fuel.

Menninger, Edwin A. Edible Nuts of the World. Stuart, FL: Horticultural Books, 1978.
A compendium of information on 1,000 species of nut-bearing plants from around the world.

Northern Nut Growers Assoc. Annual Reports, 1910–1981. Northern Nut Growers Association, 13 Broken Arrow Rd, Hamden, CT 06518.
This set of reports contains a vast amount of information on nut culture and the history of the NNGA.

Reed, C.A., and Davidson, J. The Improved Nut Trees of North America. New York: Devin-Adair, 1958.
A textbook on nut culture with solid background information.

Riotte, Louise. Nuts for the Food Gardener. Charlotte, VT: Garden Way, 1975.
Good introduction to edible nut varieties, culture, harvest and recipes.

Shumanyur, A.P., ed. Fruiting of the Siberian Stone Pine in East Siberia. Jerusalem: Israel Program for Scientific Translation, 1963.
Devoted wholly to maintenance and harvesting of wild stands of *Pinus siberica*. Information can be applied to other wild or semiwild plantings of other edible pine nut species.

Stebbins, Robert L. Growing Walnuts in the Pacific Northwest. Extension Bulletin #974. Corvallis, OR: OSU, 1979.
A brief summary of walnut culture within the region.

Thompson, B. Black Walnut for Profit. Forest Grove, OR: Timber Press, 1976.
Presents a well researched and realistic appraisal of the potential for black walnut as a timber and food crop.

Weschke, C. Growing Nuts in the North. St. Paul, MN: Webb, 1953.
Detailed information on nut tree culture and propagation by one of the North American masters. How to adapt nuts to colder climates.

Woodroof, Jasper Guy. Tree Nuts. Westport, CT: AVI, 1979.
A current world-wide overview of commercial-scale production and processing of nuts. Includes the latest cultivars. Well referenced.

ORGANIZATIONS

Northern Nut Growers Association, 13 Broken Arrow Road, Hamden CT 06518.
The NNGA contains some of the sharpest minds in the country on the subject of nut growing. Their newsletter and annual report contain much valuable information.

Nut Growers Society of Oregon and Washington, PO Box 23126, Tigard, OR 97223.
An association serving commercial nut growers.

Oregon Filbert Commission, PO Box 23126, Tigard, OR 97223.

Society of Ontario Nut Growers, RR 1, Niagara-on-the-Lake, Ontario, Canada L0S 1J0.
A Canadian version of the Northern Nut Growers Association, with a necessary emphasis on hardy nuts. They have some excellent publications on the subject.

A Wealth of Trees

TEXT AND PHOTOGRAPHS BY KAREN HAYDEN

Harvey Johnson lives with his brother Clarence on the farm their grandparents homesteaded in 1887. Their land is on Bainbridge Island in Puget Sound. Cherry and pear trees their grandmother grafted and planted almost a hundred years ago still surround the farm house.

Harvey has been planting trees on the farm since he was twelve years old. Through the years of experimenting he has extended the range of several plant species beyond their established climatic zones, including the subtropical kiwi and persimmon. "It's just a hobby," Harvey says, "like some people go to music or dance."

One of the most surprising of Harvey's plants is his kiwi. The vines stretch along sixty feet of trellis; above the top wire, tangled coils of vine curl over, or arch and trail to the ground. Their trellis is on a north-facing slope which helps to protect them from leafing too early in the spring. The kiwis begin growth sometime in April and flower in June. Bees are necessary for pollination. Harvey's two six year old female kiwis started bearing three years after they were planted, and have yielded a hundred pounds of fruit each season for the last three years. The plants are free from mildew or bug damage. Kiwis seem to have left behind the disease and insect predators they might have had in their native environment, the Yangtze Valley in China.

Harvey propagates kiwis from his established plants by layering the vines. The newly rooted vines generally are kept indoors for the first couple of winters as they have a better chance of surviving a freeze after they have some woodiness in their stems. Kiwis are vigorous and rampant growers. They set a lot of fruit which needs to be thinned in order to ripen large kiwi fruits. If the fruit is

picked before the first frost and stored in a cool room, it will keep until February. The fruit is brought to room temperature to ripen. After the plant becomes dormant it can survive temperatures as low as ten degrees.

Young persimmon trees also can tolerate temperatures as low as ten degrees, and mature trees can survive temperatures down to zero. Harvey's first two persimmon trees were the Japanese Kaki variety grafted onto American rootstock. They were planted in 1955. One of the young trees survived a freeze that dropped to zero degrees Fahrenheit the winter after it was planted. The top growth of the other tree died, but the American rootstock survived and sent up a shoot, which is now a twenty-five foot high tree. One of its branches is an improved American variety that Harvey grafted onto it. These persimmon trees stand against a south-facing wall, which aids in ripening the fruit. The trees produce an average of twenty-five pounds of persimmons a season. Harvey picks the Japanese persimmons before the first frost and stores them at room temperature. The three inch long fruits ripen around Christmas time. The American persimmon tree has fruits about half an inch to an inch long that ripen on the tree before first frost.

The persimmon trees are planted down the hill from Harvey's newest orchard which includes young pearapple, plumcot and cherryplum trees. Across the road in the garage Harvey has started eucalyptus seedlings to "see what they'll do." Outside the garage there is a two year old pomegranate tree that Harvey planted for the same reason. Scattered around the farm are other trees unusual to the region, including several varieties of fig and a twenty year old almond that occasionally bears a good crop. In one orchard there are a couple of nitrogen-fixing trees, a silk tree *(Albizia julibrissin)* and a goldenchain tree *(Laburnum anagyroides)*. Harvey commented, "I planted them in with the fruit trees and they seem to do real well." Beyond the orchard there is a small pond partially surrounded by native forest; a young Dawn Redwood overlooks the pond. After reading about this ancient species which survives from the Upper Cretaceous era 100 million years ago, Harvey decided to try one. "It's a fossil, just like me," he remarked. The six year old Dawn Redwood is now about eighteen feet tall. Delicate branches fringed with fine needles allow plenty of light to reach the ground cover.

The trees mentioned above are interspersed with trees more common to the region. In one of the original orchards are Cox Orange Pippin apple, Red Flemish pear, Green Gage and Sugar plums and Italian prunes. Near the house is a mixed grove which includes a Carmello walnut, a Daroga Red peach, two figs, a yew, elderberry, and a low trellis of Delaware grapes. Harvey's garden also contains a mixture of plants, many of them perennials—asparagus, daylilies, Egyptian multiplier onions, yucca and cardoons. This intermingling of species and propagation of perennials is basic to permanent agriculture; Harvey, however, had not heard of the word "permaculture." His farm's design was guided mainly by curiosity: "There's plenty you can do with a piece of land if you just get to it and start experimenting."

Maintaining the orchards requires as much imagination and persistence as establishing them. For example, Stellar's jays and robins consume nearly all of Harvey's cherries and nuts, so his latest idea for protecting them is to fly an owl shaped helium balloon over the trees. Frost sometimes destroys the top growth of young kiwis, persimmons and figs, but with Harvey's care, many apparently dead plants have recovered. "The main thing is to protect the roots," Harvey emphasizes. He mulches the young frost-tender varieties or brings them indoors for their first few winters. Sensitive plants need diligent attention, and some varieties won't make it.

"Plant and thin," Harvey says, "just like thoughts in the head."

Kay Hoffman

Chopaka Nut Orchard

MICHAEL PILARSKI

Bill Schildgen has spent the last 16 years developing a nut orchard which successfully integrates upper and middle-story plants. Located on the north shore of a large lake in northeastern Washington a few miles south of the British Columbia border, the elevation of Bill's land is slightly over 1,000 feet. The mountains surrounding the lake are steep and rocky and reflect heat and sunlight back into the valley where the orchards are. The lake to the south and the northern rocky slopes also reflect heat into the orchard creating an ideal microclimate for nut trees.

The upper story of the nut orchard consists of walnuts, Chinese chestnuts, heartnuts and a few Turkish tree hazels *(Corylus colurna)*. The middle story is composed mainly of filberts, with some pears, peaches and plums. The south side of the orchard is planted to peaches and nectarines to take advantage of the warm, sunny exposure. On the west side grapes have been trained along the high deer fence which surrounds the orchard. Clematis grows along the fence on the east side.

The filbert trees grow long-limbed and less productive as the walnut overstory grows in height and the shade deepens. The productivity of the filberts is maintained by periodically thinning some of the walnut trees, allowing more sunlight to reach the middle story. Bill has found that spacing the walnut trees 60 feet apart and the filbert trees 25 feet apart is ideal for both crops. Most commercial walnut orchards are planted with trees 40 feet apart.

THE NURSERY

Bill also maintains a nut tree nursery where he breeds and grows several thousand seedlings each year. Some trees are grown simply to provide seed for his own hardy rootstocks. The rootstocks are generally budded in June and July. In its first year of growth the bud shoot sometimes grows more than 10 feet. For the first few years Bill stakes the trees to keep them from blowing over. Some of his trees start bearing nuts within only a few years. For most, significant fruit and nut production is underway within 5 to 10 years.

Bill speeds up his breeding results by hand pollinating selected trees and planting the nuts which result. After one year of growth, buds from the most promising seedlings are grafted onto bearing trees. These buds can bear nuts as early as the following year. In this way the type of nut a new tree will bear can be determined years before it would normally have come into bearing. He points out, however, that for every exceptional nut cultivar he creates through breeding, he creates hundreds of others which are not worthy of continued propagation. Some of Bill's trees have over 50 varieties of nuts grafted onto them and most have at least a dozen.

Bill's nut breeding experiments have resulted in many unusual and promising varieties. His Chopaka and Cascade walnuts are twice the size of an ordinary walnut. He has successfully crossed the domestic filbert with the Turkish tree hazel, thus producing a filbert capable of growing into a tall tree. One of his current projects is to produce a successful hybrid between the English walnut and the Japanese heartnut.

ANIMALS

Within his two acre orchard, Bill has more than 2 dozen peacocks and peahens patrolling for insects. The birds eat grasshoppers and sound an alarm when rattlesnakes find their way to the orchard from the dry, rocky slopes above. He makes enough money on the peacock feathers collected during molting season to pay for the birds' feed for the whole year. Bill sells young pairs of birds, and the Schildgens dine on chestnut-stuffed peacock.

Bill has built and set out owl boxes along the edge of the orchard. The owls help control mice and other rodents which enjoy the bark of young walnut trees. He also keeps 50 to 75 hives of bees to pollinate the orchard trees and to harvest the nectar from the flowers which appear in the surrounding wildlands.

Bill Schildgen's multi-story orchard demonstrates some of the many possibilities for designing a permanent agriculture. The advantage of the orchard's topographic position has certainly contributed to its success, but it is also the result of the patient care of someone who delights in working with plants.

The Forest

Mark Musick

When the first settlers came to the Northwest, they found much of it "covered all over with timber, thick as hair on the back of a dog." Since their first interest was farming, the forests which they encountered were for them largely a nuisance, to be removed only at considerable labor. However, not many harvests passed before the pioneers learned to appreciate the true value of this nuisance, both for their own uses and as an exportable cash crop. A significant portion of the old growth timber from the Pacific Northwest became houses in pre-earthquake San Francisco. Since then, the history of many communities within the Pacific Northwest has been the story not of farming but of forestry, not of grains and vegetables but of ponderosa pine, Douglas-fir and western red cedar.

In this chapter, we expand the boundary of the forest to include the city and the orchard, mainstreet and the farm. Imagine the Pacific Northwest as a forest, a forest of fuel and food and lumber. In the clearings, the hillside patches and river ribbons, are our own homes, our cabbage patches, our wheat fields. We harvest from the forest what we need. We leave shallow footprints. We are forest dwellers.

Urban Forestry, by Gene W. Grey and Fred J. Deneke. New York: John Wiley & Sons, 1978.
This excellent book discusses the benefits of trees in urban areas and existing urban forestry programs. It also includes a chapter on how to approach city officials and planners. The useful bibliography and appendices include a list of schools offering degrees in urban forestry and organizations involved in urban forestry research.

Urban Forestry

Most cities and suburbs are home to many trees, both native and exotic. Seen from a high vantage point, these trees present the aspect of a mixed forest, predominantly hardwoods, dressed for winter with conifers. In addition to being good green company, trees add to the urban landscape shade from the summer sun, shelter from the winter wind, and a touch of quiet against the city noise.

In recent years, the long-unquestioned urban-concrete-grey assumption has been challenged by the growing practical science of urban tree use and management, called urban forestry. The new urban forestry movement fosters a cooperative spirit, involving everyone in the community in planting and protecting trees.

BIBLIOGRAPHY

Arnold, Henry F. Trees in Urban Design. New York: Van Nostrand Reinhold, 1980.
"Trees are the most exquisite and the most sparingly apportioned raw material of urban design." This is a detailed guide for utilizing trees in urban settings to best effect.

Barker, Philip A., and Durrant, M. Guy. "Space for Trees in Street Rights Of Way." *in* North America's Forests: Gateway to Opportunity. Society of American Foresters: Washington, DC, 1979.
This paper reports the requirements for minimum "treelawn" or planting strip widths in representative American cities, and recommends alternative designs for future treelawns.

Barnhart, Earl. "Tree Crops: Creating the Foundation of a Permanent Agriculture." In The Journal of the New Alchemists, edited by Nancy Jack Todd. Brattleboro, VT: The Stephen Greene Press, 1980.
Comprehensive review of the role of tree crops in rural and urban agriculture.

California Dept. of Forestry. The Hip Pocket Urban Tree Planter. Available from the Department at 1416 9th St, Sacramento, CA 95814.

Corbett, Michael N. A Better Place to Live. Emmaus, PA: Rodale Press, 1981.
Urban and suburban planning principles that make good sense. Lots of examples; real and reassuring.

Pirone, P.P. Tree Maintenance. New York: Oxford University Press, 1978.
One of the most comprehensive books on the prevention and treatment of tree diseases, injuries and insect infestations. A useful tool for urban and rural tree workers.

ORGANIZATIONS

International Society of Arboriculture
5 Lincoln Square, PO Box 71
Urbana, IL 61801

This professional society promotes trees, stimulates public action, and initiates and supports investigation of problems in tree culture. They also publish the monthly *Journal of Arboriculture* for professional tree people. Of special value for park managers and urban foresters, this magazine includes reports on research and technical information on diseases, insects and management issues. The information is not strictly biological. In the May, 1980 issue, for example, an article by Richard G. Ames entitled "The Sociology of Urban Tree Planting" examines the fact that, "In many cases sociological factors may be more important than biological factors in determining tree survival and well-being." A well-done journal.

National Arbor Day Foundation
Arbor Lodge 100
Nebraska City, NE 68410

An organization promoting the planting of trees and the celebration of Arbor Day. The foundation has begun a special "Tree City USA" program to encourage urban forestry efforts.

Tree People
12601 N Mulholland Drive
Los Angeles, CA 90210

This grassroots organization is devoted to changing Los Angeles into a greener, more livable place. They aim to create the Los Angeles Urban Forest by planting more than one million trees before the 1984 Olympics are held there.

Urban Forestry Specialist
USDA Forest Service
Box 2417
Washington, DC 20013

OTHER ACCESS

Fruition
The Plan
PO Box 872
Santa Cruz, CA 95061

Fruition promotes public access fruit and nut trees being planted on public lands such as school grounds and parks, along hiking and biking routes, throughout neighborhoods, and generally anywhere people can have access to the food these plants will supply.

"The Urban Forest" and "Trees and People Need Each Other" are two films concerning urban forestry which may help to stimulate interest in a community tree program. They can be obtained for a nominal fee from **Instructional Media Services, WSU, Pullman, WA 99164.**

Urban Resource Systems, Inc.
783 Buena Vista West
San Francisco, CA 94117

Urban Resources is an internationally-focused, non-profit organization which provides professional consultation, planning, and direct technical assistance to urban areas in developed and developing countries. Their principal goal is to assist the development of urban self-reliance strategies worldwide. Their particular emphasis is designing alternative approaches to urban resource management which directly serve the urban poor.

Isabel Wade is president of Urban Resource Systems, Inc. She is working to bring food and fuel production into cities around the world. Following is an excerpt from a two-part report on urban self-reliance which originally appeared in the September and December, 1981 issues of Development Forum, *published by U.N. Division for Economic and Social Information, Palace of Nations, CH-1200, Geneva 10, Switzerland.*

Fertile Cities

ISABEL WADE

The resilience of any given city to a food/fuel crisis in the 1980s will depend largely upon its capacity to meet at least some of its own basic food and fuel needs. While this idea may seem novel or even preposterous to some, the fact remains that through the first half of this century, most urban areas around the world produced a significant amount of food and other items required by local residents. Production was not limited to the urban fringe but included substantial yields in home gardens and market gardens within the cities themselves. The Parisian "marais" or market gardens of 100 years ago serve as an excellent example of this type of productive urban agroecosystem.

Fortunately, the developing world itself provides some of the most impressive examples of cities adopting more self-sufficient approaches in meeting food and fuel needs. Some of these cities have programmes which are more than half a century old. Each of the examples here is an exciting illustration of the possibilities for greater self-reliance in urban areas in diverse geographical, climatological, and cultural settings.

The cities of Addis Ababa, Ethiopia, and Lae, Papua New Guinea, offer two examples of urban settlements which have become more self-reliant in fuel needs by developing urban forestry programmes. The programme in Addis Ababa dates back to the turn of the century when massive plantations of *Eucalyptus globulus* were planted through a Government-sponsored programme in an effort to solve the fuelwood shortage of the nation's capital. By 1964, the urban forest of Addis reportedly had grown to 250 square kilometres (135 kilometres actually covered by trees). The eucalyptus wood is used for a multitude of purposes besides fuelwood ranging from lumber to fence posts.

Lae's urban forestry scheme, begun in 1977, was designed to produce both fuelwood and tree crops and to arrest the erosion of hillsides around the city caused by rampant cutting of trees. An initial 12 sq km site has been planted adjacent to the city, primarily with *Leucaena*. A smaller area has been planted with both leucaena and various fruit trees. A further aspect of Lae's program which helps to meet the need of the urban poor for fuelwood is a system of firewood distribution centres around the city. Set up by the city council, the centres provide off-cuts from the city's lumber mill free of charge to settlement dwellers.

Growing more food and woodfuel in urban areas requires not only a recognition of the productive potential of small and marginal spaces but also an awareness of how to maximize the food or fuel potential of vegetation that already exists and how to select appropriate vegetation with the interests of greater self-reliance in mind. For example, many trees and shrubs with multiple-use potential are found in urban areas but are not utilized to their fullest capacity. In Manila, *Leucaena* trees now grace the city's streets for shade and beautification. These trees, however, are also a source of food for humans and livestock, as well as an excellent source of fuelwood.

Street trees and parks also provide an opportunity to produce fruit and nuts for urban residents. In Bangalore, India, for example, approximately 25 percent of the street tree population are fruit trees. A co-ordinated programme of tree crop planting along city streets and in parks is one of the most effective ways to produce high yields of valuable food on limited space in urban centres.

Reisinger, Richard D., and Kubosumi, Elizabeth. A Community Tree Program: Colfax, Washington. Pullman: Dept. of Extension Forestry, 1979. Pub. No. EM4497.

Designed as a guide for the development and management of public trees in Colfax, this 51-page manual contains information that would be valuable to any community. Included are street planter building plans, sample shade tree and other municipal tree ordinances, and an extensive listing of books, organizations and periodicals.

Maryanne Caruthers-Akin

Giant European chestnut trees (Castanea sativa), planted at the turn of the century, line Bigelow Street on Seattle's Queen Anne Hill. Hundreds of people harvest the abundant crops of large, sweet chestnuts which drop each fall.

Seattle's City Forest, An Owner's Manual. City Arborist, Rm. 704, Municipal Bldg., 600 Fourth Avenue, Seattle, WA 98104.

Colorful introduction to the ways trees bring a city to life. Includes lists of recommended and non-recommended trees.

Stauffer, Ms. George Hopkins. The Urban Foresters' Notebook. Syracuse, NY: The US Forest Service and the School of Forestry, SUNY, 1980. Available from the School of Forestry, SUNY, Syracuse, NY 13210.

The editors of this Notebook select information from the body of urban forestry knowledge, package it and periodically send it free of charge to those on the Notebook mailing list. It is designed for urban forestry practitioners, researchers, and others for whom the information is relevant.

USDA Forest Service. Forest Recreation and Urban Forestry Research Publications. Rev. 1981.

An outstanding bibliography of papers available from the nine experiment stations of the Forest Service.

Trees of the Northwest

MICHAEL PILARSKI

When European explorers first gazed upon the coniferous rainforests of the maritime Northwest, trees grew almost unbroken from the ocean shore to timberline in the Cascades. This was an exceptionally rich cathedral forest, with beams of sunlight streaming through the tall columns and dense archways of the forest canopy which towered 200 to 300 feet above the forest floor. The Sitka spruce was the most common tree along the shoreline, intermixed with western red cedar, grand fir, western hemlock, Douglas-fir, and lodgepole pine.

At higher elevations the Maritime forests are dominated by mountain hemlock, Pacific silver fir, noble fir, Alaska cedar, and Douglas-fir (where it plays a pioneer role). At lower elevations the forests are dominated by western hemlock and Douglas-fir, with western red cedar prominent in wetter locations. The Maritime forest becomes more diversified the further south in Oregon one travels. Incense cedar, sugar pine, ponderosa pine, Jeffrey pine, and many other conifers make their appearance.

The northern portion of Puget Sound lies in the rain-shadow of the Olympic mountains and thus has a drier climate and a unique mix of trees. Western white pine, madrone, white birch, Oregon oak, and even Rocky Mountain juniper are found there, along with the expected Douglas-fir, western hemlock and western red cedar.

A dense undergrowth of shrubs and ferns is found in almost all the Maritime forests. Common shrubs include salal, rhododendron, hazel, Oregon grape, vine maple, ocean spray, huckleberries, and the appropriately-named devil's club. In the stream courses are found red alder, Sitka alder, cascara, wild cherry, buckthorn, native crabapple, and Oregon ash.

Although small pockets of climax deciduous forest did exist naturally on specialized sites in the region, most deciduous trees play a pioneer role in the Northwest forest ecosystem, shifting into an area following some sort of disruption such as fire, landslides, floods, and clearcutting.

Red alder is an agressive, fast growing pioneer found across the Maritime lowlands. As a result of logging and fires, extensive forests of this short-lived tree now inhabit the region. Long considered a weed by loggers, alder is now becoming recognized as a vital and valued element in the forests of our region.

Big-leaf maple is often associated with both alder groves and cedar forests. Since it is larger and longer-lived than alder, maple tends to persist in coniferous forests. The big-leaf maple is a distinguished tree with a huge spreading crown and thick branches draped with mosses, lichens and ferns.

Vine maple, and less commonly the Douglas maple, often form the understory in moist western hemlock and

BIBLIOGRAPHY

Baumgartner, David M., and Boyd, Raymond J. Tree Planting in the Inland Northwest. Pullman, WA: Washington State University, 1976.

This book gives many details on tree planting methods currently utilized in the interior Northwest. The chapter by Frank TerBush on information sources on Northwest forestry is an excellent tool for forestry workers and researchers.

Bernatzsky, A., et al. Trees in the 21st Century. Berkhamsted, England: A B Academic Publishers, 1982.

An up-to-date synthesis of the importance of trees and forests for the well-being of humanity. Written by eminent forest scientists from around the world, contents include the role of trees in the evolution of the British, European and American landscapes; the effect of trees on urban climate; urban woodlands; the effects of tree roots on structures; and planting trees for the reclamation of desert land and global tree cover.

Brockman, C. Frank. Trees of North America, A Guide to Field Identification. Racine, WI: Golden Press, 1979.

The best field guide available for tree species throughout North America. Detailed species information and distribution maps.

Douglas-fir forests. These small shrubby maples are an important source of food for wildlife.

Five species of oak are found in southwest Oregon, but only the Oregon white oak and the Kellogg oak reach into the Willamette valley. The oaks play various roles in different ecosystems, growing as climax forests, widely spaced savannas, dense shrubby thickets, or as the middle story under taller, coniferous trees. The native oaks, often in association with blue camas, once dominated open savannas in northern California, in the Rogue, Umpqua, and Willamette valleys, and in the "prairies" of southern Puget Sound.

Oaks are drought resistant and inhabit the drier parts of the otherwise wet Maritime environment. They grow as companions to the drought-hardy ponderosa pine and can survive even where this hardy pine fails in the arid Interior. This oak forests cover the relatively dry slopes of the Klamath and Siskiyou mountains, where their common associates are the ponderosa pine, madrone, Douglas-fir and incense cedar.

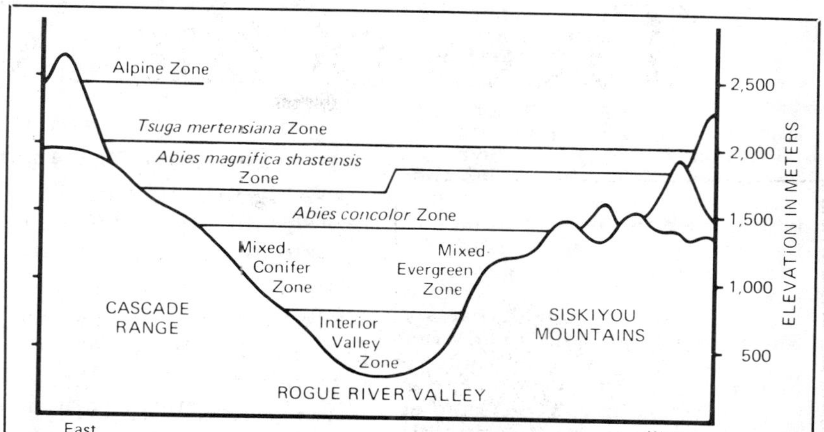

Figure 99.—Arrangement of vegetation zones in the Cascade Range and western Siskiyou Mountains of southwestern Oregon.

Franklin anc Dyrness, *Natural Vegetation of Oregon and Washington*

The Klamath Mountain region of southwest Oregon and northern California contains one of the most complex vegetation patterns in North America. The oceanward slopes of the Klamath mountains support dense coniferous forests. The renowned redwood forest reaches its northernmost limit in the Klamath mountains of Oregon. At higher elevations is a forest zone which contains the only natural habitat of the Port Orford cedar. Even higher is found the rare Brewer's weeping spruce, along with Alaska cedar, Shasta fir and mountain hemlock.

Most forests in the Klamath region have rich shrub understories. Sometimes these shrubs are predominantly deciduous, sometimes evergreen chaparral species. A number of them fix nitrogen, such as the California lilac and most other ceanothus species, the wax myrtle, and buckthorn. There are also shrubs with edible berries, such as native blackberries, salal, native gooseberries, Klamath plum, and Oregon grape.

On the drier, east side of the Cascades the change in climate is reflected in forests of ponderosa and lodgepole pine, western larch, and an Interior Douglas-fir. At higher elevations the subalpine fir, Englemann spruce, subalpine larch, whitebark pine, and mountain hemlock are found. Understory shrubs range from drought resistant species such as sagebrush, bitterbrush, squaw currant, and serviceberry to Oregon grape, snowberry, thimbleberry, bunchberry, white alder, and willows.

Ponderosa pine is one of the most common forest trees in arid regions of North America. Since the ponderosa can endure where other trees cannot, it is often found at the

Elias, T.S. <u>The Complete Trees Of North America</u>. New York: Van Nostrand Reinhold, 1980.
A detailed identification book with information on range, habitat, uses, and botanical characteristics.

Hosie, Robert C. <u>Native Trees of Canada</u>. Don Mills, Ontario: Fitzhenry & Whiteside, 1979.
A basic text for university forestry students. 180 species are covered, with a range map for each.

BLM

Fowells, H.A. <u>Silvics of Forest Trees of the United States</u>. Agricultural Handbook No 271. Washington, DC: USGPO, 1965.
Descriptions of nearly 125 trees, their life histories and ecology. Information provided for each species includes habitat, associated trees and shrubs, reproduction and growth, races and hybrids, detailed range maps, and references.

Johnson, Hugh. <u>The International Book of Trees</u>. New York: Simon & Shuster, 1973.
The single best book on trees of the world, with beautiful photographs and solid information.

The Oak/Camas Connection

"Swine were among the very first animals brought to Astoria and later taken inland to Fort Vancouver.... Versatility was the hog's outstanding virtue. No other domestic animal exhibited such facility and prowess in foraging for the necessities of life. If there was nothing to eat above ground, it would root up food from beneath the soil or even from underwater. Swine hardly ever died from starvation.

"In the Willamette Valley, two native plants, the Oregon oak *(Quercus garryana)* and the camas *(Camassia quamash)*, became especially important in the pig's diet....

"After studying stands of small trees in the mountains of northern California in 1944, botanist Carl B. Wolf noted that in normal crop years the Oregon oak produced about five hundred pounds per acre. Allowing for the larger tree crowns characteristic of the central Willamette Valley's extensive oak woodlands and oak openings this figure might be increased by 50 percent or more. When multiplied by the thousands of acres covered by the trees in the 1840s, the harvest must have totalled millions of pounds....

"The food supply offered by the oaks and camas was so great and so easily obtained that swine multiplied rapidly and escaped into the woods. By the late 1830s, numerous bands of wild hogs roamed the countryside. Since so little interest was shown at the time in the activities of these animals it is now difficult to accurately assess the effects they had on the land. There are some suggestions that they had a tremendous impact on the native economy. By eating large amounts of camas and acorns, pigs effectively reduced the Indians' main supplies of carbohydrate. Their consumption was so great in some areas that bands of Indians were driven to near starvation. On March 27, 1854, Indian Superintendent Joel Palmer wrote a letter concerning the situation facing the Calapooya who still lived along the Tualatin River: 'The Wappato, Kammas and other nutritious roots once produced abundantly in the marshes and lowlands around their principal residences, and consituting their principal means of subsistence, have, since the increase of swine, gradually diminished in quantity and must soon entirely fail.' "

-William A Bowen, *The Willamette Valley, Migration and Settlement on the Oregon Frontier*. University of Washington Press, Seattle, WA, 1978. pp 86-87.

interface of forests and grasslands. The ponderosa is long lived and its bark becomes lighter with each succeeding decade, changing from the dark brown of youth to lighter and lighter shades of red and cinnamon. A fully mature tree stands out for miles in the hot sunny air which is typical of ponderosa country.

The western larch is a deciduous conifer which grows on north-facing slopes and other moderately moist sites east of the Cascades. The larch is most widespread in southeastern British Columbia, northeastern Washington and northeastern Oregon, but its range also extends into northern Idaho and western Montana. Larch cannot tolerate shade, but grows so fast that it becomes both the colonizer and the climax species in many Interior forests.

Lodgepole pine is found throughout the Interior. Lodgepole is a most adaptable tree; its cones can survive even the heat of fires. In fact, heat facilitates the release of seeds. So along with larch, lodgepole is usually one of the first trees to sprout after the forest fires which are common in this area of low summer rainfall.

The best developed forests of the Interior are found in the highest rainfall zones on the western slopes of the Rockies. Here there is enough rainfall to support forests remarkably similar to the rainforests of the coast. Species include western red cedar, Pacific yew and western white pine as well as the expected western larch, lodgepole pine, Douglas-fir and Engelmann spruce. These trees are found mixed with two climax species, grand fir and western hemlock.

A number of deciduous broadleaf trees are also found in the Interior. Aspens, willows, cottonwoods, birches and alders are scattered over the Inland Empire, ranging from small groves along streams and natural springs to forests of birch or aspen on wetter mountain slopes. The presence of these trees is an indication of moist areas in the dry Interior. Many an old homestead is located near birch or aspen groves.

The quaking aspen is known and loved by anyone who has spent much time in the mountains of the West. Found from the Arctic Circle to northern Mexico, aspen has the widest range of any tree in North America. It takes only the slightest breeze to set aspen leaves dancing, their bright surfaces glittering against the clear blue sky.

Like the aspen, hillsides covered with birch forests are a fine sight in spring, with their shimmering leaves set against ranks of straight white trunks. The paper birch is the best known and most widespread of the birches in the northern Interior. As mentioned above, large stands are also found along the coastal lowlands of British Columbia and northern Washington. The bark peels in smooth horizontal bands, usually with pink or cream colored markings. Intolerant of shade, it is often seen in pastures and clearings. For most of its young life the paper birch has a slim, graceful elegance, but by middle age it often reaches the stout proportions of a proper forest monarch.

While forests cover higher elevations, much of the lowland Interior is treeless. Lines of trees, however, can generally be found along watercourses. Black cottonwood, water birch, willow, and the striking red-osier dogwood colonize islands and banks which are exposed as rivers and streams change their course.

The white alder and the thin-leaf alder are also important colonizers of disturbed sites in the Interior. They are water dependent, however, and appear only along stream courses. Alders are important wildlife trees; chickadees, pine siskins and other seed eating birds use the female catkins as survival food in winter and early spring.

Juniper is one of the characteristic trees of the arid zones of eastern Oregon, southern Idaho and a few locations in southeast Washington. Two junipers are found in the Interior. Rocky Mountain juniper grows as an occasional tree

Little, Elbert L. Atlas of United States Trees. Washington, DC: USGPO, 1971.
A four volume set which covers all the common and rare conifers and deciduous trees of the United States.

Peattie, Donald Culross. A Natural History of Western Trees. New York: Houghton-Mifflin, 1953.
A detailed, readable and enjoyable introduction to Northwest trees.

Sudworth, George B. Forest Trees of the Pacific Slope. New York: Dover, 1967.

Written in 1908 by a man who had first-hand knowledge and experience with most Western forests before they were despoiled. The book's best points are its exact descriptions of each tree's range, habitat, site tolerances and reproduction habits.

Will the True Cedar Please Stand Up?

Cedar #1

Port Orford cedar *(Chamaecyparis lawsoniana)* is an ancient species which was widespread in past geologic ages. It has been almost eliminated from its native haunts by logging, fire and a recently introduced fungal root disease. It is now native only in the Klamath Mountains, growing almost exclusively within 15 miles of the ocean. Its beautiful weeping appearance and wide variation in natural form have made it a favorite ornamental. Today, hundreds of named varieties are being sold all over the world. Its fragrant wood has many valuable properties and is prized by woodworkers, especially in Japan. Boat builders make frequent pilgrimages to Port Orford, Oregon in search of this ideal hull and deck lumber.

Cedar of Lebanon
Cedrus libani

Cedar #2

Alaska Cedar *(Chamaecyparis nootkatensis)*, also known as yellow cedar, is native only to the Northwest. It is one of the longest-lived trees in the world, attaining an age of 3,500 years and more. The wide spreading root system of the Alaska cedar does not need deep soil in areas where precipitation exceeds 60 inches per year. Its lightweight, close-grained, decay resistant wood was so highly valued by native peoples that coastal tribes carried it over mountain passes to trade to Interior tribes. Today, it is particularly valued for boat building and greenhouses, since the wood swells only slightly after being repeatedly soaked and dried.

Cedar #3

Incense cedar *(Calocedrus decurrens)* was listed as a *Libocedrus* until recently. The new name means "beautiful cedar." Its neat, pyramidal crown and graceful form have brought it world renown as an ornamental. Unlike other "cedars" of the Northwest, the incense cedar grows on very dry sites, often with ponderosa pine. Its drought tolerance is demonstrated by the remarkable natural stand of these trees which grows, surrounded by sagebrush, east of Bend, Oregon.

Cedar #4

Western red cedar *(Thuja plicata)*, one of the most loved trees of the Northwest, has been planted in parks and gardens throughout the world. Its wood and bark were key elements of the domestic crafts of the Indians of the Maritime region. Because of its stubborn resistance to decay, Western red cedar is the homebuilder's choice for lumber that is to be continuously exposed to the weather. Until recently, most logged cedar stands were replanted with faster growing Douglas-fir, giving rise to fears that the red cedar eventually would disappear from our forests. Fortunately, many foresters are now coming to appreciate the value of red cedar, both for its ability to grow in moist soils and for the unique properties of its wood.

THE TRUE CEDARS

None of the above is a true cedar. There are only four species of true cedar in the world, all of which are becoming increasingly common as ornamentals in the Northwest. The deodar cedar *(Cedrus deodara)*, native to the Himalayas, the Atlas cedar *(C. atlantica)* native to the Atlas mountains of North Africa, the Cyprus cedar *(C. brevifolia)* native to the island of Cyprus, and the cedar of Lebanon *(C. libani)* can be identified by their distinctive windblown silhouettes amidst the trees and rooftops of our urban landscapes.

-Michael Pilarski

Friends of the Juniper Forest

The Friends of the Juniper Forest was organized to protect a unique Northwest ecosystem. About 20 miles northeast of Pasco, Washington, is the border of a 15,000 acre area of Juniper forest and sand dunes. The dunes are the largest in the state, the western Junipers are at the northernmost limit of their range, and the forest is home to many significant forms of wildlife.

Unfortunately, the area is threatened by a proposed BLM plan to develop a 5,000 acre Off-Road-Vehicle park in the middle of the forested dunes. For more information and to help in their preservation attempt, contact **Friends of the Juniper Forest, 204 Newell Ave, Walla Walla, WA 99362.**

in the northern forests, while western juniper forms extensive parklands and savannas in the arid regions of Oregon and Idaho. The western juniper is fire sensitive and has been diminished in range in the last century. The Indians gathered western juniper berries for food, used its bark to make ropes, sandals and other objects, and carved its wood for tools.

In the Klamath mountains, between the lowland savannas and the forests at higher elevations are found extensive shrublands known as chaparral. Chaparral is made up of densely woven thickets of various shrubs with stiff branches and leathery leaves. Chaparral includes many species, such as manzanita, ceanothus, mountain mahogany, canyon live oak, tan oak, chinquapin, chamise, and buck brush. Because of its dense growth, chaparral is very fire prone.

As a result of poor logging practices and natural and humanmade fires, chaparral is increasing in the Klamath region. Chaparral's advance indicates environmental abuse and degradation, and is a visible reminder of the need for reforestation and ecosystem rehabilitation.

Plant a Tree...Grow Tomorrow

MICHAEL SOULE

The band was playing Sousa as it marched down Main Street, just behind the group of small children dressed like Paul Bunyan and Johnny Appleseed carrying seedlings and small shovels. All but a handful of the town's 5,000 inhabitants lined the street for the parade. There was an old fashioned log truck with a single tree on it that took up the whole bed, a van for senior citizens which was decorated in green, pre-schoolers costumed as leaves strung together, and the town's fire truck. The four princesses of the parade rode on the back seats of convertibles loaned by Bob's Motors. It could have been any parade on any Fourth of July in almost any small town in America, but this celebration carried a special meaning.

This was the Oakridge Tree Planting Festival, the only tree planting festival in Oregon. The signs read, "Plant a Tree. Grow Tomorrow!" and people talked about meeting, after the parade was over, on Cougar Mountain, where a clearcut was waiting to be planted. There was excitement about the tree planting, but it was only one of the many activities which made up the celebration. There was also a pancake breakfast, square dance, golf tournament, parade, teen dance, community rummage sale and more.

Oakridge is a small town in the Cascades, 41 miles east of Eugene, Oregon, just across the river from Westfir. Its major industry is trees: cutting, hauling and lumbering. In each of the past 30 years, the town has let go of its daily routine for a weekend in May to celebrate trees and the community life that trees have given them.

The group of Cub Scouts stood at the top of the ridge with shovels in hand and bags of tree seedlings slung around their waists. After brief instructions and tree planting tips from a local ranger, the children were off. They worked in pairs, one digging, one readying the two-year-old Douglas-fir seedling, and both putting it into the ground. "Don't take the trees out too early." The day was cold, the ground steep, the shovels too big, the bundles of trees too heavy, yet in only a couple of hours the entire three acres was planted with seedlings. The work didn't seem that difficult after all.

The day was fading. The Forest Service ranger picked up stray shovels and collected lost jackets and buckets. The scouts were gone, off to see the last of the auction and the bowling tournament. Looking across to other hills, trees covered the earth as far as one could see. If your eyes were good enough, you might be able to see beyond these trees, beyond this forest, to places where there are no trees, where there is no shade, where the sun is hot and the ground is hard. People live there, too. You might be able to see the blowing sands of the Sahara, the flash floods caused by deforestation in India, or the vast areas of tropical rain forest which are being cleared in Brazil and the Philippines. If we could see these places, we might have a better understanding of how much we depend upon trees. Even if your eyes cannot see as far as the Sahara or the tropical rain forests, perhaps you could still take one weekend in May and have a celebration for the trees. Just drive down to Oakridge and join them in their tree planting festival or start one in your own town. Everyone is welcome. Remember—Plant a tree. Grow tomorrow.

Stalking the Savory Native Truffle

Truffles, flavorful underground fungi, have been a gourmet food in Europe for centuries. The pungent flavor of the truffle is much sought after as a seasoning for gourmet dishes. The small potato-like fruits have a fragrance which has been characterized variously as cheesy, spicy, fruity, garlicky and even fishy. Their monetary value may be $40 per pound or more.

Recently researchers have discovered that truffles are widespread in the Pacific Northwest. Dr. James Trappe, mycologist at Oregon State University, believes that in many parts of the Northwest truffles may be as common as above-ground mushrooms. Eight new truffles species were discovered in northwestern Oregon in 1980 alone.

John Stamets

Dr James M Trappe, Principal Mycologist with the USDA's Forest Sciences Laboratory in Corvallis, Oregon, introduces a class to the art of truffle hunting at the Mycomedia Conference on Orcas Island, Washington in October, 1980.

Researchers at Oregon State University also have discovered an important connection between truffles, trees and animals. Certain small mammals eat a large quantity of truffles. They are attracted by the truffles' pungent odor, and either dig them up to carry home or tunnel underground from truffle to truffle. Spores from the truffles pass through the digestive system of these animals and the spores are left to inoculate seedlings of the host species. This is the only known way the spores of many of these highly adapted fungi can be dispersed.

All known truffles function in mycorrhizal associations with plants. Some trees require the truffle associate for establishment and survival. In the early 1800s, once the association between truffles and its mycorrhizal host species, the oak, became known in Europe, fields of oaks were planted specifically for truffle production. In the Pacific Northwest, truffles are a relatively untapped source of wild food.

Truffles are not easy to find without knowing where and how to look for them. Further information on truffles and truffling may be obtained from **The North American Truffling Society, 720 Manner Way, Lebanon, OR 97355.**

Seedlings of oak and filbert inoculated with the famous Perigord truffles *(Tuber melanosporum)* of southern Europe are available from Mushroompeople, PO Box 158, Inverness, CA 94937. Mushroompeople also distributes a book on modern truffle cultivation entitled *Truffle, the Black Diamond*, by F. Picart ($10.75).

Forest Mycorrhizae

Some bacteria and fungi live in symbiotic association with roots of higher plants. These mycorrhizae (a term which literally means "fungus roots") were first observed on forest trees in the late 19th century.

Much of the current knowledge of the workings and significance of mycorrhizal associations comes from research in forest tree nurseries where the growth of the treated and untreated stock was studied. The results were so conclusive that today inoculation of reforestation conifer seedlings with mycorrhizal fungi is being practiced as a complement or alternative to chemical fertilization. Inoculating seedling roots with site-specific strains of fungi has also been found to increase seedling survival, especially on harsh sites.

Anderson, Robert L., and Cordell, Charles E. How to Recognize and Quantify Ectomycorrhizae on Conifers. 1979. USDA Forest Service, 1720 Peachtree Road NW, Room 816, Atlanta, GA 30309.
Pamphlet which describes the various forms of mycorrhizae fungi, with a description of nursery and field sampling techniques. Excellent color identification photos.

Molina, Randy. Ectomycorrhizal Inoculation of Containerized Western Conifer Seedlings. PNW-357. Portland, OR: PNW Forest and Range Exp. Stat. 1980.
Evaluation of inoculation rates of 15 fungi on five container-grown species, with a brief discussion of methods. Thirteen references cited.

Trappe, James M. "Selection of Fungi for Ectomycorrhizal Inoculation in Nurseries." In Annual Review of Phytopathology 15 (1977) 203-222.
Dr. Trappe emphasizes that "good nursery management requires good mycorrhiza management." This paper reviews the history and potential of mycorrhiza utilization (including the inoculation of nursery stock with truffles and other valued fungi). 127 references.

Waring, R.H., ed. Forests: Fresh Perspectives from Ecosystem Analysis. Corvallis, OR: Oregon State University Press, 1980.
Proceedings of a symposium on forest and stream ecology which emphasize the newly recognized importance of soil biology in forest management.

Raphael, Ray. Tree Talk, The People and Politics of Timber. Covelo, CA: Island Press, 1981. 287 pp, $12.00.

The interwoven issues of modern forestry are discussed with a direct and human emphasis, with examples of old and new ways of living more harmoniously with our forests. An important book for those who want a new perspective on people, land and trees.

Forest Planning

Forest Planning, PO Box 3479, Eugene, OR 97403. Monthly, $10/year.

Journal of the Nationwide Forest Planning Clearinghouse, this publication emphasizes the public's role in forest planning. News and techniques are presented with a balanced perspective.

Cascade Holistic Economic Consultants. The Citizens' Guide to Forest Planning. CHEC, Box 3479, Eugene, OR 97403.

The U.S. Forest Service is developing a master plan allocating the resources of all national forests. Produced by the editors of Forest Planning Magazine, this citizens' guide is a tool kit for people wishing to take an active role in the planning process. It concludes with a chapter on how concerned citizens can be effective in proposing alternatives which encourage environmentally sound approaches to forestry.

Tree Talk

Forest policy is fraught with controversy. Conflicts rage over wilderness preservation, clearcutting, and the use of herbicides. While differing points of view may seem beyond reconciliation, Ray Raphael has brought together diverse voices for forestry reform in an important new book entitled Tree Talk. *Much of* Tree Talk *consists of interviews with people who live and work in the forests. What emerges is a creative and challenging call for a more holistic approach to forestry. Below are some excerpts from this valuable new book.*

RAY RAPHAEL

Nature, if left alone, will certainly grow trees. The impressive coniferous forests that European-Americans discovered in the Pacific Northwest less than two centuries ago bear witness to nature's bounty. But the ways of nature are slow; to some even tedious. Since we want more timber from fewer forests, we grow anxious to get on with the business of growing trees. We substitute genetic engineering for evolutionary processes; we bypass natural stages of forest succession; we even cut the timber before it has reached its optimum productivity. By intensively managing the forests, we hope to straighten out the kinks in the meandering ribbon of time. The problem, however, is that we don't always know what we are doing. In the name of making improvements, we often create new obstacles.

So what should we do? Should we sit back and wait for nature alone to heal the scars that we ourselves have created? Or should we grab destiny by the wings and hitch a ride through time by altering the genetic selection of trees, eliminating unwanted brush with herbicides, cutting the adolescent timber as soon as we can find a market for it? Fortunately, these are not the only alternatives. There are ways of tinkering with nature that complement, rather than contradict, natural processes.

Holistic forestry, as it is sometimes called by proponents, aspires to work *with* natural laws rather than *against* them. Reacting to the excesses of industrial manipulation, self-proclaimed holistic foresters attempt to manage land according to the basic ecological principles of balance and diversity. Instead of simplifying the ecosystem by eliminating unwanted elements, they prefer to utilize all elements to the best advantage of the entire forest. Like industrial foresters, they are interested in the production of timber; unlike industrial foresters, they treat the forest as a complex, interdependent system with a life of its own.

The prime concern in holistic forestry is growing trees. But holistic foresters are not interested in just planting trees and spraying them with various chemicals. Instead, holistic foresters offer a sweeping program for environmental repair. Since much of our forested land has already been severely damaged, holistic foresters feel we must take steps to heal the ground that supports the trees. We cannot expect the land to produce to its full potential, for instance, unless we can halt the erosion that washes away the topsoil. We must also replenish depleted soil nutrients, but this doesn't require chemical fertilizers. "Green-manure" trees, such as alder or *Ceanothus*, fertilize the ground as they grow.

Since holistic foresters believe in balance and diversity, they try to maintain a variety of species and ages in the forest at all times. They are reluctant to remove the entire forest canopy at once, since this will alter the microclimate and drastically change the ecology of a site. When they do harvest the trees, they devote their attention to a careful and complete utilization of all the material they remove; they try to find some use for the slash, the low-quality lots, and the species of lesser commercial value.

Holistic forestry treats each specific site according to its own needs. The ecosystem is maintained in its basic form, but is modified here and there to bring it into harmony with human needs.

In some spots, for instance, a brush cover might be maintained to stabilize the soil or to replenish valuable nutrients, but just a few feet away, where the soil is in better shape, the brush might be cut back to allow room for conifer seedlings. The premise is that each place in the forest is absolutely unique, even though there are basic ecological rules which must be adhered to if the forest as a whole is to prosper.

BUD McCRARY
LUMBERMAN, BIG CREEK LUMBER COMPANY

"Our family started lumbering commercially in 1946. When my brother, my dad, and I got out of the service, we formed a partnership with my uncle. We didn't have much land; we were buying timber from other people. We built a sawmill and moved around to the different places.

"The way we practice good forestry on these steep hills is with labor-intensive forestry. We do things the hard way. We're not trying to modernize as much as we should, I guess, but we seem to be surviving. We feel it's important to give people jobs in this area. There's just a shortage of work for people, and if everybody tries to economize on labor, you run out of jobs for people and you've got a bunch of machines and nobody around to run them."

(cont. on next page)

PERIODICALS

Agroforestry Review, PO Box 666, Winters, CA 95694.
The journal of the International Tree Crops Institute U.S.A. High quality review of the tree crops literature and developments. Recommended.

Forestcover, RR 3, Queen's Bay, Nelson, BC V1L 5P6.
A periodical for tree-planters and other forest workers in British Columbia.

International Tree Crops Journal and The Journal of World Forest Resource Management
A B Academic Publishers
PO Box 97
Berkhamsted, Herts, HP4 2PX England
The first of these journals is scholarly, with a primary emphasis on agroforestry for developing nations. The second is new (begun in 1982), and focuses on all aspects of the management of forests as a global resource. Articles will include detailed studies of forestry policy, and the connections between the nature of forest utilization and the health of the biosphere. The editor, Alan Grainger, is also director of the International Tree Crops Institute.

ORGANIZATIONS

Agri-Silviculture Institute, c/o Paul J Marks, PO Box 4166, Palm Springs, CA 92263.
The Institute's focus is on tree crops research and on involving young people in alternative agriculture projects. Their emphasis is on the arid and semi-arid regions of the world.

Agroforestry Associates, PO Box 666, Winters, CA 95694.
Complete agricultural forestry service for Northern California, including design, nursery stock and management of new plantings. Catalog available.

American Forestry Association, 1319 18th St. NW, Washington, DC 20036.
One of North America's oldest conservation organizations, the American Forestry Association helped establish forest management as an art and science. Their goals are to replant and preserve the nation's forests and rangelands. They provide information and assistance for private woodlot and forest owners, and sponsor the National Register of Big Trees. They also publish a monthly journal, American Forests.

Forever Forests, PO Box 212, Redwood Valley, CA 95470.
Forever Forests is helping create protected forest preserves by providing incentives to landowners to put land into perpetual trusts. The incentives include free or low cost tree planting and erosion control by teams of Forever Forest experts and volunteers.

Friends of the Trees, Star Rt, Box 74, Oroville, WA 98844.
The Friends of the Trees Society is a network of tree enthusiasts and an information service on tree crops for the Pacific Northwest. A grass-roots movement encouraging wide-spread reafforestation and local self-sufficiency.

International Tree Crops Institute, Convent Lane, Bocking, Braintree, Essex, England.
The international headquarters for ITCI maintains a network of cooperating tree societies and researchers around the world. ITCI promotes the development and implementation of agroforestry and publishes a quarterly journal.

International Tree Crops Institute USA, Western Office, PO Box 666, Winters, CA 95694. Appalachian Branch, Rt. 1, Gravel Switch, KY 40328.
Both US Branches of ITCI are devoted to research and the exchange of information on tree crops. Both have nurseries which supply seedlings, cuttings and scion wood to the general public. The Western Branch is a good source of information on windbreaks. Their journal, The Agroforestry Review, is a good way to keep in touch with the tree crops movement.

Men of the Trees, Crawley Down, Sussex, England.
Richard St Barbe Baker started Men of the Trees in 1922. It continues to be one of the most effective organizations in promoting the cause of forests around the world.

Oregon Small Woodland Association, 4033 SW Canyon Road, Portland, OR 97221.
The objective of the OSWA is to promote and carry out programs which increase the productivity and health of the membership's forest lands. The ongoing use of newsletters, talks, presentations, demonstrations and tours is central to achieving this goal.

Society of American Foresters, 5400 Grosvenor Lane, Washington, DC 20014.
Professional foresters' association, with an emphasis on forestry science, technology and education. Includes an Urban Forestry Working Group to support and promote urban forestry. They publish the Journal of Forestry monthly and Forest Science quarterly.

Horse Logging

Karen Hayden

"You don't want a lot of payments to make if you're in horse logging. You need to have a little bit of money in the bank. And you need to have an awful lot of experience behind you, not only in logging, but in horse psychology, horse sense."

A horse doesn't weigh as much as a cat. It doesn't push around the topsoil. It doesn't spin its tracks, create erosion channels, or significantly compact the earth. It doesn't bump into residual trees. All in all, a horse is gentler on the forest than a logging tractor or a wheeled skidder. And a horse consumes renewable resources—hay and grain—rather than unrenewable fossil fuels. A horse is not made out of metals that must be extracted from the earth. It is not manufactured in factories that dump polluting by-products into the air and water. Environmentally speaking, a horse is preferable to a cat in every respect. And it is also cheaper to buy.

For thinning small timber or harvesting poles, a horse can hold its own; but try as they might, even the mightiest draft horses can't tug on a large log as well as a cat. Horse logging is slow and tedious work. When the timber is small and scattered throughout the forest, environmental benefits can offset the relative inefficiency of horse logging. But when the timber is large and most or all of the trees are to be harvested, the horse will be competing out of its league. Horse logging, in short, is a valuable tool, but it is not a panacea. Like any good tool, draft horses should be used only on the jobs for which they are best suited.

JOAQUIN BAMFORD
HORSE LOGGER

Joaquin: "I've always believed that if you want to look into something, you go to whoever knows most about it. I heard about this fellow Al Merrill. He was about sixty-five at the time, and he was still out there logging. I went out to meet him and went around with him, helping him out when I could, watching him and asking him questions. Of course, the first thing he said to me was: 'If you want to go into horse logging, you might as well figure on starving to death.'

Forest Ecology & Management

<u>Forest Ecology and Management</u>
Associated Scientific Publishers
P.O. Box 211
1000 AE
Amsterdam, The Netherlands

An international journal concerned with forest science and conservation, and in particular the application of ecological knowledge to forest management. The scope of the journal includes all forest ecosystems of the world. Technical, but useful to the serious forester.

Franklin, Jerry F., et al. <u>Ecological Characteristics of Old-Growth Douglas Fir Forests</u>. USDA FS Gen Tech Rep PNW-118. Portland: PNW For & Range Exp Stn, 1981.

This report helps establish the scientific basis for preserving old-growth forests. The authors provide management options to aid people concerned with saving old-growth forests for the future.

Kittridge, Joseph. <u>Forest Influences: The Effects of Woody Vegetation on Climate, Water and Soil</u>. New York: Dover, 1973.

First published in 1948, this book remains one of the most comprehensive, balanced coverages of the field. The author examines forest influences in great detail: filtering of light, air temperature, wind velocity, atmospheric moisture, precipitation, soil moisture, soil temperature, decomposition of forest litter, interception, transpiration and evaporation, runoff and stream flow, floods and erosion.

Minckler, L.S. <u>Woodland Ecology, Environmental Forestry for the Small Owner</u>. Syracuse: Syracuse University Press, 1980.

How to maximize woodland production while maintaining a healthy forest ecosystem. Written for Northeast hardwood forests, but applicable to the Northwest.

National Forest Products Association, Private Woodlands Committee. <u>America Grows on Trees: The Promise of Private Nonindustrial Woodlands</u>. Washington, DC: National Forest Products Association, 1980.

Approximately 4 million landowners hold 58% of US forest land. With demand for timber projected to double in the next 50 years, these lands have become a prime target for more intensive management and utilization. This report is the timber industry's blueprint for increased utilization of private forests. It outlines a joint government/industry program for research, education and financial incentives to stimulate production and reforestation on private lands.

Thomas, Jack Ward, ed. <u>Wildlife Habitat in Managed Forest—The Blue Mountains of Oregon and Washington</u>. USDA Forest Service, Agricultural Handbook No. 553. Washington, DC: USGPO, 1979.

An analysis of habitat types (e.g. riparian zones, edges, snags and cliffs) and how they might best be managed to preserve and enhance wildlife.

"So we set out to get our finances all squared away before we started. You don't want a lot of payments to make if you're in horse logging. You need to have a little bit of money in the bank. And you need to have an awful lot of experience behind you, not only in logging, but in horse psychology, horse sense. What to do when a situation arises. How to move. How to be light on your feet. You have to be able to perceive what's going to happen to you fifteen, twenty feet down the road. You have to be able to jump out of the way, stop them, turn them. Think what the horses are thinking: whether they are spooked or scared; whether they'll stand for you; whether they'll back up for you if you want them to. If somebody else is out there, make sure you're never in a bind in case they happen to say a word that means for the horses to move ahead or move back.

"The thing about horses is you can't get in a hurry. No way. Whenever you start getting excited and worried about getting enough volume in, then something's going to happen. You can't hurry the horses. You just have to take one step at a time. It's not like machine logging. When you own machinery and you have big payments (I mean they don't give that machinery away), the logger has to rip and tear and go right after it. There is no being really careful. He's got to get in ten loads a day, or whatever it takes to make those payments.

"In horse logging you have to look at the timber before you can go and say, 'Yeah, I can log that.' The best type is second growth, timber that ranges from fifteen years to forty or fifty years. You want it small, but not too small. We usually won't go into timber that is less than twelve inches on the butt.

"And you need to look at the land. You want a downhill pull, if at all possible. You look at your terrain. You find out what the soil is like: whether it's loose soil, whether it's on a steep hillside, whether there will be erosion where you drag the timber down. You have to look at it as a conservation practice."

MECA WAWONA — NEW GROWTH FORESTRY COLLECTIVE

"A vision, a dream of ours, is to develop a system of land-based foresters, a system of stewardship. It takes two directions. One is on public lands. An example is the Ohai system they have in New Zealand. They ask nine adult members to sign a long-term lease—some are thirty-year leases, some are lifetime leases—for stewardship and occupancy of a certain amount of acreage in their national forests. Those people, in return for their use of the forest and their occupancy there, do certain types of management or caretaking activities. The system hasn't been going on for that long, so I don't know how it's working out. But it's something that we've given a great deal of thought to in terms of the National Forests here: a group of people, whether it's a handful of families or a large communal group, having stewardship contracts on a number of acres, large acreage, from hundreds of acres to thousands of acres.

"They could do a number of things there. In terms of maintenance activities, they could do everything from collecting seeds, growing trees, planting trees, thinning, and precommercial logging, to logging and milling. In the more remote rural areas, they could do the milling right there and truck the lumber down to the towns. Where there are large mills in the area, they could either transport the logs out or have trucks come up and take them out. They could also be doing hardwoods, which require greater understanding of how to cut them, how to dry them, and how to stack them. They could also help maintain fisheries and do monitoring. They could do the maintenance and monitoring for work and research set up by university people. They could also maintain hostels, whereby people could come into the area by trails and participate in a number of work-oriented forest activities, just for the work experience. It would move away from 'Winnebago' recreation, toward learning about the forest and having a real vacation.

"Those communities would be doing their own training of young people in their work, their own schooling, homesteading, food production, shelter production, and those kinds of things, along with the stewardship of the forest resources."

Forest Work

The Northwest Forest Workers' Association (NWFWA) is a trade group of cooperatively structured, democratically managed companies which has added the human resource as a factor in the complex reforestation equation, bringing with it an emphasis on human capital and labor-intensive forestry practices. The state of the art is being developed by its two dozen coop members representing a thousand workers, most of whom live in Oregon, although there are also small chapters in Washington and California.

The first coop was born at Eugene, Oregon, still the center of the reforestation industry, in 1971. It has spawned many others, peaking in the boom year of 1978 with the emergence of most of the present cooperatives. These perform contracts throughout the region, primarily on public lands.

Treeplanting, which requires crews of ten to twenty workers, follows timber harvesting by about eighteen months. The decline in the timber cut recently has led the coops to diversify into smaller, more locally-based groups, who perform contracts in the other areas of reforestation work. These include timber stand exams, stocking surveys, stand maintenance such as thinning and protection from animals and competing brush species, cone collection, and trail building and maintenance, as well as other activities necessary to growing a healthy forest crop.

The development of the Northwest's forest resource is yet to be completed. Satisfying an international need for fiber will be combined with providing for the regional need of a stable ecosystem that nurtures the people who cultivate the forest crop. In this work the Northwest's reforestation cooperatives have an important and special role that is worth the observation and participation of anyone dedicated to a future which is abundant.

TOM CLEVELAND

The address of the Northwest Forest Workers' Association is 795 Willamette, Eugene, OR 97401.

Community Forestry

a new FAO program

"Agricultural planning in China is fully integrated with forestry, animal husbandry and fisheries, so that maximum benefit can be derived from the land and water resources of the country....The following systems are employed: Types of 'taungya plantation' with inter-row cropping of such crops as melons, cassava, groundnut, ginger and soyabeans; Raising of fodder crops and grazing under forests; Growing of non-timber forests using food, fodder, medicinal and oil trees and other economic crops such as walnut, chestnut, fig, camphor, tea oil, tung oil and bamboo; 'Four around' forestry, around houses, villages, along roads and waterways using such fast-growing trees as poplars, willow, pines, firs, eucalyptus, and other types of vegetation such as bamboo; forestry farms with the primary objective of timber production, which often have subsidiary activities yielding minor products such as medicinal plants, mushrooms or basket fibers."

FAO, Forestry for Local Community Development

People in poor countries around the world are redefining forest policy to recognize social and cultural goals as well as economic needs. This new movement has come to be called community forestry, and it is now being fostered and promoted by the Forestry Department of the Food and Agriculture Organization of the United Nations.

Community forestry is bringing new hope to poor people in many parts of the world. Following are excerpts from two FAO reports which describe their new programs, Forestry for Local Community Development *and* Forestry for Rural Communities.

Forestry for community development must reflect the needs, problems and aspirations of local people as seen through their eyes. To be truly appropriate its strategy will vary according to community and place. In underlining social, rather than strictly commercial needs, community forestry calls for new policies and attitudes. In particular, the division between agriculture and forestry which seems so clear as between industrial forestry and intensive agriculture, must be abandoned. The preoccupation of the rural poor rests not in whether they are farmers or foresters; they can be, and usually are, both. There is no place for rivalry between the two: the issue is how forestry and agriculture can be combined to raise nutrition and living standards among the rural poor.

Shiitake - "new" crop for small farmers

MIKE MAKI

FAO

For centuries in Japan, the Shiitake mushroom *(Lentinus edodes)* has been taken from its wild habitat and cultured on logs. Today, over 180,000 Japanese growers produce shiitake on a small scale, mostly on family farms. Much of the annual shiitake crop, harvested spring and fall, is dried for year-round home use and for export. It is the second most important mushroom crop in the world, after the familiar "button" or "champignon" mushroom, *Agaricus bisporus*. Shiitake is prized not only for its firm, meaty flesh and rich flavor, but also for its medicinal properties.

While shiitake grows on decaying oak logs in its native habitat, it has been successfully grown on many other hardwoods. Shiitake grows well and fruits abundantly on chestnut, especially on the Japanese chestnut *(Castanea crenata)*, and on oak and alder. Logs for culturing shiitake are usually cut three to six feet long and four to six inches in diameter. They are then stacked under a forest canopy, where they will be moist and sheltered. The logs are inoculated by drilling holes and inserting plugs of mycelial culture. Fruiting begins 18 to 24 months after inoculation, and continues every spring and fall for five years or more.

Shiitake culture is ideal for small-scale production. Spawn and information on shiitake production is available from The Kinoko Company, PO Box 6425, Oakland, CA 94621, Kurtzman's Mushroom Specialties, 815 South Harbour Way #12, Richmond, CA 94804, and Mushroompeople, PO Box 158W, Inverness, CA 94937.

AGRISILVICULTURE

In agrisilviculture, the land supports both forest trees and agricultural crops, together or in rotation. At its simplest level, which is little more than an improved version of shifting cultivation, trees are grown on fallow land. The difference is that the trees, which are valuable species, are planted or sown deliberately instead of leaving the land to be recolonized by surrounding forest vegetation.

The trees are usually species that grow quickly above the crops. They have deep root systems to avoid root damage by the cultivators. The trees have to be able to withstand competition from the crops for light, water and nutrients. At the same time, the crops should not exhaust the soil. On the contrary, plants such as legumes which help to enrich it, in this case with nitrogen, are preferred.

Agrisilviculture, with its mixture of crops, may offend the eye of foresters accustomed to the tidy and regular appearance of plantations, but the evidence shows that it suits the environment, maintains soil fertility and combats the twin evils of erosion and leaching.

MULTIPLE PRODUCT FORESTRY

Local productivity can be improved simply by taking better advantage of the forest environment itself. Known as 'multiple product forestry,' the aim is to increase the yield of the many commodities, other than timber, that forests can provide. This approach can be particularly successful in the case of forest communities that already have some experience of obtaining from their surroundings items such as fruit, nuts, honey, game and fish, and herbal medicines.

In India and Pakistan some 80 percent of the people are believed to rely exclusively on medicinal plants to heal themselves. The use of such natural medicines is rapidly increasing elsewhere. Apart from encouraging local populations, especially people living in remote areas, to grow medicinal plants for their own use, good prospects exist for selling them. Honey is another product which can be both an essential item in local consumption and an increasingly popular article on a world-wide scale. Once again, it may be a question of introducing beekeeping to the rural community or improving traditional honey production.

Multiple product forestry, by diverting attention away from the timber value of trees, provides a considerable service to forest dwellers who, for too long, have failed to benefit adequately from commercial logging operations. It offers immediate, if often modest, improvements in local supplies of food or income without making any great demand on the people. By its essentially piecemeal approach, multiple product forestry can rarely provide a long-term solution to present-day pressures on rural and forest communities. Its successes, however, can pave the way for other systems that provide a key to enhancing the productivity of forests and marginal lands.

FAO

The Forestry Department of the Food and Agriculture Organization (FAO) of the United Nations is spearheading development of community forestry programs. Publications include:

Forestry for Local Community Development, FAO Forestry Paper 7.

A detailed report on the rapid deforestation and impoverishment of rural areas of the world, with analysis of new forest policies which have the potential to respond quickly and effectively to the needs of hungry people and exploited landscapes. The paper includes seventeen detailed case studies which evaluate community forestry programs in Asia, Africa and Latin America. They cover a broad range of approaches and include many inspiring examples.

Forestry for Rural Communities, FAO, 1978.

An excellent introduction to community forestry concepts. Included are descriptions of Multiple product forestry (the use of forest products such as honey, game meat and fruits, as well as wood), Small-scale forestry (the cultivation of individual trees for fruit or nuts), Arboriculture (intensive cultivation of individual trees for fruit or nuts), Agrisilviculture (the combination of forestry and crop production), and Silvipasture (the controlled grazing of forest vegetation).

OTHER FAO PUBLICATIONS

FAO publishes two regular journals. CERES ($12/year) is a bimonthly review of world agriculture. UNASYLVA ($8/year) is a quarterly journal of forestry.

FAO also publishes a series of nearly 200 Technical Papers on forestry, soil and water conservation, crop production, livestock, food and nutrition, and irrigation. Information on these and the reports listed above are available from the following address:

Publications Division, Food and Agriculture Organization of the United Nations, Via delle Terme di Caracalla, 001000 Rome, Italy.

Coppicing For Fuel

Coppicing, or the repeated harvest of sprouts which have grown from tree stumps, is an ancient practice which is currently enjoying a revival. The art and science of coppicing have become quite sophisticated over the centuries, and several species can be planted specifically for this purpose, including hazel, red alder, chestnut and poplar.

For the small woodlot owner or manager, the advantages of coppicing are many. Proper planning, planting and rotations can insure a virtually perpetual harvest of firewood for home or commercial use. The manager can see a return from a piece of land planted to trees in as few as five years, thus making tree-planting more economically feasible. Because sprouts are relatively thin when harvested, the labor of splitting firewood is greatly reduced.

Research is also underway to determine the value of coppiced trees as a source of alcohol and methanol feedstocks, and as green-chop forage or silage for ruminant livestock, whose intestinal flora enable them to utilize the food energy in these high-cellulose materials. In Europe, coppiced woods are also used for a wide variety of traditional crafts. Thus, the products of a piece of coppiced land may have several markets in the future.

Planting a coppice, then, may be one of the most rewarding things a landowner can do. Listed below and on the following page are some references to provide guidance and inspiration. For species selection, see the **Plant Species Index** on pages 151-170, where species especially suitable for coppicing are noted.

BIBLIOGRAPHY

Lambert, F. Tools and Devices for Coppice Crafts. (Young Farmers' Booklet # 31). Available for £1.50 (about $4.00) from **The Centre for Alternative Technology, Machynlleth, Powys, Wales, U.K.**
Good introduction to the ways of putting coppice wood to work in making fences, tools, and garden accessories.

Site-Specific Forestry

MIKE MAKI

Site-specific forestry is an approach to forest management which takes into account existing species and age mixtures, local microclimatic conditions, terrain and hydrology. Conventional monoculture forestry relies on slash-burning and herbicides to try to maintain woodlands at the pioneer stage of succession, allowing only one tree, usually Douglas-fir, to grow to maturity. Site-specific forestry, on the other hand, requires more sensitive harvest and management techniques, and more human labor. While net profits may be somewhat reduced in a site-specific management system, better utilization of forest products such as posts, poles, firewood, pulpwood and small sawlogs is made possible, and higher quality timber is also produced.

Hand clearing of weedy vegetation from around young conifer seedlings, called manual release, rather than using aerial herbicide sprays to kill competition, is an important site-specific forestry technique. In many cases, the skilled eye and weeding tool or chainsaw of a person on the ground turns out to be more cost effective than aerial sprays. Claudio Katz, a forest and watershed rehabilitation contractor from Arcata, California, explains some of the advantages of manual release in his 1981 paper, *An Economic and Silvicultural Analysis Between Alternative Forest Vegetation Management Practices and Aerially Applied Phenoxy Herbicides for Conifer Release:*

> Manual release of conifers in the Pacific Northwest began in 1977 with experimental work clearing brush from around conifers. Until recently the most common method used was total clearing of hardwood and brush species, using chainsaws and other cutting equipment. Current manual release practices combine the release operation with a tree thinning operation. Brush is cut only around the crop trees in an upward projected cylinder or half-circle. In some cases, certain species of brush and hardwoods are exempted from cutting, and in one case, the crop trees selected for release were hardwoods, in the absence of healthy Douglas-fir or pine trees. The decreased costs associated with improved contract specifications are a direct result of increased skills and labor productivity and the demand for work opportunities.
>
> An advantage of manual release is that it can be combined with tree thinning at no or only slightly additional cost. Manual release can also be more sensitive than sprays in the buffer zones along streams, near residences, and in the vicinity of crops. It also eliminates the problem of toxic sprays drifting onto land not intended for application.

Because of its reliance on local conditions and careful workers, site-specific forestry provides opportunities to experiment with creative management techniques. Wildlife managers have shown that planting certain herb, legume and shrub species provides wildlife browse which deer and other grazing animals usually prefer to conifer seedlings. Coast trailing blackberry *(Rubus ursinus)* , for example, has been widely planted in some clearcut areas. The further enhancement of blackberry production by using select cultivars could benefit both wildlife and berry pickers. On suitable sites, fruit and nut trees may also be planted by land owners in forest clearings created after logging.

Studies are now underway to determine if grazing sheep on newly replanted forestland may help control brush while tree seedlings become established. Flocks of sheep have traditionally been moved into mountain meadows and logged-off lands during the summer when the dry lowland pastures could not provide sufficient forage, but until recently grazing sheep had not been considered as a commercial management tool. Since sheep prefer the brushy shrub species to the young conifer trees, however, using them as an alternative to herbicides is now attracting interest among forest managers.

For the sheep to be effective, they must be turned in while the shrubs are young and just beginning to sprout. At this stage, the sheep's grazing can kill or set back the shrubs. Later browsing may actually increase foliage production of established shrubby species, just as pruning stimulates new growth in trees and bushes.

The latter technique could be useful to land owners who wish to increase the productive capacity of a plot of forestland while they are waiting for the trees to mature. Sheep turned in at the right time could increase production of a middle-story crop such as huckleberries or blueberries. Good management of any native or domestic animals in a woodlot is crucial, as uncontrolled browsing can seriously deform young crop trees.

Control of vegetation by sheep may be combined with other income-producing activities, such as bringing honeybees to recently logged lands during the herb and shrub stages of succession. This is now being profitably done on some clearcut lands. These intermediate crops could provide a financial return from the land until commercial forest trees overshadow the shrub and ground layers, and when combined with manual release, could also provide employment opportunities from lands in early stages of reforestation.

All site-specific forestry management techniques require people who are willing to diversify their economic activities and take advantage of seasonal forest work. They may become involved as tree planters, thinners, fuelwood harvesters, sheep herders or cone gatherers, as well as the more familiar forest occupations of logger, equipment operator, truck driver or mill worker. Their contribution to a more intensive, diversified forestry could mean more employment opportunities as well as better forest management.

Modern industrial timber managment sees the reduction of employees as a way to decrease overhead and increase profits. In the future, however, increased employment opportunities will be an essential condition for, and a major advantage of, intensive locally based, responsible woodlot management.

National Academy of Science. Firewood Crops, Shrub and Tree Species for Energy Production. Washington, DC: National Academy of Science, 1980.

Half of the timber cut in the world is used as fuel for cooking and heating. Although most of the 60 species listed in this book are not adapted to this region, the book provides an inspiring example of the potential for tree crops to serve humanity.

Stanford, Geoffrey. "Coppicing Your Home Woodlot." In The Next Whole Earth Catalog, edited by Stewart Brand. Sausalito, CA: POINT, 1981.

Excellent introduction to coppicing as a highly productive method of sustained wood production, with bibliography. "You can do all your cooking for a family of four forever with the wood you can cut from a tract of well-managed coppice that measures less than 65 ft. on the side."

Stewart, Philip. "Coppicing With Standards." In CoEvolution Quarterly, No. 30, Summer 1981.

"Standards" to grow to maturity and provide sawlogs in a mixed wood where other trees (usually other species) are coppiced. Ancient practice of multiple use brought up to date.

Habitat Improvement

Campbell, Dan L, and Evans, James. Establishing Native Forbs to Reduce Black-Tail Deer Browsing Damage to Douglas-Fir. 1978. U.S. Fish and Wildlife Service, Forestry Sciences Laboratory, Olympia, WA 98502.

Report on a six-year study which "demonstrated that prompt establishment of highly palatable native forbs reduced summer browsing on planted seedlings to the point that black-tail deer were no longer a factor limiting Douglas-fir regeneration."

——, and Johnson, Larry E. Guide for Collecting and Seeding Native Forbs for Douglas-fir Clearcuts. Wildlife Leaflet 513. 1981. U.S. Fish and Wildlife Service, Colorado State Univ., Ft. Collins, CO 80523.

Techniques for identifying, collecting and establishing seed of six native forbs which provide superior wildlife browse—catsear, hawksbeard, fleabane, hawkweed, phacelia, and redstem fireweed.

Alder

Red Alder *(Alnus rubra)* is found in a narrow region along the Pacific Coast from California to southeast Alaska. Farmers and woodspeople west of the Cascades know red alder's ability to colonize disturbed or abandoned land. Red alder seedlings quickly cover road cuts and clearcuts, and crowd the edges of fields and streams. It is a strong competitor with Douglas-fir, the major commercial timber tree.

Red alder obtains nitrogen from the air through colonies of symbiotic actinomycetes which form nodules on the tree's orange roots. This helps it survive on exposed, nutrient-poor ground. Within a few years the soil surface is covered with leaf litter and twigs. The soil becomes enriched with nitrogen and organic matter, thus preparing the way for succeeding species such as Douglas-fir, western red cedar, Sitka spruce or hemlock.

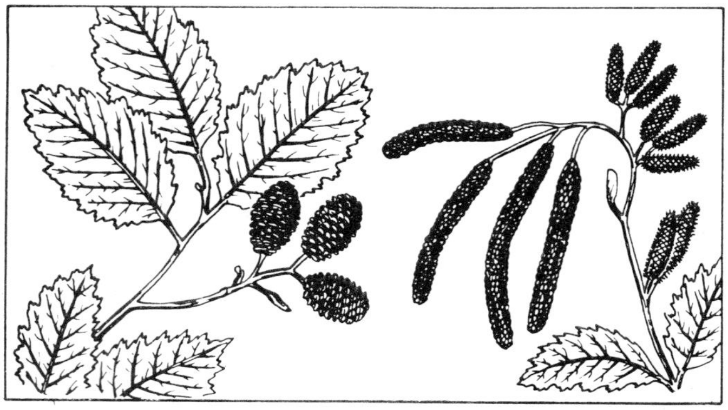

Trees of Washington

Red alder is also a good biomass energy crop, as it produces an abundance of usable fuel wood in less than 20 years. In Japan, one species of alder *(A. inokumae)* is grown in fuelwood plantations on poor sites. After harvest, the enriched land is often used for farmland for a few years before being replanted to conifers.[1] In the Pacific Northwest, a few test plantations of alder are being established. One of them, near Centralia, Washington, is on land reclaimed after coal mining.

When used in a mixed stand or in short successional rotation with Douglas-fir, red alder helps fir resist parasitic colonization by *Poria weirri*, a fungus which attacks both living and dead fir trees. Recent experiments by the U.S. Forest Service show that planting red alder after Douglas-fir trees have been harvested can greatly reduce or eliminate the threat of this disease.[2]

U.S.Forest Service silviculturist Dean DeBell nicely sums up the virtues of alder when he states, "Although most foresters know that alder grows fast and improves soil conditions, they may not be aware that the energy that must be invested to grow a given amount of wood is lower for alder than for most other species. Alder also offers forest managers a lot of flexibility because it grows rapidly over a wide range of site conditions. It is suitable for management systems ranging from short coppice (stump sprouting) rotations to sawlog production. It can be managed to supply nitrogen to conifer forests, improve forest soil, rehabilitate disturbed sites, and produce wood."[3]

[1] Biology of Alder, p. 197.
[2] Ibid. pp. 195, 247.
[3] "Let's Harness the Energy of Red Alder," p. 1.

Hardwoods

While conventional forestry concentrates on reforestation and production with evergreen conifers, many Northwest forest lands are covered with deciduous hardwoods. These hardy, soil-building trees have been considered worthless, but now their utility and value are being recognized by mill and land owners.

Markets for furniture-grade hardwood lumber from red alder, Western maple, Oregon ash, and Oregon white oak, have opened new possibilities for these naturally occurring species. The Maritime region of the Pacific Northwest is also well-suited to non-native hardwoods. The growth of such valuable hardwood species as walnut, chestnut, and oak may equal or exceed those grown in their native habitats when grown on a suitable site. The flourishing shade and fruit tree nursery industry indicates that growing hardwoods in the Northwest may be an idea worthy of the woodlot owner's consideration.

Northwest Hardwood Association
1220 SW Morrison
Portland, OR 97205

The Northwest Hardwood Association is the only organization dedicated exclusively to promoting and marketing alder and maple. Members include land owners, logging and trucking companies, wholesalers and manufacturers. They've published the Western Alder Species Book which describes the availability, quality control, grading requirements and uses of alder. They can also provide an "Alder and Maple Products Buyers' Guide," which lists over 50 Northwest member companies and the hardwood products they manufacture.

BIBLIOGRAPHY

Bergstrom, Dorothy. "Let's Harness the Energy of Red Alder." *in* Forestry Research West, USDA Forest Service, 1979.

DeBell, Dean S. "Future Potential for Use of Symbiotic Nitrogen Fixation in Forest Management." *in* Gordon, J.C., et al., eds. Symbiotic N-Fixation in the Management of Temperate Forests. Available from the Forest Research Laboratory, Oregon State University, Corvallis, OR 97331.

Delver, P. and Post, A. "Influence of Alder Hedges on the Nutrition of Apple Trees." Plant and Soil, 28:2 (1968), 325-336.

Miller, Richard E., and Murray, Marshall D. "The Effects of Red Alder on Growth of Douglas-Fir." *in* Utilization and Management of Alder, USDA Forest Service Gen. Tech. Report PNW-70, 1978.

____. "Fertilizer vs. Red Alder for Adding Nitrogen to Douglas-Fir Forests of the PNW." *in* Gordon, J.C., et al., eds. Symbiotic N-Fixation in the Management of Temperate Forests. Available from the Forest Research Laboratory, Oregon State University, Corvallis, OR 97331.

Trappe, J.M., et al., eds. Biology of Alder. Portland, OR: PNW Forest and Range Experiment Station, 1968.

Animals

A healthy, well-integrated agriculture includes animal life. Wild and domestic animals are closely interconnected with plants, other animals and the physical environment. One of the roles of the human animal is to protect, nurture and manage these interactions in a way that keeps the environment healthy, or moves it toward health, and produces usable surplus. Respect for the integrity of other animals is an important part of the system.

We begin this chapter by discussing the place of wildlife in sustainable ecosystems. We then focus on the importance of breed selection in domestic livestock management, highlighting unusual, endangered breeds or species which deserve more recognition. We conclude with some ideas about aquaculture and mariculture—growing and gathering fish and vegetables in ponds and in the sea.

It is not within the scope of this book to list all the functions livestock animals traditionally perform, or to discuss the many management decisions each animal's care requires. Please refer to the bibliographic and access references for information about choosing and caring for animals appropriate to your situation.

Wildlife

SEGO JACKSON

Animals on the farm can be roughly divided into two groups: wildlife and livestock. Wildlife includes animals living in their natural environment, from ladybugs and robins to garter snakes and bears.

Wildlife in and of itself has value. Wild animals are an essential part of the biosphere of which we are, or should be, the gentle caretakers. They have been a source of symbols, legends and spiritual inspiration since the dawn of human culture.

Of course, wild animals can also damage domestic crops and livestock. Deer and rabbits love young orchard trees, gophers go for gardens, birds crave cherries, coyotes kill sheep. Thus, encouraging wildlife must always be balanced with protecting crops from them, discouraging those wild animals who might do more harm than good, and trapping those who won't give up. You'll probably have more success, however, protecting domestic crops than you will trying to eliminate wild animals from the area.

Many species of wildlife are also considered game animals. Game consists of animals that live in the wild and are harvested by hunting or trapping. Larger game animals such as deer and elk provide meat and other useful products for some families. Game populations are usually low as a result of limited food and habitat, and competition for territory. Improving wildlife habitat also benefits most game animals, while adding beauty and diversity to the landscape.

In addition to their intrinsic value, wild animals, especially birds and bats, can play a great role in pest management. So if you enjoy listening to songbirds, marveling at hummingbirds, admiring a browsing deer or knowing that your insect population is being kept in check by the many birds flitting around you, you will probably want to find ways to make your surroundings more hospitable to wild animals and encourage their presence.

WILDLIFE MANAGEMENT

Each wildlife species has specific habitat needs, but all require food, water, and shelter. Food must be of the proper kind and must be available in sufficient quantities in each season and during each stage in the animal's life. Water may be needed in the form of open ponds, seepages, dew drops or succulent plants. Shelter may be needed for roosting or nesting sites or as protection from weather or predation. Other specific conditions may also need to be provided, such as roosting sites, sites for hibernation, or the gravel stream bottom with clear running water required by salmon for successfully laying and hatching eggs.

Invite Wildlife to Your Backyard

The Backyard Wildlife Program, a project of the National Wildlife Foundation, encourages and aids all interested people to transform their backyards into "mini-refuges" for wild animals. To this end, the Foundation has published an excellent booklet entitled "Invite Wildlife to Your Garden," which gives plans and advice on designing and planting a yard to attract the greatest variety of wildlife species. Especially inspiring are the color illustrations of a quarter-acre backyard over a period of 30 years, as it turns from a lawn fringed with young trees and shrubs into a small clearing within a dense wilderness, home to birds, bats, frogs, rabbits and others.

If you would like to have your backyard certified as a Backyard Wildlife Habitat, request information and application form from the **National Wildlife Foundation, 1412 16th St NW, Washington, DC 20036.** "Invite Wildlife to Your Backyard" was reprinted from the April-May 1973 issue of National Wildlife Magazine, and is available from the Foundation for $.25.

BIBLIOGRAPHY

Allen, Durward L. The Farmer and Wildlife. 1977. Wildlife Management Institute, Wire Building, 1000 Vermont Avenue NW, Washington, DC 20005. $1.00
Strategies for enhancing the environment for wildlife.

Chiavetta, Kenneth J., ed. Wildlife Review. An Indexing and Abstracting Publication of Wildlife Literature. No. 180. Washington, DC: US Dept of the Interior, Fish and Wildlife Service, 1981.
This guide to publications is a valuable tool for people who wish to improve wildlife habitat in their vicinity. The extensive listings are well organized in several dozen categories. Author, geographical and subject indexes are included. Many entries are from other countries, making it useful worldwide.

Edminister, Frank C. American Game Birds of Field and Forest: The Habits, Ecology and Management. New York: Charles Scribner's Sons, 1954.
Very detailed, excellent text.

Leedy, D., et al. Planning for Wildlife in Cities and Suburbs. Washington, DC: US Department of the Interior, 1978.

That's Incredible!

Dr. Noel Vietmeyer of the National Academy of Sciences provided some fascinating examples of wildlife "farming" in his paper, "Underexploited Plant and Animal Resources for Developing Country Agriculture" (in Background Papers for Innovative Biological Technologies for Lesser Developed Countries. U.S., Congress, House, Committee on Foreign Affairs, 97th Congress, 1st session, 1981, pp. 485-511). According to Vietmeyer, "some natural products now virtually unknown are likely to become mainstays of world agriculture." Examples include winged beans, *Leucaena* (a tree legume), Chilean toads, guinea pigs, and crocodiles.

Perhaps the most unusual example of wildlife farming cited by Dr. Vietmeyer is the butterfly farming operation in the north of Papua New Guinea.

Butterflies may seem exotic livestock to us, but even in the remotest jungle, a villager knows and understands their habits, location and lifestyle. ...To a Papua New Guinean, the strange thing is that people are willing to pay for a butterfly. And pay well they do. Ounce for ounce, exotic butterflies are more valuable than cattle. And worldwide demand for butterflies is rising.

The New Guineans plant flowering shrubs such as hibiscus, flame-of-the-forest and aristolochia to attract the adult butterflies. Within half-acre circlets of these flowers are planted leafy plants that the caterpillars feed on. The combination provides a complete habitat for butterflies from the beginning to the end of their life cycle, which keeps the farmer's flock at home without fences.

In addition to their ease of management, Dr. Vietmeyer points out the important ecological benefits of butterfly farming—

Perhaps the most striking feature of the program is that it is actually conserving, and even increasing, the numbers of butterflies. Basically, it is an exciting, pioneering conservation project because it develops a tremendous economic incentive to preserve populations and habitats....Because of this, conservation organizations are becoming excited by the program, seeing it as a model that could be duplicated to help save endangered exotic butterflies everywhere.

If you would like to explore the possibility of butterfly farming in the Northwest, see Robert Michael Pyle's Watching Washington Butterflies. The book, part of the Audubon Society's Trailside Series, is an interpretive guide to 134 native butterfly species, including most of those found in Washington, Oregon, Idaho, and British Columbia. Included is a section on the design and establishment of butterfly gardens.

Watching Washington Butterflies, by Robert Michael Pyle (1974), is available for $4.50 from the Seattle Audubon Society, 619 Joshua Green, Seattle, WA 98101.

Food & Shelter for Wildlife & You

One of the most exciting wildlife plant catalogs we've seen is from Kester's Wild Game Food Nurseries in Omro, Wisconsin. Their slant is towards hunters who want to enhance game populations, but the recommendations and suggestions they give promote sustainable agriculture wholeheartedly. As they note, "The planting of natural foods is conservation in the highest degree."

Kester's carries books as well as seeds and plants of such species as wild rice *(Zizania aquatica)*, wapato duck potato *(Sagittaria latifolia)*, and an intriguing perennial grain named Agrotriticum W-21. The grain was created by crossbreeding wheat *(Triticum)* with perennial tall wheatgrass *(Agropyron)*. (See page 50 for more details.)

The Kester's catalog contains a wealth of information on techniques and plants for increasing wildlife habitat. The following quote exemplifies the advice and attitude found in the Kester's catalog.

An ideal, useful and productive food and shelter planting should be started by plowing and working up a strip of land, preferably close to water, wooded areas or other wild areas. This strip can be as wide or as long as you care to make it. We suggest a minimum of 50 to 75 feet wide and 200 feet in length or as long as you can make it. In the Northern areas which are frequented by howling blizzards of snow or freezing rain, it is an absolute necessity to start the planting with two, three or more rows of conifers, such as Pines, Spruce, Cedar or Fir planted down the center of the strip of worked soil.

On each side of the rows of conifers plant a row or two of food producing shrubs such as Russian Olive, High Bush Cranberry, Autumn Olive, Multiflora Rosa, Arnot Bristly Locust, Pin Oak, Mulberry, Hazelnut, etc. In among these shrubs plant some vines of False Bittersweet, American Bittersweet and Wild Grapes. These vines will climb up among the trees and shrubs to make a tangled shelter as well as produce berries which stay on the vines long into the winter to provide food.

On each side of the shrubs plant a strip of perennial food and cover grasses and legumes, such as Cicer Milkvetch, Lespedezas, Sweet Clover, Switchgrass, Birdsfoot Trefoil, Partridge Peas and Jerusalem Artichoke, or a mixture of these varieties. These varieties provide both food and permanent nesting areas as well as succulent foliage for Mammals, Upland Game Birds and other Wildlife.

The remaining strips of soil on either side of the permanent planting should be planted by sowing or drilling some of each of the following annual varieties, especially the varieties such as Duckwheat, Dove Proso Millet, Hegari, Milo, Mung Beans, Sesbania, Corn, Buckwheat, Hull-less Barley, and Canary seed which have the ability to hold their seed long into the winter to provide available food. Other good food producing annuals are Pearl Millet, Browntop Millet, German or Foxtail Millet, Common Vetch, Proso and Japanese Millet.

Kester's Wild Game Food Nurseries, Inc.
PO Box V, Omro, WI 54963
Catalog free.

Martin, A.C., et al. American Wildlife Plants: A Guide to Wildlife Food Habits. New York: Dover, 1951.
A guide to the food habits of birds and mammals in the United States. An excellent reference work.

McAtee, W.L. Wildfowl Food Plants: Their Value, Propagation and Management. Ames, IA: Collegiate Press, 1939.
The main plant species valuable for wildfowl food are discussed along with much valuable information on how to establish them.

Picart, Francois. Escargots from Your Garden to Your Soil. 1550 Ridley Avenue, Santa Rosa, CA 95401.
A Frenchman's guide to raising this French delicacy. Says Mr. Picart, "Snails don't have a good image here. They're not a pest in France because everyone eats them as soon as they grow." Picart purchased the first stock for his California snail ranch from an artichoke farmer overrun with snails for the paltry price of 10 cents per pound. Now he papers Santa Rosa with posters: "Wanted Alive, Common Garden Snail. 25 cents per pound reward." Snails, anyone?

Terres, John K. Songbirds in Your Garden. New York: Hawthorn Books, 1977.
Attract songbirds by enhancing habitat; gain beauty and biological control.

Thomas, Jack Ward, ed. Wildlife Habitat in Managed Forests-The Blue Mountains of Oregon and Washington. USDA Forest Service Agricultural Handbook No. 553. Washington DC: USGPO, 1979.
An analysis of habitat types (e.g., riparian zones, edges, snags, cliffs) and how they might best be managed to preserve and enhance wildlife. Includes practical forest management guidelines for wildlife habitat preservation.

USDA Forest Service. Wildlife Habitats in Managed Rangelands-The Great Basin of Southeastern Oregon. Portland, OR: PNW Forest & Range Experiment Station.
Series of reports for range managers on wildlife habitat improvement. Although specific to southwestern Oregon, the principles are generally applicable to all managed rangeland.
Reports available thus far include Plant Communities and Their Importance to Wildlife (GTR PNW-120); Riparian Zones (GTR PNW-80); Edges (GTR PNW-85); Geomorphic and Edaphic Habitats (GTR PNW-99); Manmade Habitats (GTR PNW-86); and Native Trout (GTR PNW-84).

Wildlife cannot be managed in the same way that domesticated livestock is managed; it can only be encouraged to live on or near the site. Existing wildlife habitat can be protected by leaving snags or old hollow trees standing. Small areas throughout a farm can be fenced from livestock and allowed to become overgrown.

Nesting boxes can be provided for squirrels, song birds and owls. An open nesting box in the loft of a barn can serve as a home for owls. Boxes can also be specially designed for bat roosts (see below). These and other wildlife houses may be attached to outbuildings, trees or specially placed posts. Hollow logs placed in trees or partially buried in the ground also provide good nesting and den sites.

Water should be open and available, especially near special plantings. Many ground birds need water every four to six hours. Creating a pond will greatly increase the number of animals who visit the farm or garden.

Bats Need Friends

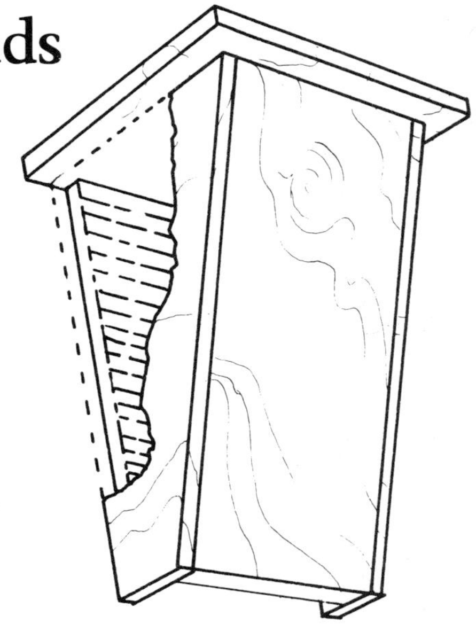

Bat box dimensions are as follows:

Front— 6 X 13-3/16"
Sides— 2-1/4 X 12-11/16 X 6 X 13-1/4"
Back— 6 X 11"
Top— 7-3/4 X 8-1/2"
Entry reducer— 1-1/2 X 6"

Bats are unique flying mammals essential to a healthy ecosystem. Some species eat up to half their weight nightly and colonies of bats consume tons of mosquitoes, gnats, mayflies, moths and beetles during the summer foraging season.

Many species of bats are declining. Three bat species are on the United States Endangered Species List, two others have been proposed, and still others are probably in need of this protection. Putting a bat on the list does not insure its survival, but does provide protection from some habitat alteration. On the home scale, building bat boxes to attract bat populations can put these helpful, misunderstood creatures to work for you.

Bat boxes have been used successfully in Europe, but are new to the United States. They are simple to make and use inexpensive materials. One design is shown here. The exact shape and size of boxes are probably not important except for the width of the entry space. This should not exceed one inch, and 3/4 inch is ideal.

Bat boxes should be fastened securely to a tree trunk or the side of a building 12 to 15 feet off the ground. Boxes should be placed where they receive morning sun and afternoon shade, and most bats seem to prefer sites relatively protected from the wind. It is important to note that bats can live only where adequate food supplies exist. Most colonial bats are found near swamps, rivers, ponds and other areas where insect populations are high. Boxes located more than 1/2 mile from such habitats probably will not be successful. Bats are extremely shy and intolerant of disturbance. While incidence of rabies in bats is low (contrary to popular belief), bats, as other wildlife, may carry rabies. Healthy bats are difficult to catch, so those which can be approached should be handled with great caution.

For more information about bats, including other bat box plans, contact The Center for Action on Endangered Species, 175 West Main Street, Ayer, MA 01432, or see "Build Bat Boxes" in Mother Earth News, March/April, 1981.

Bees

MICHAEL PILARSKI

Beekeeping fits well into a diversified agroecosystem. Insects, primarily bees, are the only means of pollination for many crops. Colonies of bees nearby will increase yields where such crops are grown. Along with their important role in pollination, bees provide honey, pollen, royal jelly, beeswax, and a valuable substance called propolis, which is used in producing an antibiotic.

Beekeeping is already an important industry in the Northwest, with over 300,000 hives in Oregon, Washington, Idaho and British Columbia. Although the bloom in some areas is fairly well utilized, there is still room for expansion. Landowners can consciously increase the amount of bee forage on their properties by "beescaping" or planting a carefully selected combination of flowering plants.

The best beescapes are composed of a mixture of farmland, meadows and open woodland which contain an abundance of legumes, herbs, wildflowers, fruits and berries. People who have studied the economics of beescaping say that it is not economically sound to plant farmland to plants whose sole purpose is bee forage. Many of the best bee plants, however, also have value for other purposes. For example, fruit and berry producing plants are excellent sources of nectar and pollen. Black locust trees are heavy nectar producers and are potentially valuable sources of hardwood. Many culinary and medicinal herbs, such as anise hyssop, borage, thyme, lavender, rosemary, sage (*Salvia* spp.), lemon balm, catnip, and several other mints are also favored by bees. Many of these herbs have blue flowers, the most effective color for attracting bees.

Bees will range for a two mile radius from the hive. However, since they prefer shorter hauls, production can be much higher if sufficient forage is closer, particularly when the weather is unsettled.

Beescaping requires plant diversity, since bees need to have a succession of bloom throughout the season. Bees begin working early in the spring and continue until the last of the hardy flowers have withered in early winter. Pollen is important throughout the year because it is the food used for rearing brood. Nectar is the source of honey. Some bee plants produce pollen only, while others produce both pollen and nectar.

There are three main periods of bloom to consider in designing beescapes:

1. EARLY SPRING BLOOM. Availability of early spring bloom can affect honey yields for the whole year. The hive is weak after the winter and needs to build up strength. The success of early spring brood rearing also determines how well the hive will be able to utilize nectar during the main honey flow. Trees, especially willows and maples,

BEEBLIOGRAPHY

Aebi, Ormond and Harry. The Art and Adventure of Beekeeping. Santa Cruz: Unity Press, 1973.

Folk wisdom from long study and observation is shared in this personal and information-filled narrative. The Aebis are listed in the Guiness Book of World Records for harvesting 420 lbs. of honey from one hive.

——. Mastering the Art of Beekeeping. Santa Cruz: Unity Press, 1979.

The Aebi brothers (their name means "bees" in Latin) share more of their decades of experience.

Cosper, Nancy. You Can Can With Honey. Portland, OR: Self-Published, 1981.

A useful book by a regular winner of local made-with-honey bake-offs. Available for $2 from the author at 2270 NW Irving, Portland, OR 97210.

Dadant, C., and Grout, A.L., et al. First Lessons in Bee-Keeping. NY: Charles Scribner's Sons, 1982.

An introductory text by two master beekeepers.

——. The Hive and the Honey Bee: The Classic and Definitive Book on Beekeeping. Hamilton, IL: Dadant & Sons, 1975.

This book lives up to its subtitle. It is the direct successor to one written by Lorenzo Lorraine Langstroth, published in 1853 under the title Langstroth on the Hive and the Honey-bee, which made available to the world his invention of the top-opening, movable-frame hive which made modern beekeeping possible.

are often the principal source of spring food. In some areas bees rely heavily on the sweet exudations of cottonwood buds.

2. SUMMER BLOOM. The main nectar flow occurs in the long, hot days of late spring and summer when hives are strong. Given fine weather and a plentiful supply of nectar plants, a hive of bees can produce up to 40 pounds of honey per week. Since one hive will produce 70 to 120 pounds of honey in one year, the bulk of the year's crop can be made in just 2 or 3 weeks of ideal conditions. Some of the best summer-blooming honey plants are black locust, sagebrush, buckwheat, anise hyssop *(Agastache foeniculum)*, white clover, alfalfa, sweet clover and many other legumes and mints. Anise hyssop is a special favorite of bees and has one of the longest bloom periods in the mint family.

3. AUTUMN BLOOM. Sufficient autumn bloom is necessary for the bees to go into the winter in strong condition, reducing the chance of winter kill. Some late-blooming bee plants include the asters, garden hyssop *(Hyssop officianalis)*, heartsease, borage, buckwheat, goldenrod, and wild sunflower.

Beekeeping can be practiced almost everywhere in the Northwest, but the Interior region is generally more favorable than the Maritime because there are fewer rainy days and because the hotter temperatures induce higher nectar production. There is often a shortage of bloom in the Maritime lowlands during the main nectar production time, so some beekeepers move their bees to the Interior or to the mountains in summer to take advantage of the fireweed, huckleberry, heather and other plants prevalent in clearcut, burnt over and alpine areas.

Howes, F.N. Plants and Beekeeping. London: Faber and Faber, 1979.
An account of wild and cultivated plants valuable for honey production in the British Isles. When Plants and Beekeeping was first published over thirty years ago it was widely acclaimed and no other book has since taken its place. Although written with British beekeepers in mind, it is of value to beekeepers throughout temperate regions since so many of the plants discussed are widely planted. Pellett's book (below) contains more species, but Howes' book goes into more intricate detail on the merits and demerits of the major honey plants. The book contains chapters on establishing bee pasturage, bee gardens and apiary hedges, and windbreaks. This edition contains a supplementary bibliography which will lead the reader to the most prominent literature which has been published on the subject since the book was first written.

Pellett, Frank C. American Honey Plants. New York: Charles Scribner's Sons, 1982.
A helpful book if you are planning to plant bee pasture or locate hives. Pellett's seed company specializes in bee plants.

Root, A.I., et al. ABC and XYZ of Bee Culture. Medina, OH: A.I. Root Co., 1970.
Standard reference work for all beekeepers.

BULLETINS

Beekeeping. PNW Bulletin #79. Cooperative Extension Service: Pullman, WA, 1974
An introductory manual with sufficiently detailed information to answer most of the questions of the amateur beekeeper. Includes a bibliography with 36 entries.

Beekeeping for Beginners. Home and Garden Bulletin #158. USDA, Washington, DC 1974.
Includes plans for home construction of a beehive.

How to Reduce Poisoning of Bees From Pesticides. Bulletin #3473. Cooperative Extension Service: Pullman, WA, 1976.
Even if you don't use pesticides, your neighbor might, much to the detriment of your bees. This booklet gives several strategies for protecting your bees.

Pollination of Agricultural Crops by Bees. Published for the USDA Agricultural Research Service and the National Science Foundation. Available from the US Dept. of Commerce, National Technical Information Service, Springfield, VA 22161.

Selecting and Operating Beekeeping Equipment. Farmers' Bulletin #2204. USDA: Washington, DC, 1969.
Describes equipment for medium to large scale honey production.

PERIODICALS

American Bee Journal
Hamilton, IL 62341

Canadian Beekeeping
Orono, Ontaria L0B 1M0

Gleanings in Bee Culture
PO Box 706
Medina, OH 44258

BEE SUPPLIES

A. I. Root Company
PO Box 706
Medina, OH 44258

Dadant and Sons
51 S. Second Street
Hamilton, IL 62341

Diamond International
Apiary Dept.
Box 1070
Chico, CA 95927

Hodgsons Bee Supplies
7925 13th Avenue
New Westminister, BC V3L 4Y6

Hubbard Apiaries
PO Box 160
Onsted, MI 49265

R.G. Bee Equipment Sales
6200 No. 4 Rd.
Richmond, BC V6Y 2S9

Animals at Work

SEGO JACKSON

Livestock includes all domestic animals that people introduce to the land. They must be carefully managed by feeding, breeding, sheltering and harvesting.

Domestic animals are more than merely a source of eggs, milk and meat. Fed from the products of the garden or farm, domestic animals are vital links in the cycle of soil fertility. Animals have the capacity to convert plant materials inedible by humans into food and fiber, while their manure is one of the most valuable renewable resources for agriculture. When properly used as fertilizer, animal manures increase crop yields and contribute to sustained fertility in the soil. Composted manures are high in stable organic matter, essential for humus formation.

One of the problems that commercial farmers have in converting to soil-conserving methods is the lack of economic justification for including soil building crops in long-range rotations. Farmers know that grasses and legumes protect against erosion and build soil humus, but they cannot justify the income lost from taking lands out of production. By reintroducing and properly managing and marketing livestock and their products, effective use can be made of grasses and legumes in crop rotations, thus insuring sustained fertility and productivity.

DRAFT ANIMALS

Animals can perform chores around the farm which would otherwise have to be done by hand or with machinery. Large draft animals such as mules, horses and oxen (castrated cattle) can work land which is impossible or unsafe for tractors, such as steep land or soils with poor drainage. Llamas and dogs are good pack animals and can be used for other light draft work.

For beginners interested in using large draft animals, oxen work at a slower pace than horses, and are easier to work with. Because oxen can be developed from any cattle breed, they are also less expensive to start with. Unlike costly draft horses, oxen can be butchered for meat if the experience of working with them is unsatisfactory without great financial loss.

Both oxen and draft horses can be maintained on quality summer pasture during extended idle periods, but they must have feed supplements when they are working. Draft animals can be "fueled" by farm-produced feeds. Working with draft animals and raising their feed at home saves the cash outlay needed for fossil fuel to run machinery or for animal feed raised off the farm. The work received from these animals often more than repays the farmer for raising the feed.

BIBLIOGRAPHY

Belanger, Jerome, D. The Homesteader's Handbook to Raising Small Livestock. Emmaus, PA: Rodale Press, 1974.
Often-used guide for raising rabbits, chickens, turkeys, goats, sheep and hogs.

Cullison, Arthur. Feeds and Feeding. Englewood Cliffs, NJ: Reston, 1979.
New edition of the standard text on animal nutrition.

Ensminger, M.E. The Stockman's Handbook. Danville, IL: Interstate, 1970.
"Currently, we are producing manure (exclusive of bedding) at the rate of 1.8 billion tons annually. That is sufficient manure to add nearly one ton for each year to every acre of the total land area (1.9 billion acres) of the United States."
This is the most comprehensive text available for planning and managing commercial livestock. Includes detailed sections on breeding, pastures, animal care, financial management, marketing and access to additional sources of information.

Fraser, Alister, and Thear, Katie, eds. Small Farmer's Guide to Raising Livestock and Poultry. New York: Arco, 1980.
The best, most detailed overview of small-scale livestock husbandry.

Gates, Norman. Constructing an Effective Anticoyote Electric Fence. USDA Leaflet No. 565. Washington, DC: USGPO, 1978.
Details on New Zealand fencing system effective for coyote control.

Holderread, Dave. The Book of Geese. Hen House Publications, PO Box 492, Corvallis, OR 97330. 1981.
The only thorough book on geese around.

Levy, Juliette de Bairacli. Herbal Handbook for Farm and Stable. Emmaus, PA: Rodale Press, 1976.
Practical experience with a unique point of view in this popular and delightful book.

MacKenzie, David. Goat Husbandry. Salem, NH: Faber and Faber, 1981.
A classic goat reference. Can be eccentric but is quite useful.

Martin, George A. Fences, Gates and Bridges: A Practical Manual. Brattleboro, VT: Stephen Greene Press, 1974.
Reprint of 1887 edition. Good chapters on gates and fastenings and on fencing over gullies and streams.

Miller, Lynn. The Work Horse Handbook. Reedsport, OR: Mill Press, 1981. $12.95.

This is a great book for anyone interested in working horses. While mostly an instruction manual for first-timers, it also contains much useful information for experienced teamsters.

The subject matter includes the horse itself, feeding and housing, harnesses and equipment, and a chapter on the relative value of horses vs. tractors on the farm. Perhaps the most valuable parts of the book, for beginners, are the many detailed pictures of the harness and hitching gear. The only subjects not thoroughly covered in The Work Horse Handbook are pastures and fencing, both essential for anyone keeping horses.

Spaulding, C.E., DVM. A Veterinary Guide for Animal Owners. Emmaus, PA: Rodale Press, 1976.

General overview of veterinary care for livestock and pets, emphasis on proper management and prevention. The tone is straightforward with good indications on when owners can deal with problems themselves and when it is necessary to call a vet. Like any encyclopedia, this volume offers a good introduction to a wide variety of livestock, but needs to be balanced with more detailed texts devoted to management of specific species.

Telleen, Maurice. The Draft Horse Primer. Emmaus, PA: Rodale Press, 1977.

The basics of farming with horses and mules, plus information on finding and repairing equipment.

Templeton, George S. Domestic Rabbit Production. Danville, IL: The Interstate Printers and Publishers, 1968.

PERIODICALS

Backyard Poultry
Countryside Publications
312 Portland Road
Waterloo, WI 53594

Backyard Poultry is a folksy periodical designed for the small flock owner and rare breed fancier. It is one of a series of similarly designed magazines published by the editors of Countryside, including Rabbits, Dairy Goat Guide, sheep!, and National Stock Dog. All are inexpensive, informal hubs of information exchange concerning the many farm animals of their concern.

Black Sheep Newsletter
28068 Ham Road
Eugene, OR 97405

The Black Sheep Newsletter is dedicated to the black sheep and other fibered animals. A refreshing alternative to the slick "industry" style magazine, this quarterly publication is simultaneously a pleasure to page through as well as informative to read. All of the articles are written by people who have personal experience with sheep in a small-scale management situation, so the Black Sheep Newsletter is people-to-people, about sheep.

Every spring, the Newsletter publishes the "Annual Black Sheep Directory," in order to facilitate information exchange among sheep raisers.

Dairy Goat Journal, PO Box 1908, Scottsdale, AZ 85252

Detailed information on dairy and range goats.

The Draft Horse Journal
Rte 3
Waverly, IA 60577

Reliable source of information on farming with horses.

Journal of Applied Rabbit Research
OSU Rabbit Research Center
Corvallis, OR 97331

Reports on research underway to develop rabbits as a major livestock enterprise for small farms.

Llama Newsletter, Box 135, Athena, OR 97813

Llamas are becoming popular as pack animals because they train easily and are adapted to rocky mountain trails. North America's largest llama herds are in Oregon. Andy Tillman, editor of this newsletter, is author of Speechless Brothers, a book on llama training.

Small Farmer's Journal (see p. 52)

Features horse farming and livestock management.

Garden Way Books

There are ten titles in the Garden Way series of books on livestock raising. All are excellent introductions to small-scale livestock husbandry. Especially refreshing is the absence of cute, empty reassurances about the supposed simplicity of growing farm animals. All farm animals demand sensitive, informed, close attention.

Garden Way Publishing
Charlotte, VT 05445

Keeping Livestock Healthy by Dr. Bruce Haynes. An essential veterinary guide emphasizing preventative medicine. Covers feeding, shelter, reproduction, and diseases of horses, cows, pigs, sheep, and goats. Field guide edition. $12.95.

Small Scale Pig Raising by Dirk Van Loon. Produce the most meat with the smallest investment. Complete details on everything from buying a pig to smoking your own bacon. $7.95.

Raising the Home Duck Flock by Dave Holderread. Includes choosing breeds, building low-cost housing, growing feed, incubation of eggs, butchering, recipes, and more. $6.95.

Raising Milk Goats the Modern Way by Jerry Belanger. Complete information on raising, feeding, housing, grooming, kidding, breeding, milking. $5.95.

The Family Cow by Dirk Van Loon. Covers buying a cow, housing, feeding, milking, health care, breeding, calving, land use. $7.95.

Raising a Calf for Beef by Phyllis Hobson. Enjoy superior quality beef at low cost. Includes complete care, feeding, and housing needs. Plus, step-by-step butchering directions. $5.95.

Raising Sheep the Modern Way by Paula Simmons. Over 100 "how-to" illustrations covering feeding, breeding, care, shearing. Plus, how to make a profit. $6.95.

Raising Poultry the Modern Way by Leonard S. Mercia. Complete "how-to" for fresh, additive-free meat and eggs. $6.95.

Raising Rabbits the Modern Way by Bob Bennett. For home and semi-commercial producers of rabbits. Easy-to-read handbook covers breed selection, housing feeding, disease prevention, marketing and much more. $5.95.

Raising Your Own Turkeys by Leonard Mercia. Complete handbook covers selection, housing, feeding, butchering and dressing turkeys. Extensive coverage of disease control. Plus equipment, supplies and source lists. Plans for a low-cost turkey shelter. $5.95.

The Loss of Rare Breeds

The loss of old seed varieties due to trends in commercial marketing and plant patenting has received a lot of attention. People are less aware that many livestock breeds are also in danger of extinction. These breeds represent thousands of years of selective breeding under specific environmental conditions. Once extinct, their genetics cannot be recreated.

Many of these once-common breeds have been replaced by new ones developed for large-scale farm operations. Yet many of the rare breeds are more appropriate for a small farm system. For example, the Irish Dexter, the smallest breed of cattle in the U.S., was bred in resource-poor Ireland to be a high meat and milk producer. Irish Dexters require less acreage and feed and are able to consume rougher forage. They are not favored by commercial beef breeders because of their size and mediocre performance on good grasslands and in the feedlot.

Buff Orpington chickens, one of the most popular egg and meat producers in the U.S. earlier in this century, are increasingly difficult to obtain. Their tendency to become broody made them unpopular when the poultry industry began expanding.

These breeds, and hundreds of others, will only be saved through the establishment of rare breed centers (domestic livestock zoos, in essence) and through the efforts of private individuals. Those who choose to maintain an endangered breed will get much pleasure from these unusual animals and from the research necessary to locate stock. Rare breeds can be profitable as well. After the initial cash outlay for breeding stock, the offspring can be sold to others interested in protecting the breed, or to farmers interested in the animals for their special qualities.

FOR MORE INFORMATION

Alderson, Lawrence. The Chance to Survive, Rare Breeds in a Changing World. Brattleboro, VT: The Stephen Greene Press, 1978.

A beautifully illustrated book about the extinction of domesticated breeds of livestock in the British Isles. Britain has lost close to 25 native breeds this century. One hundred ten breeds remain today, and over half of these are in danger of extinction.

Whitlock, Ralph. Rare Breeds: The Vulnerable Survivors. New York: Van Nostrand Reinhold, 1981.

Whitlock, a founder of England's Rare Breeds Survivial Trust, points out that in the race for more efficient production, better conversion rates, early maturity, and other apparent improvements in livestock, many advantages of the original breeds have been lost. According to Whitlock, we must strive to preserve a genetic reservoir of these primitive breeds from which needed traits may be drawn in the future.

American Minor Breeds Conservancy, c/o Ridgway Shinn, Box 225, Hardwick, MA 01037. $1.00 for inquiries.

Rare Breeds Survival Trust, 127 Abbots Road, Abbots, Langley, Herts, ENGLAND WD5 0BJ

Society for the Preservation of Poultry Antiquities, James K. Rice, Rt. 3, Greenwood, WI 54437

Fjord Horses *Work*

The Norwegian Fjord Horse is to work horses what the Volvo is to jumbo American sedans—compact, sturdy, and fuel efficient. The breed appeared 4,000 years ago when the direct descendants of European prehistoric breeds migrated from Asia to Norway. By 2000 BC the Fjord Horse had been domesticated. In 1850, the Norwegian government embarked on a breeding program, using only the best stallions as traveling studs for Norway's many mares. Through strict breeding standards, the Norwegians have produced quick, sure-footed animals as comfortable on a steep, icy slope as on pasture. In size fjord horses range from 13.3 to 14.3 hands, often smaller than the 14.2 division between horses and ponies. Adult weights range from 1,000 to 1,200 pounds. They are quiet of temperament and stout of spirit.

The Norwegian Fjord Horse is beginning to attract the interest of small landholders across the United States, where it recently has become a registered breed. In 1977, associations of people interested in the breed were established both in Canada and in the United States. Both associations maintain stud and registry books and issue newsletters for membership communication. They emphasize the breeding of a well-rounded horse for both working and riding.

Writing in Harrowsmith (No. 41) Fjord Horse owner Pamela de Alth talks of the work her gelding Roo does to earn his oats: "Every year Roo brings in five cords of firewood, as well as any lumber logs we might need for construction projects, does all the garden work, harrows almost half an acre in small fruit, mows and rakes an acre of hay and does various small hauling jobs."

Roo is kept on pasture from June through October. In addition, he requires about 120 bales of hay, a little grain during periods of heavy wintertime work (about one quart per day), and a few roots each day from November through April. If Roo lived south of his northerly Nova Scotia home, his feed requirements would be even less. Writes Ms. de Alth, "As far as we are concerned, there is not a better animal one could have working for a family farm. It is a wonder that Fjordings are not better known on this continent, but the formation of the two associations should help to change that. We heartily agree with one association pamphlet that asserts, 'The Fjord horse, stout and strong and Norwegian in ancestry, is our Lord's masterpiece among domestic animals.' "

The Canadian Fjord Horse Association
PO Box 495
Sundre, Alberta T0M 1X0

Norwegian Fjord Horse Association
RR #1, Box 370
Round Lake, IL 60073

Dogs & Sheep Can Mix

Sheep losses from coyotes, wolves and wild dogs vary throughout the U.S., but in a recent year over one million sheep were lost, despite $30 million worth of predator controls. Looking for an effective solution to the problem, scientists at the New England Farming Center at Hampshire College, MA, the U.S. Sheep Experiment Station, Dubois, ID, and Colorado State University, Fort Collins, are working with guarding dogs to protect sheep from coyotes and other canines.

Guarding dogs have well defined traits: 1. They do not chase or kill sheep. 2. They follow sheep around both day and night, unsupervised. 3. They protect the flock from predators. Guarding dogs are not herd dogs. They are big, 100 pound, placid, aloof dogs that work alone and respond slowly to commands. In a fight they are quick and effective.

The most distinct breeds of guarding dogs are Anatolian Shepherd Dogs, Great Pyrenees, Maremmas, Komondors, Kuvasz and Shar Planinetz.

What do the dogs do to protect the sheep? First, the guarding dogs are large and they bark. This is enough to keep many predators away. Secondly, the dogs relate to other dogs. A potential sheep eater stays around to sniff and the larger guarding dog establishes dominance over it. Third, guarding dogs respond negatively to any new stimuli, anomalies in their routine. They will bark at, threaten, and attack predators of their sheep. The dogs do not chase after the predator if it flees; they remain with the herd.

Problems can occur, usually when the dogs are young and rather too playful for sheep. Also some dogs do not have the instinct it takes to be a guarding dog. Generally, though, they have worked well.

Researchers in the U.S. have been testing the effectiveness of a number of breeds of guarding dogs, and are trying to determine the best method of training. While there are traditional ways of raising these dogs (having puppies suckle ewes), tests by the New England Farm Center have shown that the behavior of the guarding dog is primarily unlearned.

In field tests for the New England Farm Center, 70 dogs were leased to farmers throughout the U.S., and so far, most of the dogs have been quite effective. The rest either need another year maturing, or are not cut out to be guarding dogs.

Excerpted from The IPM Practitioner, Volume 3, Number 9, September 1981.

The Livestock Guarding Dog Project is a non-profit, cooperative venture set up between livestock owners and Hampshire College's New England Farm Center. Their primary goal is to develop good guarding dogs which farmers can rely on to help protect their livestock from predators. The project has imported a number of dog breeds from Europe, carefully selecting individuals with a background of guarding livestock. As of fall, 1981, the Project had placed over 300 dogs on farms and ranches across the United States, for the nominal fee of $1 for the first year and $50 per year after the dog becomes effective. In addition to paying the fee, farmers are expected to provide for the physical needs of the dogs and to complete an annual questionnaire on the dogs' behavior.

This project is part of the work of the New England Farm Center, which publishes a newsletter (The Farm Center Newsletter) and other publications of general interest to the farmer and of special interest to the guard dog owner.

The New England Farm Center
731 West Street
Amherst, MA 01002

FOR MORE INFORMATION

The Guardian
540 Palm Avenue
Penngrove, CA 94951

The Guardian describes itself as a publication designed to meet those needs particular to the owners of livestock guardian dogs. "Informative articles, training tips and firsthand experiences provide an abundance of material useful to the sheep producer and/or breeder of livestock." Included in The Guardian is a directory of breeders of guard dogs, listings of breed clubs, and pertinent advertising. The staff maintains a referral service for anyone seeking local information about guard dogs.

Coppinger, Lorna, and Coppinger, Raymond. Livestock Guarding Dogs for U.S. Agriculture. 1978. Livestock Dog Project, Chestnut Hill Road-RFD, Montague, MA 01351.

Written by the directors of the Livestock Dog Project, this book is the most comprehensive source of information about livestock guard dogs.

ORGANIZATIONS

Anatolian Shepherd Club
R.C. Ballard
3620 Alpine Blvd.
Alpine, CA 92001

Komondor Club of America
B. Schendelman
19900 Hiawatha Street
Chatsworth, CA 91311

Great Pyrenees Club of America
Whitney Coombs
3119 Valley Road
North Millington, NJ 07946

Kuvasz Club of America
Aino Andres
18 S. Terrace Avenue
Mt. Vernon, NY 10550

North American Jugoslav Shepherd Dog Club
Marjorie Nadeau
247 E. Neece
Long Beach, CA 90805

Water Culture

MIKE MAKI

The abundance of life surrounding a body of water, even a small pond or marsh, helps one understand water's productive potential and longstanding importance for human civilizations. With the depletion of common fishing grounds, many people are now turning toward raising fish and wetland plants in private ponds and other improved aquatic systems. Good opportunities exist for increased production through introduction of especially productive varieties and polyculture of aquatic organisms.

The best overall strategy for establishing a self-sustaining aquaculture is to maintain and enhance native fisheries and, where necessary, to rehabilitate degraded watersheds. Many streams and rivers which were once home to salmon, trout and steelhead are now blocked by dams, stripped of streambank vegetation, and choked with sediment by poor logging and farming practices. As a result, salmon, trout and steelhead fisheries have declined sharply in recent years. In response to this situation, stream rehabilitation projects are now underway throughout the region.

Salmonids

Salmonids (săm'-ə-nids) are salmon, trout and steelhead of the *Salmonidae* family. Anadramous salmonids are fish which live in both freshwater and marine environments. The movement to restore their habitat and bring back the once-awesome salmonid runs is one of the most exciting and important community efforts underway in the Northwest. The following books include information on the state of our fisheries, and their vital role in the future of our region.

Brown, Bruce. Mountain in the Clouds: A Search for the Wild Salmon. New York: Simon and Schuster, September 1982.
History, ecology and current plight of Olympic Peninsula salmon. Lots of detailed salmon observation.

Netboy, Anthony. The Columbia River Salmon and Steelhead Trout: Their Fight for Survival. Seattle: University of Washington Press, 1980.
History of fishing on the Columbia and the impact of dam construction. The book provides details on fish biology and on efforts being made to restore tributary spawning streams.

Stream Rehabilitation

Studies in eastern Oregon have demonstrated that the restoration of streambank vegetation provides rapid benefits. In several pilot projects year-round flows returned to streams that once ran dry each summer. Wildlife soon returned, and fish were once again able to spawn in the clear, cool waters protected by overhanging trees and shrubs.

As a result of these experimental projects, in 1981 the Oregon Legislature enacted SB 397, a law which provides tax incentives to enable landowners to invest in stream and streambank restoration. The new law is administered by the Oregon Department of Fish and Wildlife (see Oregon Salmon and Trout Enhancement Program, below).

The Oregon programs are just one example of stream rehabilitation projects in the region. Several others are listed below.

Adopt-a-Stream, Snohomish County Dept. of Public Works, County Administration Bldg., Everett, WA 98201.
Unique program which supports grass-roots efforts to save spawning streams. Community groups assist in surveying stream conditions and the county provides small grants, materials, and technical support for rehabilitation projects.

British Columbia Salmonid Enhancement Program, Fisheries and Oceans, 1090 W Pender Street, Vancouver, BC V6E 2P1.
Actively working with landowners and local communities in the restoration of streams for improved spawning habitat. This is the most comprehensive program in the region.

Robert C. Watts

Northwest Watershed Research Center, PO Box 2700, Boise, ID 83701.
The US Forest Service has undertaken a 20-year project to restore the South Fork of the Salmon River. As part of the project 400 miles of old logging roads have been closed and river banks revegetated.

Oregon Salmon and Trout Enhancement Program, Oregon Dept. of Fish and Wildlife, 506 SW Mill Street, Portland, OR 97201.
Assistance provided for the placement of hatchboxes and for stream improvement projects. Also source of information on Oregon's new Riparian and Instream Habitat Tax Incentive Programs.

Helm, Michael. "Salmon Enhancement." In Raise the Stakes, The Planet Drum Review, Spring 1982, pp 8-9.
Description of efforts to enhance salmon runs in the Mattole River of northern California, with an emphasis on the project's effect in uniting the local community.

Reiser, D.W., and Bjornn, T.C. Habitat Requirements of Anadromous Salmonids. PNW-96. Portland, OR: PNW Forest and Range Experiment Station, 1979.
Detailed analysis of water and riparian habitat conditions required for natural fish reproduction.

Juelson, Thomas C. Suggestions for Streambank Revegetation in Western Washington. Report No. 13. 1980. Washington State Game Dept., Olympia, WA 98504.
Discussion of the value of streambank revegetation, with a listing of 19 suitable plant species, propagation techniques, and nursery stock sources.

PERIODICAL

Pacific Watershed Enhancement, PO Box 96, Honeydew, CA 95545.
Excellent review of research and projects aimed at the restoration of native fisheries and local communities.

"Salmonid enhancement projects are restoring more than our fish. They are restoring our sense of community and belonging together."

Amity

The Amity Foundation is a publicly supported non-profit corporation in Eugene, Oregon, engaged in research and development in agriculture and aquaculture. Along with their book Fish Farming in Your Solar Greenhouse ($5.00), they publish a series of useful and informative Technical Bulletins as part of their Small Farm Water Management Project. Titles and 1982 prices are: Some Useful References in Aquaculture, $.50; Making Your Pond Water Clear, $.75; Keeping Water in Your Pond, $.75; Growing Fish in Cages, $1.00; Controlling Aquatic Weeds Without Chemicals, $1.00; Wild Rice Fact Sheet, $.75. The entire series is available for $3.50. Include $1.00 for postage and handling with any order from **Amity Foundation, PO Box 11048, Eugene, OR 97440.**

BIBLIOGRAPHY

Bardach, J.E., et al. Aquaculture: The Farming and Husbandry of Freshwater and Marine Organisms. New York: John Wiley & Sons, 1972.
In-depth treatment of aquaculture.

Harbell, Steve C, and Carkner, Richard. Trout Farming in Washington. Extension Bulletin EB 756. Cooperative Extension Service: Pullman, WA, 1981.
This pamphlet describes the proper trout pond and gives the basics for growing this popular food and game fish.

Heritage, Bill. Ponds and Water Gardens. Poole, Dorset, England: Blandford Press, 1981.
Start small and grow from there. This is a good introduction to small pond design and maintenance.

POND CULTURE

Ponds and other water storage structures can be of great value in restoring natural conditions by retaining water in the surrounding soil and by providing a supply of irrigation water for plants used in watershed rehabilitation. Building ponds also results in increased wildlife diversity.

Algae, plankton, and tiny crustaceans are the basic foods for fish and other higher forms of pond life. For algae-feeding fish species, occasional fertilization of the pond with animal manure or similar high-nitrogen food will induce growth of plankton and algae, but abrupt population surges of these tiny organisms may cause nutrient competition and oxygen depletion. For this reason, population and oxygen levels should be closely monitored. Good monitoring equipment is essential for intensive aquaculture.

Aquatic polyculture has been carried on for centuries in various parts of the world. In northern Europe, ducks, fish, and crayfish are raised together in freshwater ponds. The ducks eat the algae and small fish, and deposit manure which promotes further growth of algae and aquatic plants. Crayfish, considered a delicacy by Europeans, eat these plants, as do other herbivorous fish. These are in turn eaten by predatory fish such as bass. People then harvest the fish, ducks, and crayfish for food. This type of polyculture is a managed imitation of natural ecosystems.

Aquaculturists generally recommend stocking ponds with native species whenever possible. The ecological implications should always be considered when non-native species are introduced. Carp, for example, are non-native, bottom-feeding vegetarian fish which can survive in many of the region's warmer waters. They flourish in irrigation canals in California, and several species are used in polyculture in Asia, Europe and Africa. However, their feeding activity stirs up sediment, ruining the habitat for trout and other native fish. So despite its potential aquacultural value, the introduction of carp for commercial purposes is banned in most parts of the United States.

The common Pacific Northwest crayfish *(Pacifasticus leniusculus)* is an edible crustacean which flourishes in slowly moving water. Crayfish eat a wide variety of foods, from fish wastes to pond weeds. Crayfish are planted in California and Louisiana in flooded rice fields for insect and weed control and harvested after the fields are drained. Our native species is now being planted widely in Europe, to replace the native European crayfish which were wiped out by a fungal epidemic.

Among freshwater fish, trout seem to have the greatest potential in the Northwest. Trout farming is already a big industry in the region, with over 90% of the rainbow trout in the U.S. produced in the Magic Valley of southern Idaho. Active programs for superior stock selection have been underway for over 50 years, and much information is available on trout biology and ecology.

The use of home-produced feeds for pond fish has been

extensively and successfully experimented with, using ingredients such as low-grade beef and beef liver, wheat, fish meal, whey, bran and greens. Earthworms are an excellent fish food. They can be grown near the pond and fed whole to the fish, or mixed into a balanced blend.

For the small pond manager, a good feeding strategy is to allow fish to search the pond themselves for most of their food, providing supplemental feedings once or twice a day. In a large-scale pond culture, the best approach may be to release limited numbers of desired species, such as crayfish, trout, catfish, bullfrogs or ducks, and allow time for natural population dynamics to establish a balance.

AQUATIC PLANTS

Freshwater shorelines are productive natural areas which support many useful plants and animals. Cattail *(Typhus latifolia)* is one wetland plant with a variety of uses. Wild food gatherers know the delicious taste of its young shoots and pollen heads. Researchers in Minnesota and New York are investigating cattails for their potential as human and animal food, and as a source of biomass energy. Starchy cattail storage roots, or rhizomes, can yield up to 30 dry tons per acre. They contain more protein on a dry weight basis than corn or rice, and more fat and minerals than corn, rice or wheat. Even if the roots are intensively harvested, there are usually enough small rhizomes left in the ground to reseed a stand.

Another useful freshwater plant is duckweed *(Lemnaceae)*. These tiny floating plants act as bio-filters by feeding on algae and organic waste. Duckweed is a favorite food of wild ducks (hence its name), but recent experiments show the potential value of harvesting and feeding this quick-growing, high protein crop to domestic animals.

Although most algae have only recently been considered valuable by humans, an exception is the spirulina algae *(Spirulina plantensis)*, which has been gathered for food and fertilizer in some parts of the world for centuries. Three areas which are particularly rich in this high protein freshwater organism are Lake Chad in Africa, Lake Texcoco in Mexico, and Upper Klamath Lake in southern Oregon. Large scale harvesting of algae from upper Klamath Lake, originally begun to control pollution, now provides valuable animal feed and fertilizer. Deliberate culture of freshwater algae is still largely experimental in this country, but the practice is growing quickly in Japan and other parts of the world.

MARICULTURE

The intertidal zone is particularly rich and in many parts of the world it is extensively harvested for human food. At times this coastal zone is flooded by ocean tides, and at other times it is dry and exposed to direct sunlight. This ecological edge is well populated by many species of algae, sea vegetables and shellfish.

Hills, Christopher, and Nakamura, Hiroshi. Food From Sunlight. Boulder Creek, CA: University of the Trees Press, 1978.
The classic study of spirulina algae. Gives historical, biological, and modern cultural information, including reports from Japan and Mexico.

Kachtik, Don; Davis, Jim; and Gerlow, Art. Crawfish Production in Orange County. Extension Service, Texas A&M, College Station, TX 77843.
Guidelines for commercial production, with details on pond size, management, and costs.

Krauss, Robert W., ed. The Marine Plant Biomass of the Pacific Northwest Coast. Corvallis, OR: Oregon State University Press, 1977.
This hefty volume reports on a 1977 symposium sponsored by Oregon State University to consider the potential of coastal marine plants.

Logsdon, Gene. Getting Food from Water. Emmaus, PA: Rodale, 1978.
In his usual thorough and practical manner, Logsdon introduces many ways of obtaining, enhancing, and developing aquaculture systems.

Madlener, Judith Cooper. The Sea Vegetable Book. New York: Clarkson N. Potter, 1977.
Perhaps the best English language book on gathering and using sea vegetables.

McConnaughey, Evelyn. Sea Vegetable Recipes from the Oregon Coast. 2 vols, 1980. McConnaughey Recipes, 1653 Fairmount Blvd, Eugene, OR 97403.
A collection of recipes from around the world and from Evelyn's kitchen which offer delicious and nutritious ways to use the bountiful vegetables from the sea.

Ponds for Water Supply and Recreation. Agriculture Handbook #387. USDA: Washington, DC, 1971.
A detailed manual for pond construction, maintenance and utilization.

Stephenson, W.A. Seaweed in Agriculture & Horticulture. Edited by B. & G. Rateaver. Pauma Valley, CA: Rateaver, 1974.
Documents the traditional uses of seaweed in northern Europe, the development of seaweed extracts, and their many modern uses.

Waaland, J. Robert. Common Seaweeds of the Pacific Coast. Seattle: Pacific Search Press, 1977.
A field guide. Dr. Waaland is a researcher and instructor at the University of Washington.

Raft Culture

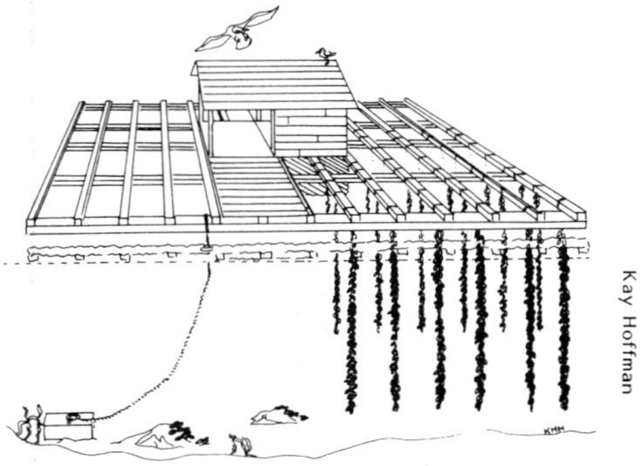

Using rafts and floating pens for raising fish and shellfish can be highly productive. The first use of nets for "growing" marine shellfish in Europe is attributed to a shipwrecked Irish sailor who suspended a net between two poles to catch sea birds. Later he discovered that mussels had attached themselves to the net when it was submerged at high tide. Nori, a seaweed of the *Porphyra* genus, is grown on suspended nets in the shallow waters of coastal Japan and can be combined with raft culture of fish and mussels.

Many varieties of nori grow naturally in Puget Sound, but development of raft culture is just beginning. San Juan Sea Farms on Lopez Island is currently establishing both rope culture for mussels and nets for nori production. A Pacific Northwest Nori Growers' Association has also formed to develop what could become a major new industry in the region.

ORGANIZATIONS

Marine Research and Development Center
Department of Natural Resources
Mail Stop EX-12
Olympia, WA 98504
The Washington State Department of Natural Resources manages hundreds of acres of State-owned tidelands, so their interest in aquaculture comes as no surprise. They are considering the development of a Nori demonstration farm.

Pacific Northwest Nori Growers' Assocation
Gary Jenks
18954 Arrowwood Avenue
Lake Oswego, OR 97034
A networking organization of nori growers which sponsors workshops and exchanges of information among its members.

In the Northwest, harvesting of intertidal life for personal consumption is permitted in many places, including many public-access beaches. Harvest limits and regulatory agencies have been established to protect these areas from depletion. Of the native shellfish, one of the most abundant and least utilized is the blue mussel *(Mytilus edulis)*. In many parts of the world these mussels are harvested both from the wild and from domestic plantings. They are just now gaining well-deserved acceptance in the Northwest as a delicious sea food.

From late spring to autumn, a toxic algae bloom called red tide occasionally makes shellfish poisonous to humans. Red tide is caused by small red algae, *Gonyaulax catenella*, which grow in huge colonies under certain conditions. The shellfish themselves are unharmed by red tide, and in fact flourish on the tiny algae. Commercially grown oysters, mussels and other filter-feeding shellfish are best harvested in winter, after the danger from red tide has passed.

SEAWEED

Many seaweed species thrive in the cool waters of the Pacific Coast. Most need rocks or other solid structures to which they can attach themselves. In Scotland, coastal villagers once put stones on sandy intertidal bottomlands to act as seaweed holdfasts. The stones were placed in rows so wagons could pass between them, making for easy harvesting at low tide. In Japan, sophisticated techniques for seaweed culture have evolved over the centuries.

Many people in the Northwest are beginning to include seaweeds in their diet. Evelyn McConnaughey has collected much useful information about eating seaweed in her book, Sea Vegetable Recipes from the Oregon Coast. She says:

> *There are hundreds of species of marine algae (seaweeds) here, many of which are similar or identical to those used for centuries as an important and highly prized part of the diets of other sea coast people. Lest you reject eating seaweeds if all you've seen are those washed up, decaying and swarming with beach hoppers, consider the difference between fresh garden vegetables and those consigned to a compost heap some days ago. Observing seaweeds exposed on rocky shores at low tide one can begin to imagine their beauty, variety and color under water, supported by the nourishing ocean.*

Another important use of seaweed has been for fertilizer. The carbohydrates in these aquatic plants break down readily in the soil and promote the growth of soil microorganisms. They are also a rich source of minerals. On the Aran Islands off the northern Scotland coast, where the soil is little more than bare rock, potatoes are grown using a heavy mulch of seaweed and sand. This is called the "lazy bed" method. Many coastal people in the Pacific Northwest also gather seaweed and eelgrass for winter garden mulch and fertilizer.

Community

John Stamets

No book about sustainable agriculture would be complete without a discussion of people and communities. In the transition to a permanent, stable, self-reliant agriculture, relationships among people and between people and land will be transformed. We open this chapter with some guidelines to one part of the process essential for such a transformation—sharing nature with children. For only if our children share our goals and dreams will substantial change take place.

We then go on to feature groups which share the belief that as long as land is seen as a commodity to be bought and sold, its true value as the source of life and sustenance is forgotten. We highlight several regional, national and international groups who are saving land in ways ranging from forming land trusts to planting forests which will halt spreading deserts. Groups analyzing, evaluating and developing more equitable political systems are also featured. And we close with an introduction to the Tilth community, a network of local chapters which actively promotes and demonstrates the possibility of a sustainable, abundant agriculture for the Pacific Northwest.

Sharing Nature With Children

Joseph Cornell has written a book, Sharing Nature With Children, *which provides valuable insights for adults and children alike. Cornell emphasizes the value of having fun while in nature. His book contains 42 games which portray the natural world as a friend and teacher. Although Cornell's book focuses on young people, the human spirit of everyone is nurtured by the sensitivity and awareness he talks about (no matter how old). Following are some essential principles of outdoor activities from this book.*

JOSEPH CORNELL

I would like to share with you five tenets of outdoor teaching that have helped me work with children's lively energies-channeling them away from mischief, and toward more constructive, and ultimately satisfying, pursuits. Underlying these principles are basic attitudes of respect for children and reverence for nature—attitudes to which they will surely respond.

1. Teach less, and share more. Besides telling children the bare facts of nature ("This is a mountain hemlock tree"), I like to tell them about my inner feelings in the presence of that hemlock tree. When we share our own ideas and feelings, it encourages children to explore, respectfully, their own feelings and perceptions. A wonderful mutual trust and friendship develops between the adult and the child.

2. Be receptive. Receptivity means listening, and being aware. It is one of the most richly rewarding attitudes you can cultivate while working with children. The outdoors brings out a spontaneous enthusiasm in the child that you can skillfully direct toward learning.

Joseph Bharat Cornell. Sharing Nature With Children. 1979. Ananda Publications, 14618 Tyler Foote Road, Nevada City, CA 95959. $4.95.
Delightful collection of games to share with children of all ages.

Universal Children's Garden, Grand Central, PO Box 2698, New York, NY 10163.
Universal Children's Garden is a plan to create gardens around the world, gardens especially for children. It is sponsored by Planetary Citizens, a non-governmental, non-profit educational organization co-founded by Donald Keys, editor-writer Norman Cousins, and U Thant, former Secretary General of the United Nations. Their goal is a Universal Children's Garden in every nation's capital, and two thousand chapters around the world by the year 2000. They publish a newsletter, In The Garden, which summarizes their work.

Children of the Green Earth

In Oregon, a non-profit group called Children of the Green Earth seeks to join an enhanced awareness of nature with a sense of planetary responsibility. Planting and caring for trees is a primary focus of their activities. Founding members of the group include Richard St. Barbe Baker, Dorothy Maclean and Rene Dubos.

Children working with their hands in the earth experience something that is carried with them throughout their lives. Planting a tree can be much more than just digging a hole and placing a seedling in it. It can be a whole experience in understanding tools and their uses, in understanding the particular qualities of the tree being planted and of the site it is being planted on.

The work of Children of the Green Earth helps children develop a new way of interacting with each other and the earth—a new way of integrating their sense of community with their own personal development. What is achieved in our lifetime to further permanent agriculture and a responsible relationship with the earth can only be sustained if we involve children, thus assuring that our work will be continued.

For more information, contact Children of the Green Earth, PO Box 751, Portland, OR 97207.

Be sensitive. Every question, every comment, every joyful exclamation is an opportunity to communicate. Respond to the child's present mood and feelings. Expand your child's interests by teaching along the grain of his or her own curiosity. When you respect their thoughts, you will find your time with children flowing easily and happily.

Be alert also to what nature is doing around you at the present moment. Something exciting or interesting is almost always happening. Your lesson plan will be written for you minute by minute if you tune in with sensitive attention.

3. Focus the child's attention without delay. Set the tone of the outing right at the start. Involve everyone as much as you can, by asking questions and pointing out interesting sights and sounds. Some children are not used to watching nature closely, so find things that interest them, and lead them bit by bit into the spirit of keen observation. Let them feel that their finds are interesting to you, too.

4. Look and experience first; talk later. At times nature's spectacles will seize the child in rapt attention: a newly-emerged dragonfly pumping blood into tender unfolding wings, a lone deer grazing in a forest clearing. But even if those special sights are lacking, the child can have an experience of wonder by just watching quite ordinary things with close attention. Children have a marvelous capacity for absorbing themselves in whatever they're looking at. Your children will gain a far better understanding of things outside themselves by becoming one with them than they will from second-hand talk. Children seldom forget a direct experience.

5. A sense of joy should permeate the experience, whether in the form of gaiety or calm attentiveness. Children are naturally drawn to learning if you can keep the spirit of the occasion happy and enthusiastic. Remember that your own enthusiasm is contagious, and that is perhaps your greatest asset as a teacher.

Lee, Allyson; Malool, Susan; and McPhail, Barbara. School Gardens. 1978. Lane County Housing and Community Development, 170 E 11th Street, Eugene, OR 97401.
A condensed guide to organic gardening, with particular emphasis on school projects. An outstanding resource for the innovative teacher.

Margolin, Malcolm. The Earth Manual. Boston: Houghton Mifflin, 1975.
Sensitively written guide to preserving, restoring and enjoying wild land without taming it. Good book for working with children.

Soule, Michael. Treesong. Portland: Children of the Green Earth.
A brief collection of writings about trees, from poems to planting instructions.

Sharing Nature With Teachers

Clearing—Nature and Learning in the Pacific Northwest
PO Box 751
Portland, OR 97207
Ideas, activities and resources which help readers explore more deeply their relationship with the natural world, and share this with children. Especially geared to teachers, whether in a "school" or not.

School Gardens

Energy, Food and You. Supervisor, Environmental Education Programs, Northwest Section, 2000 NE Perkins Way, Seattle, WA 98104.
A teacher activity guide and teacher training component covering energy, population, nutrition, natural resources, and agriculture. K through 12.

L.I.F.E. (Learning In Familiar Environments). Washington State Conservation Commission, PV 11, Olympia, WA 98504.
A teacher activity guide covering soil, agriculture, environment, conservation, water and natural resources. K through 5.

Project Learning Tree. Supervisor, Science and Environmental Education Programs, Office of the Superintendent of Public Instruction, 7510 Armstrong St SW, Olympia, WA 98504.
A teacher activity guide with a teacher training component which discusses environmental awareness, forest roles, natural resource management, life support systems, and life styles. K through 12.

The World Around You, National Council of State Garden Clubs, 4401 Magnolia Avenue, St. Louis, MO 63110.
An environmental education curriculum with study guides on public lands, conservation and wildlife enhancement.

Land in Trust

If our future truly is to be abundant, we must have land on which to grow our crops. Land trusts, non-profit organizations established to preserve property for environmental or community purposes, are becoming an increasingly common method of providing people with access to land. Land trusts challenge the basic set of assumptions that shape the use of land in this country, often enabling agricultural land to remain agricultural in the face of seemingly inconquerable economic and social disincentives.

Whether they call themselves "trusts," "conservancies," "foundations," or "reliances," members of the land trust movement share one unifying characteristic: a commitment to the preservation of land resources (natural, aesthetic, cultural, economic or social). They are organized to preserve land for the public good through private efforts.

While a few land trusts date back several decades, the emergence of local private land conservation groups on a widespread basis is a relatively recent phenomenon. Since 1975, local programs have been organized at the rate of 20 per year and now total over 400 nationwide. They have so far secured over 675,000 acres of open space and resource land in all parts of the United States.

Land owned by a land trust is not for sale. Trust land is taken forever ("in perpetuity," in land trust parlance) from the cycle of speculation. And since it's removed from the marketplace, it no longer is subject to the development pressures that turn marshland into marinas, farmland into parking lots and residential neighborhoods into condominiums.

Anne Maggs, author of the essay below, is a land trust advisor and treasurer of the Oregon Community Land Trust. An expanded version of this essay appeared in the June, 1982 issue of RAIN Magazine.

Types of Land Trusts

ANNE MAGGS

One well known type of land trust restricts land from all human impact. The Nature Conservancy is an example of this type of environmental trust, preserving wilderness areas in their natural states. (See sidebar.)

Emerging from the environmental land trust movement is the farm trust movement. The American Farmland Trust is an example of a trust established to save prime farm land. They are concerned with public education, land protection, and policy development as well as with providing technical assistance to help save farms from urbanization.

ORGANIZATIONS

American Farmland Trust, 1717 Massachusetts Avenue NW, Washington, DC 20036.

American Farmland Trust is a private, non-profit organization committed to the protection of farmland and farming opportunities. Their work encompasses public education, policy development and land conservancy. They publish a bimonthly newsletter, Farmland, which successfully balances general guidelines with technical details. They can also provide copies of a booklet for use at public meetings and forums, Farming in the Shadow of Suburbia: Case Studies in Agricultural Land Use Conflicts (single copies free, 10¢ each for multiple copies).

Land Trust Exchange, 3 Joy St, Boston, MA 02108.

The Land Trust Exchange is a national clearinghouse of land preservation programs and references. They publish a bimonthly newsletter, Exchange, as well as conference proceedings and a National Directory of land-saving organizations.

The Nature Conservancy

The Nature Conservancy is a private, non-profit organization dedicated to the identification and protection of ecologically significant natural lands. It maintains the largest system of privately owned natural area preserves in the world, with 193,000 acres (194 sites) of protected natural areas in the Pacific Northwest, and over 1,586,000 acres of preserves in the United States.

The Conservancy has established Natural Heritage Programs in many areas in the Northwest to conduct an ongoing inventory of natural areas and habitats of rare and endangered plant and animal species.

If you wish to contact them and assist them in this valuable work, addresses for further information are:

Idaho Nature Conservancy, 2015 Sunrise Rim Rd, Boise, ID 83705.

Montana Nature Conservancy, POBox 258, Helena, MT 59601.

Oregon Nature Conservancy, 1234 NW 25th Ave, Portland, OR 97210.

Washington Natural Heritage Program, 3111 Seminar Bldg, SE 3-109, Evergreen State College, Olympia, WA 98505.

Community land trusts (CLT's) go beyond the preservation of natural resources to the preservation of the cultural, economic and social resources of the land. CLT's are primarily focused on land for housing and productive uses. They work in both rural and urban settings to preserve family farms and revitalize neighborhoods.

Legally, a land trust is a group of people who have joined together in the form of a non-profit corporation, and because the group is formed for purposes that benefit the public interest, it is eligible to apply for tax exempt status under Section 501(c)(3) of the Internal Revenue Code. This status can serve as a double-edged sword: it breathes life into the organization by allowing tax-deductible donations while at the same time choking it with IRS restrictions and standards.

Once a group is incorporated as a non-profit corporation (done at the state level) it has a wide variety of tools at its disposal to actively seek acquisition of land and its resources. These tools include charitable contributions, conservation easements, bargain sales and limited development rights.

LAND TRUST TOOLS

The **charitable contribution** deduction in the Internal Revenue Code allows taxpayers to donate land, interests in land, cash and other resources to tax exempt organizations and subtract these gifts from their taxable income.

A **conservation easement** places a use restriction on the title to land. The easement reduces the value of the land from its speculated use to its actual use. The trusting of the easement legally binds the owner or future owner to the restricted use (for example, agriculture) in perpetuity.

In the case of a **bargain sale**, the owner conveys to the trust the full title at a price below its fair market value. The owner can then claim the difference between the sale price and the fair market value as the charitable contribution. This technique is often used in the pre-acquisition of land for public agencies. **Limited development** always involves compromises and some loss of open space. The role of the land trust is to persuade the developer to cluster the development in order to preserve the special characteristics of the land (e.g. an ecosystem, or prime agricultural soil).

Association with a land trust, be it a small local group or a regionally-based one, offers homesteaders and communal residents the benefits of shared resources such as technical assistance, news of other land trusts, updates on regulations and laws affecting land use and (most importantly) the support that comes from a shared land ethic. To be part of the land trust movement is to actively address our need to assure the preservation of land and its resources for ourselves and those who will come after us.

BIBLIOGRAPHY

Bolton, Charles. A Guide for Recycling Vacant Property in Your Neighborhood. San Francisco: Trust for Public Land, 1979. Available from the Superintendent of Documents, USGPO, Washington, DC 20402.
Includes a lengthy list of resources available to help turn vacant urban property into homes and gardens.

Geisler, Charles, and Matthei, Charles, eds. The Community Land Trust Handbook. Emmaus, PA: Rodale Press, 1982 (forthcoming).
This 300-page book promises to fill a long-time need for basic and technical information on community land trusts (CLT's). According to the authors, the book describes the basic CLT model, includes 20 photo-interviews with CLT members and organizers nationwide, and contains technical information on organizing CLT's. The Community Land Trust Handbook is a project of the Institute for Community Economics, 151 Montague City Road, Greenfield, MA 01301.

Hoose, Phillip M. Building an Ark, Tools for the Preservation of Natural Diversity Through Land Protection. Covelo, CA: Island Press, 1981.
A guide for the protection of special natural environments through various legal means. Detailed and helpful to anyone involved with land preservation.

"Reinhabiting the Land." A special issue of RAIN Magazine, 2270 NW Irving, Portland, OR 97210. June, 1982. $1.50.
A nicely done 24-page magazine with articles and access information on land trusts, owner-builders and building codes, and ways of developing and maintaining land-based communities. A good job by fine folks.

Swann, Robert. The Community Land Trust: A Guide to a New Model for Land Tenure in America. Available from The Institute for Community Economics, 151 Montague City Road, Greenfield, MA 01301.
The long-standing first source of information about land preservation through the establishment of land trusts.

Land Trusts

Common Earth Land Access and Stewardship Society, 1842 E Georgia St. Vancouver, BC.

Common Earth was formed to learn about and promote stewardship of land, and to alter traditional conceptions of land stewardship and management. They wish to develop a network of community land trusts.

Evergreen Land Trust, 747 16th Ave E, Seattle, WA 98112.

A community land trust which holds (at this Summer 1982 writing) 2 urban and 2 rural properties on which communities live. They have developed excellent model Use Agreements (between land trust and residents) and Trust Agreements (between land trust and donor). Copies are available for $2.

The Montana Land Reliance, PO Box 355, Helena, MT 59601.

The Montana Land Reliance develops and implements programs to protect significant lands by working directly with private landowners. Their work includes studies, research programs, land acquisition and negotiating conservation easements.

Northern California Land Trust, 1505 Milvia St, Berkeley, CA 94709.

Northern California Land Trust is a regional service organization and educational institution dedicated to forming community land trusts.

Oregon Community Land Trust, c/o Dept. of Landscape Architecture, 409 AG Hall, OSU, Corvallis, OR 97330.

The Oregon Community Land Trust is a non-profit corporation founded to promote mutually sustaining relationships between people and land. Their main activities include education and land management services.

Oregon Women's Land Trust, POBox 1692, Roseburg, OR 97470.

OWL was formed to give women access to land and to the skills needed to live well in the country. They currently own one 147-acre piece of land. OWL needs people and financial support.

The Trust for Public Land, 82 Second St, San Francisco, CA 94105.

The Trust for Public Land is a private, non-profit conservation organization specializing in acquiring land for public use, both urban and rural. Excellent technical resources.

Turtle Island Land Stewardship Society, Manson's Landing, BC V0P 1K0.

Turtle Island is a charitable society working to develop research and teach ecological stewardship of land; promote land-based communities; and pursue spiritual growth and planetary consciouness.

Land Preservation

Fletcher, W. Wendell, and Little, Charles E. The American Cropland Crisis: Why U.S. Farmland is Being Lost and How Citizens and Governments are Trying to Save What is Left. Bethesda, MD: American Land Forum, 1982.

"The American cropland crisis does not lie in the current ability of U.S. agriculture to produce: it has never been more productive. Rather, it is a political crisis...."

This is the best current report on strategies for agricultural land preservation. Available from American Land Forum (below right).

The National Agricultural Lands Study. Amherst, MA: Regional Science Research Institute, 1980. Available from the Superintendent of Documents, USGPO, Washington, DC 20402.

An excellent source of a multitude of thoroughly-researched publications concerned with the preservation of agricultural land.

Sampson, R. Neil. Farmland or Wasteland. Emmaus, PA: Rodale Press, 1981.

"Every day, 26 square miles of U.S. topsoil wash and blow away....At that rate, it could take less than 100 years for us to lose every single acre of cropland now farmed in the U.S. Then our once-fertile farmland will be little but barren wasteland." Farmland or Wasteland presents a passionate, well-documented statement on the current farmland crisis—causes, effects and plausible solutions—by the publishers of Organic Gardening and The New Farm.

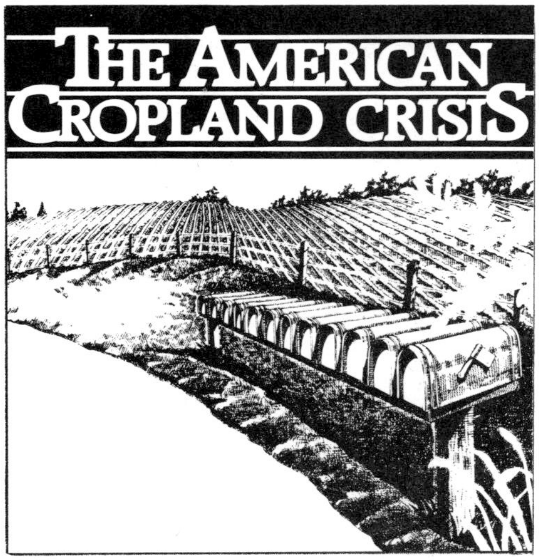

Steiner, Frederick. Ecological Planning for Farmlands Preservation: A Sourcebook for Educators and Planners. Pullman, WA: WSU Department of Horticulture and Landscape Architecture, 1980.

The experience of citizens in Whitman County, Washington to preserve farmland is outlined as a model for ecological planning. Excellent sourcebook, with an extensive bibliography.

A film, "The Only Essential Industry...Farmland Preservation," was produced as part of the project. It is available from Radio-Television Services, WSU, Pullman, WA 99164.

Who Owns the Land? ESCS #70. Washington, DC: USDA, 1979.

The report of a demographic study of land ownership. "Half of the privately owned land is owned by only 1 percent of the landowners."

Wolf, Peter. Land in America: Its Value, Use and Control. New York: Random House, 1981.

Comprehensive review of the history of land speculation and current legal battles as land is seen less as a private commodity to be exploited and more as a resource to be preserved. Wolf analyzes the impacts of current planning efforts, with suggestions for making them more effective.

ORGANIZATIONS

American Land Forum, 5410 Grosvenor Lane, Bethesda, MD 20814.

Formed during the mid-1970's, the American Land Forum sponsors conferences, special studies, and research projects pertaining to land conservation and use. They have published The American Cropland Crisis (see accompanying bibliography) and publish the quarterly American Land Forum magazine.

The Urban Land Institute, 1090 Vermont Avenue NW, Washington, DC 20005.

Source of information on land use decision-making. They publish Land Use Digest, a newsletter which provides access to current research.

Politics—Bibliography

Belden, Joe, et al. New Directions in Farm, Land and Food Policies. 1979. Conference on Alternative State and Local Policies, 1901 Que St, NW, Washington, DC 20009.

Anthology of strategies covering farmland preservation, soil conservation, market development, access to land, and city food plans. Excellent resource for anyone engaged in public policy making.

Goldschmidt, Walter. As You Sow: Three Studies in the Social Consequences of Agribusiness. Montclair, NJ: Allanheld, Osmun & Co., 1978. Originally published in 1947.

The majority of this book is a reprint of the classic study of two comparable towns in California, Arvin and Dinuba. One was surrounded by family farms and one by giant agricultural enterprises. The study's clear conclusion was that the quality of rural life was seriously eroded when neighbors were corporations. A new closing essay, "Agribusiness and Political Power," describes the events that have taken place since the landmark Arvin-Dinuba study, including the extensive efforts made by agribusiness interests to sabotage and suppress the information.

Goodwyn, Lawrence. The Populist Movement. New York: Oxford University Press, 1978.

One hundred years ago the agrarian revolt grew into the largest movement for economic democracy in US history. This history of the movement provides useful background for today's continuing struggle for social change.

Hart, E. Richard. The Future of Agriculture in the Rocky Mountains. Salt Lake City: Westwater Press, 1980.

This book is the result of a symposium held at Sun Valley, Idaho, to discuss the future of agriculture in the northern Rocky Mountain states. The tone of the book is progressive but written in a style that makes it acceptable to many policy makers. The four-chapter coverage on the Sagebrush Rebellion offers many insights into the politics of that movement.

Illich, Ivan. Shadow Work. Salem, NH: Marion Boyars, 1981.

Behind the scenes in the industrial world, where there is a salaried specialist for every social function, the daily work of subsistence goes on as before. Illich's analysis brings light to the inherent contradictions of contemporary culture.

Lappé, Frances Moore, and Collins, Joseph. Food First: Beyond the Myth of Scarcity. New York, Ballantine, 1979.

Lappé and Collins document their claim that every nation has the capacity to feed itself. They describe the true causes of world hunger and propose effective solutions.

Loomis, Mildred J. Alternative Americas. New York, Universe Books, 1982.

A personal account of numerous movements for social change written by a leading activist. Emphasis is on the heritage of Ralph Borsodi and other pioneers in areas such as land trusts, intentional communities, cooperative businesses, holistic health, and organic agriculture.

Perelman, Michael. Farming for Profit in a Hungry World: Capital and the Crisis in Agriculture. Montclair, NJ: Allanheld, Osmun & Co., 1977.

Perelman analyzes the forces behind the destruction of small-scale, self-sufficient farming in a hard-hitting critique of contemporary agriculture.

Sherman, Jay, and Webb, Lee, eds. Assisting Beginning Farmers: New Programs and Responses. Washington, DC: The Agriculture Project of the Conference on Alternative State and Local Policies, 1980.

By describing programs that work, this briefing book is designed to aid and inspire state legislators, farmers, farm leaders and citizen activists across the continent who are struggling to preserve and strengthen the family farm.

Swaczy, Barbara. Getting People Together in Rural America. 1979. Northern Rockies Action Group, 9 Placer Street, Helena, MT 59601.

A basic text (40 pages) for individuals and groups interested in organizing in rural areas. Its step-by-step analyses of how rural areas function, how to bring people together and keep them together, and how to learn and share necessary skills are straightforward and practical.

Vogeler, Ingolf. The Myth of the Family Farm: Agribusiness Dominance of U.S. Agriculture. Boulder, CO: Westview Press, 1981.

Dedicated to "a new Populism in our lifetime," this book describes the subsidized growth of agribusiness, its nurturing and fattening by governmental policies. The losers are the general public, the consumers of the products of agribusiness, and rural America. Vogeler advocates that "...farmers and other workers need to rediscover their radical roots and finally forge a unity to establish a truly democratic political and economic society."

PERIODICALS

Common Sense
The Kootenay Environmental Institute Ltd.
Box 2, Galena Bay
Via Revelstoke, BC V0E 1X0

Environmental news for Canadians, including information about agriculture.

The Ecologist, Journal of the Post-Industrial Age
Worthyvale Manor Farm
Camelford, Cornwall
England PL329TT

Worldwide outlook on the environment, including articles on agriculture and tree crops.

Food Monitor
350 Broadway, Suite 209
New York, NY 10013

Published by World Hunger Year, Food Monitor provides insight into the causes of hunger around the world, with a very human perspective.

Not Man Apart
124 Spear Street
San Francisco, CA 94105

The journal of the Friends of the Earth. Reports from around the nation and the world on how industry is depleting the environment for profit and what people are doing to try to stop them. Friends of the Earth actively promotes organic farming and works to limit pesticide use.

Nutrition Action
Center for Science in the Public Interest
1755 S Street NW
Washington, DC 20009

Published 10 times per year, this journal reports on a broad range of food issues, with special emphasis on politics and science.

Small Farm Advocate
PO Box 405
Walthill, NE 68067

Quarterly newsletter surveying legal and administrative issues affecting small and low income farmers.

Community Organizations

ComeUnity Network, c/o Alpha Farm, Deadwood, OR 97430.
Regional network of intentional communities in the Pacific Northwest.

Idaho Citizens Coalition
605 W Franklin
Boise, Idaho 83702
A coalition concerned with agriculture and resource development in the Snake and Columbia river basins, especially regarding agribusiness expansion and exploitation of natural resources.

Institute for Food & Development Policy
2588 Mission Street
San Francisco, CA 94110
The search for an answer to the question of why there is hunger in a world of plenty motivates the work of this research and education center. Co-founded by Frances Moore Lappé author of Diet for a Small Planet, the institute has published several valuable books, including Food First: Beyond the Myth of Scarcity, Aid as Obstacle, and World Hunger: Ten Myths.

National Land for People, 2348 N Cornelia, Fresno, CA 93711.
Best known for its efforts to achieve enforcement of federal irrigation laws. Currently promoting small-scale farming and developing an arid land demonstration farm.

North Olympia Living Lightly Association (NOLLA), Box 1073, Sequim, WA 98382.
The first of what has become a regional network of groups sharing skills and building local communities. Regular workshops and newsletters.

Northcoast Environmental Center
1091 H St
Arcata, CA 95521
Center for exchange of information on forest practices and other environmental issues in northern California. Extensive library. They publish "Econews," a hard hitting newspaper covering the wide range of environmental concerns confronting California.

Oregon Food Action Coalition
1414 Kincaid
Eugene, OR 97401
Composed of many public action groups working to share information and promote high quality food, small family farms and a public policy that will insure a healthy agriculture.

Oregon Gleaning Consortium, c/o Margret Grant, Mid Willamette Valley Community Action Agency, 1275 Chemeka Street NE, Salem, OR 97301.
Coordinates 16 gleaning projects statewide, publishes yearly statistics on gleaning, and coordinates an annual spring training program.

Oregon Food Share, 718 W Burnside, Portland, OR 97209.
This is the first program in the nation to link urban and rural food outlets into a statewide distribution network for donations of surplus food. They also provide technical assistance and training for emergency food agencies and distribution centers throughout the state.

Rural America, 1346 Connecticut Avenue NW, Washington, DC 20036.
A national membership and advocacy organization for people in rural areas and small towns. Their excellent newspaper, ruralamerica, reports on government and corporate policies, with in-depth investigations into crucial rural issues.

Save Our Public Lands, Inc., Box 844, Boise, ID 83701.
People opposed to the exploitation of public lands by commercial interests.

Small Towns Institute
PO Box 517
Ellensburg, WA 98926
As our population continues to decentralize, small towns will once again become important community centers. The Small Towns Institute fosters this change in communities throughout North America. They publish a bimonthly magazine which includes articles and news on community, technical, social and economic programs, with an eye toward informing people of projects that may work in their own areas.

Pesticides

National Rural Health Council. Pesticides and You. Rural America, 1346 Connecticut Avenue NW, Washington, DC 20036.
This publication is subtitled "A Guide for Farmworkers, Small Farmers and Rural Communities." It contains information both for working with and around pesticides, and information about the politics of pesticides.

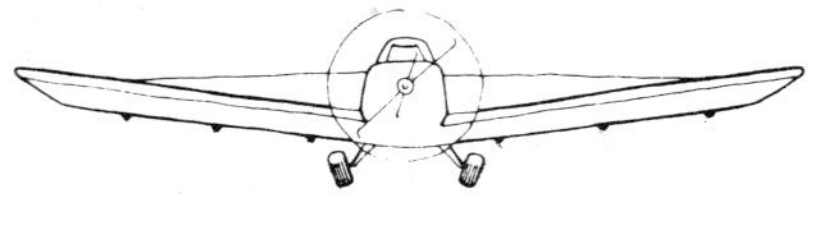

Van den Bosch, Robert. The Pesticide Conspiracy. Garden City, NY: Doubleday, 1978.
Authoritative report on the deadly cycle of pesticide use promoted by chemical companies.

Northwest Coalition for Alternatives to Pesticides (NCAP)
PO Box 375
Eugene, OR 97440
Active network of groups opposing pesticide sprays. Sponsors independent research on alternatives and publishes a quarterly news magazine. These groups have been instrumental in affecting spray policies throughout the region.

Pesticide Action Network International
c/o Inst. for Food & Development Policy
2588 Mission Street
San Francisco, CA 94110
International network established to halt the overseas marketing of hazardous pesticides which are banned in this country. These pesticides are "responsible for at least 10,000 deaths in the third world each year." The network is working to stop such practices and develop alternatives.

Western Washington Toxics Coalition
4512 University Way NE
Seattle, WA 98105
Western Washington Toxics Coalition, organized in late 1980, is a regionally-based coalition of groups and individuals concerned about the increasing overuse and misuse of synthetic pesticides and other toxic substances.

The Green Front

MICHAEL PILARSKI

Many of the Earth's ecosystems have been degraded by human activities. The Earth's forests have been reduced 60-65% during recorded history, while desert area has doubled. Soil erosion, resource depletion and environmental degradation are found throughout the world. The work of rebuilding our Earth is a challenge to all.

Rehabilitating ecosystems is a complex task even under the best of circumstances. War, oppression, drought and poverty complicate possible solutions. Present patterns of land ownership and governance discourage many people from attempting land reclamation work.

All is not doom and gloom, however. Over the years a global movement for earth-healing has been forming. Examples can be found in every country, though few of the people involved think in terms of being part of an international movement.

"The Green Front" is a title particularly appropriate for this earth-healing movement. The term was first coined in 1951 when Richard St. Barbe Baker initiated an international conference to discuss the alarming spread of the Sahara desert. He went on to work with native people, planting a "green front" of trees to halt the desert's spread.

Today the Green Front is given form by the combined activities of thousands of people, planting trees, planting seeds, and relating to nature first hand. There are many ways each person can contribute to the success of the international Green Front and the creation of a sustainable agriculture. Search for ways in which you can help. Chances are you will not have to look far... since the Green Front starts in your backyard.

Father of the Trees

Richard St. Barbe Baker was one of the world's most eloquent and influential defenders of the trees. "You can gauge a country's wealth," he said, "its real wealth, by its tree cover." By that measure, he found much of the world rapidly becoming impoverished.

St. Barbe's efforts to enlist trees in the battle against desertification began in 1922. Working with the Kikuyu people in Kenya, he convinced them to plant trees as a defense against the advancing desert. As a result of his work the Men of the Trees Society was formed, and the people named St. Barbe 'Father of the Trees.' *Twahamwe*—work together—became their password.

John Watkins

In the spring of 1982 St. Barbe made his last tour on behalf of the trees. He became ill while in northern California and spent his last month in the Grove of Understanding, a magnificent redwood grove that he had helped preserve over forty years before. While there his concern for the future of the earth continued to grow. Don't stop with just saving single trees, he said. Continue until you save whole watersheds and, ultimately, the entire biosphere. In early June St. Barbe was taken to Saskatoon, Saskatchewan, where he had homesteaded just after the turn of the century. It was there that he died, leaving others to carry on the task of healing the earth.

St Barbe's **Men of the Trees Society (Crawley Down, Sussex, England)** publishes a quarterly journal which will continue to be a vital source of information on international reforestation efforts. In North America, St Barbe's work is being carried on by **Children of the Green Earth, PO Box 751, Portland, OR 97207** (see page 140).

Civilian Conservation Corps

Ecosystem impoverishment in the North America was given almost no attention until Hugh Hammond Bennet began writing about soil erosion in the 1920's. In 1928, Bennet and WR Chaplin published the pamphlet *Soil Erosion, A National Menace*. It was brought before Congress and caused enough of a stir that ten erosion research stations were set up by 1930. The findings of these stations, plus the desperate unemployment situation created by the Depression, prompted President Franklin D. Roosevelt to establish the Civilian Conservation Corps (CCC) in March of 1933. In its first year the CCC improved 950,000 acres of forest land, built 420,000 erosion control check-dams, and planted 100,000 acres of trees.

In 1934 the first of the Dust Bowl storms swept across the country, alerting people to the impending crisis. To respond to the situation, the Soil Conservation Service (SCS) was formed in early 1935. By the middle of 1936, the SCS was operating 147 demonstration projects, 454 CCC camps, 48 plant and seed nurseries, and 25 experiment stations.

One of the most ambitious projects the SCS initiated was the Great Plains Shelterbelt, 100 miles wide and 1800 miles long. Although this project was never completed much was accomplished, especially in the northern plains states. As an example, in 1980 there were 67,000 miles of windbreaks in North Dakota, as compared to 43,000 miles in the rest of the U.S. combined. Canada also planted large areas to shelterbelts in its portion of the Great Plains during this time.

By the time World War II began, the three million CCC workers had made an important contribution to land rehabilitation. But the business of fighting a war siphoned people off. After the war, land rehabilitation was given little attention. In the 1970's, Federal policies led to increased U.S. agricultural exports, and the quality of the nation's farmland has suffered as a result. The SCS now calculates that national soil erosion losses are once again at Dust Bowl levels.

With unemployment and soil erosion rates approaching those of the Great Depression, the CCC Alumni Association is petitioning congress to restart the CCC.

For more information, contact the National Association of CCC Alumni, 7900 Sudley Road, Suite 413, Manassas, VA 22110.

The Chipko Movement

The first Chipko Movement was born around 250 years ago. A Vishnoi woman named Amrita Devi and several hundred fellow Vishnoi attempted to save their village forest by hugging the trees as the Maharajah's axemen approached. ("Chipko" means "hug" or "cling to".) The axemen killed many people while felling the trees, and when the Maharajah heard of their sacrifice he declared their forest a natural preserve. Today that forest is one of the few green sanctuaries in what has become a desolate desert.

In modern times, the Chipko Movement was reborn in the Alakhnanda valley, a remote mountain area of the Himalayas. In the late 1950's, an organization called the Dasholi Gram Swarajya Sangh (Dasholi Society for Village Self-Rule) was formed to improve the position of the many poor and landless peasants. Cooperative action was stressed, and members formed small cooperative industries using local forest resources. During the 1960's highways were constructed into the region, making possible increased exploitation of local forests by the government and large timber corporations. Many steep slopes were clearcut, leading to increasingly disastrous floods which swept away villages and caused much death and destruction in the valley.

The villagers protested the clearcuttings, but the government's insensitive response was to refuse the village cooperatives the permits needed to cut down the few trees they required for their enterprises. Meanwhile, large corporations were allowed to continue their clearcutting.

In 1973, the villagers began the non-violent disruption of forest cutting by hugging trees in front of the loggers. Press coverage and public rallies eventually forced the government to investigate the forest situation in the Alakhnanda river watershed. As a result, the government put a ten-year ban on all tree cutting in an area of over 450 square miles.

Between 1975 and 1978, the villagers planted over 100,000 trees to reforest 1200 acres, demonstrating what can be done by people working together, without government aid, to improve local conditions.

The Chipko Movement has begun to spread to other parts of India. One of the leaders of the Chipko Movement has written a booklet entitled "Chipko, A Novel Movement For Establishment of Cordial Relationship Between Man and Nature." To obtain a copy send a dollar, or whatever you can to: Sunderlal Bahuguna, Chipko, PO Silyara Pin 249155, Tehri - Garhwal, UP, India.

Tree People

In 1970, a report issued by the US Forest Service indicated that each year at least 40,000 trees in the forests surrounding Los Angeles were dying from air pollution. The report stimulated 15-year-old Andy Lipkis to found Tree People to replace the dying trees with new seedlings. Since then, over 50,000 people have planted over 200,000 trees in the Los Angeles area. Many environmental education programs have also been carried out. "All our programs," Andy says, "have been oriented to show how we cause environmental problems and how we as individuals can solve them. Many of our daily actions are causing the forests to die. We have to accept that responsibility and do something about it."

In 1981, Andy announced Tree People's new goal, "to plant one million trees in Los Angeles before the 1984 Olympics to be held in Los Angeles. We are intending to involve the entire community in planting trees as a way to help improve air quality, conserve energy, and produce food locally. The trees will be both smog- and drought-tolerant, so they won't require the importation of additional water in the LA area."

For more information on the Tree People program, write to Tree People, 12601 N Mullholland Drive, Los Angeles, CA 90210.

PERIODICALS

Development Forum, DESI/DPI, Palais des Nations, CH-1211 Geneva 10, Switzerland.

Thorough, incisive review of international development trends, with quality information and a diversity of perspectives. 10 issues per year.

Ecoforum, Environmental Liaison Center, PO Box 72461, Nairobi, Kenya.

Highly respected quarterly newsletter on the environment, international development, and human settlements.

Green Deserts, Rougham, Bury Street, Edmonds, Suffolk LP30 9LY, England.

Articles on innovative solutions to the worldwide deforestation crisis.

Helping out in the Outdoors, 16812 36th Avenue W, Lynnwood, WA 98036.

A directory of volunteer jobs and internships in state and national parks and forests in the United States. For persons who like the outdoors and want to help protect and improve it. Volunteer jobs include campground hosts, trail workers, botanists, wilderness rangers, photographers, caretakers, biologists, interpretive naturalists.

One Earth, Findhorn Foundation, The Park, Forres, Moray, Scotland IV36 OTZ.

A lively exchange of insights from a worldwide spiritual network working to unite people, plants, and the natural world.

ORGANIZATIONS

International Voluntary Services
1717 Massachusetts Avenue NW,
Suite 605
Washington, DC 20036

A private service which places American volunteers with agricultural skills in third world countries.

Tree of Life, Bou Saada Trust
c/o Coutts & Company, 440 Strand
Duncannon Branch, London

An organization which works with rural people in North Africa and Greece to combat desertification by planting trees. Bou Saada is the name of an area in Algeria where Wendy Campbell Purdie planted the first part of what has become a green wall of trees all across Algeria.

Tilth People

Kari Berger

Breaking ground for Seattle Tilth's Urban Agriculture Center. The Center features demonstration gardens, a solar greenhouse, compost bins, trellised fruit trees, and a resource library for urban gardeners. Workshops are offered frequently on topics such as winter gardening, herb growing, and backyard livestock.

For more information on their programs, write to the Seattle Tilth Urban Agriculture Center, 4649 Sunnyside North, Seattle, WA 98103.

In 1974, Gigi Coe, Becky and Woody Deryckx and Mark Musick formed Tilth to sponsor the Northwest Conference on Alternative Agriculture. Held in Ellensburg, Washington in November of that year, the goal of the conference was to bring together the diverse groups and individuals who had been working independently to create a new agriculture—small farmers, urban gardeners, land reform advocates, environmentalists, and food system activists. Nearly 800 people attended the three-day conference, and interest ran high in workshops and discussions on all phases of crop and livestock production, seed exchanges, and alternative energy sources.

That first Tilth conference was an exciting, unique event. It certainly achieved its objective of establishing communication among people throughout the region working to strengthen small farms, establish alternative distribution networks, and expand urban gardening programs.

In February of 1975, the first issue of what has become the quarterly journal *Tilth, Biological Agriculture in the Northwest*, was published to carry on the information exchange initiated at the Ellensburg conference. Tilth continued as an informal network for two and a half years. In August of 1977 Tilth was officially incorporated as a non-profit research and educational association "to support and promote biologically sound and socially equitable agriculture."

TILTH CHAPTERS

Tilth was established as a regional membership organization. Over the years, members in many communities have gotten together to form local chapters which allow people to get involved locally while being part of a larger, regional community. The first Tilth chapters were formed in the spring of 1978. Today there are over a dozen local chapters around the Northwest.

Tilth chapters are as diverse as the region itself. Each chapter develops its own goals and programs, so styles and projects vary widely. Yet they all share a common desire to work with others in supporting local agriculture. In some communities this has meant forming a farmers' market, while in others the focus has been on getting to know neighbors and exchanging skills.

One of the best ways to describe the day-to-day work of Tilth is to summarize some of the ongoing projects of a few Tilth chapters.

Tilth—People Working Together

A living agriculture involves everyone. Getting involved in local agriculture, whether on a personal or community level, is the best way to improve the quality of your own life and the quality of life for others.

In the Pacific Northwest, Tilth is a network of people active in growing food, saving farmland, developing local markets, improving forest practices, and doing whatever they can to contribute to regional agriculture and forestry. In several communities Tilth members have joined together to form Tilth chapters.

Tilth chapters are autonomous organizations. They share the regional association's non-profit status, choose their own names, and define their own geographic boundaries. They also establish their own goals and programs to respond to the unique needs of their communities. To learn about a Tilth chapter in your area, write to the regional Tilth office at the address below.

Tilth, 13217 Mattson Road, Arlington, WA 98223.

Each Tilth chapter is encouraged to develop its own programs and create its own identity. Rogue Tilth has adopted a logo which is used to identify participating members at the local farmers' market.

Tilth Producers' Cooperative

The Tilth Producers' Cooperative (TPC) is a sister organization of Tilth which supports and promotes commercial organic farming in Washington state. Formed in 1978, TPC has undertaken many projects to encourage local agriculture. A potato storage and packing facility and a 30,000 gallon per year alcohol fuel plant have been built to serve organic farmers in northwest Washington. A comprehensive Organic Certification Program has also been established to serve farmers throughout the state. For more information on this and other organic certification programs, see page 67.

Southwest Washington Tilth was the first chapter to form. They coordinate working weekends on local farms and sponsor workshops in cooperation with the Extension Service and The Evergreen State College Organic Farm in Olympia.

North Idaho Tilth's major project is an annual tree sale. They specialize in locally-adapted trees, shrubs and vines which provide high quality fruits, nuts, timber, and wildlife habitat. They are also compiling an inventory of plants adapted to their region.

Nooksack Tilth, in Whatcom County, Washington, hosts workshops, has organized a group fertilizer purchase, and has members active with the County Weed Control Board to advocate IPM practices.

Methow Valley Tilth is active in a research program on alternatives for economic development in their area. They seek to increase employment opportunities through a revitalized local agriculture.

Palouse Tilth works with schools of agriculture at the University of Idaho in Moscow and at Washington State University in Pullman. They also cooperate in the annual tree sale with North Idaho Tilth.

Good Earth Tilth is located in the Wenatchee, Washington area. Tilth members are active in the local farmers' market and the chapter has sponsored several educational programs, including a major conference on Integrated Pest Management for commercial fruit orchards.

Seattle Tilth was Tilth's first urban chapter. They have established a lively Urban Agriculture Center in Seattle's Wallingford neighborhood, with demonstration vegetable and fruit gardens, a solar greenhouse, and an extensive reference library. They also publish a frequent newsletter and sponsor a workshop series.

Willamette Valley Tilth is a large and diverse chapter made up of farmers, homesteaders and home gardeners. They have sponsored annual conferences in cooperation with Oregon State University and coordinate an active Working Weekends on Organic Farms program.

Rogue Tilth in southwestern Oregon promotes local agriculture in a variety of ways. Members are actively involved with the local farmers' market, and the chapter coordinates a testing project to identify locally-adapted vegetable varieties with commercial potential.

REGIONAL TILTH

The regional Tilth office concentrates on the exchange of ideas and information. In addition to the quarterly journal, Tilth published the popular *Winter Gardening in the Maritime Northwest*, by Binda Colebrook, and this book, *The Future Is Abundant.* In 1980, the Tilth Information Service (TIS) formed to make selected books more widely available. TIS periodically publishes catalogs which combine articles, reviews and access to regional resources.

If you find this book interesting and useful, you would probably enjoy being a member of Tilth. Tilth members receive the quarterly journal, *Tilth, Biological Agriculture in the Northwest*, Tilth Information Service catalogs, and mailings announcing workshops, publications and conferences of special interest. To join Tilth and receive information on local chapters use the form on page 192 or write to **Tilth, 13217 Mattson Rd, Arlington, WA 98223.**

Plant Species Index

SHERY LITWIN

The Plant Species Index arose out of my own need for a streamlined way to choose plants in designing for small landholdings. The main criteria I used in selecting plants for the Index is usefulness. A plant which performs one function outstandingly, or a variety of functions tolerably, has been chosen over a less useful or less multi-purpose plant. I tried to include plants with a well-rounded variety of uses, and also to include species mentioned elsewhere in this book. In all cases where I had to choose between equally valuable native and non-native species, I chose native species. In retrospect, I see that I have omitted some valuable species. I hope to be able to add them to future editions of the Index.

This edition lists 330 plant species. I am not advocating the planting of all of these species in any situation where they would flourish. These are species that you may want to plant if they are not already present, or to encourage or manage if they are. Some of the species listed are invasive. Before introducing them, please consider seriously and thoroughly the management and ecological problems they may present to you and your community in the future.

The factor which often shapes the choice of species in a diverse planting is the way the plant occupies space. Accordingly, I have grouped the plants by the physical form they most commonly assume. An exception is bamboo. Bamboo life styles and characteristics as a group are so distinct that I have listed them together in their own section.

Once you have selected a category, the charts enable you to compare other physical characteristics, habitat tolerances, and potential uses. Knowing the characteristics and limitations of your site, and keeping in mind what you want a plant to do, you can use the key to select one or several plants that most closely fit what you and your land need. The Comments point out additional interesting and salient characteristics of the plant. Don't, however, depend on the Comments for details. They are meant only to provide highlights, and I have given priority to warnings and cautions.

This Index is designed to be used only as an aid in making species selections. Detailed information on cultural requirements and uses can be found in other references, or by talking with experienced nursery people and plant growers in your area. Many of the references I have used were not written with the Pacific Northwest's conditions in mind, and there is much semantic variation between references. Interpretation was necessary to standardize the chart and make comparisons possible. (This was especially true in the Habitat Tolerance section.) I believe that I have made reasonable judgments in all cases.

This is my first attempt to create a useful and comprehensive Plant Species Index for the Pacific Northwest. I hope that you will write in with improvements, additions, corrections and modifications. I work as a design consultant in western Washington and Oregon, and may be contacted in care of Tilth.

The following people contributed information about the species listed, and in some cases undertook considerable research on several species: John Paul Bund, Randall S. Buck, Binda Colebrook, Lynn Coody, Michael Dolan, Fred Euphrat, Larry Geno, Leo Givens, Janice Masterjohn, Pamela Melcher, Miles Merwin, Michael Pilarski, David Pilz, Dinah Ross, William Ross, John Thomas, and Gail Work. Jim Orjala contributed much of the bamboo information. Lee Jacobson, who has made a special inquiry into plant heights, contributed many of the height entries. Lee Jacobson, Larry Korn, Robert Kourik, Mark Musick, and Barbara Snyder provided valuable critical evaluation of the form and content of the Index, and Jennifer Neighbors helped edit this introduction. The Tilth Association funded my research.

Plant Species Index

Discussion of Terms

Scientific names were taken from *Hortus Third*, a few names being synonymous second choices when the synonym was more in keeping with Pacific Northwest references. **Common names** were chosen on the basis of their predominance in the major local texts.

Tree refers to a woody perennial with a single, erect stem or trunk and a well-developed crown or leaf canopy. **Shrub** refers to a several-stemmed woody perennial which, unlike a tree, branches from the base. **Vine** refers to a slender and flexible-stemmed climbing or trailing plant.

The **Height at Maturity** designations are variable since some species grow to very different heights depending on genetics and conditions. Where only one height is given, it indicates the plant's mature height under ideal conditions. Some species, especially those which are commonly domesticated, have varieties which differ greatly in height. When a species is commonly grown on a dwarfing rootstock, such as Pear (*Pyrus communis*), the height of the dwarfed form has been included.

Climatic Zone refers to the maps and designations in the Sunset Western Garden Book and the Sunset New Western Garden Book. These excellent references feature detailed cultural information on many plants listed in this Index, as well as a good, useful climatic zone system.

It is important to remember that many of the plants' **Adaptations/Tolerances** are interdependent. For example, a species listed as adapted to Climatic Zones 5-7, with a lowland wet/lowland moist/saltwater margins habitat tolerance may have a shade tolerance of full sun/partial shade. That species may be tolerant of full sun in Zone 5 (a seacoast habitat where conditions are humid and cloudy) and need partial shade in a Zone 7 lowland moist habitat.

Information for the **pH Tolerance** column was almost all taken from a paper by James Duke, head of Economic Botany at the USDA Beltsville Agricultural Research Center. I wish to stress that this designation refers to the maximum range of soil pH within which the plant will grow, not the optimum pH range.

I have interpreted the **Uses** broadly. For example, a tree species may coppice productively in the wetter of the climatic zones, but may grow too slowly in the dryer of the zones to be ideal for coppicing there. In this case, I have listed the species under **Coppicing.**

There are three designations for **Medicinal** uses of the Index—one under Edible, and two under Other Uses. Information on plants which are medicinal for hoofed stock, either ingested or used externally, comes largely from Juliette de Bairacli Levy. Information for the other medicinal designations (ingested and external for humans) comes from a number of references, including ethnobotanical reports from native Pacific Northwest tribal people and American folk remedies. I wish to stress that most of the medicinal entries have not been tested or confirmed by scientific research, and in many cases have not yet been checked against the species' medicinal use in other cultures. Medicinal designations are therefore indications that a species may have medicinal value (and therefore is also potentially toxic).

In the **Wildlife** section of the key, **Insectary plant** means a plant which provides habitat and/or forage for beneficial insects. **Beneficial insects** are insects which are predators and/or parasites of pest insects.

Coppicing indicates that the plant will sprout from cut stumps.

Under **Other Uses**, **Biocontrol** means that the plant may be used to control pests. Besides indicating that a species may function as an insectary plant, this may also mean that the species produces insecticides, repels insects in its immediate area, produces insect growth inhibitors, inhibits neighboring plant growth, or provides forage or habitat for beneficial bird populations. **Oils**, in the designation "Chemicals, oils, cosmetics" refers to non-edible (for instance lubricant) uses. **Landscaping** describes the use of species which, either singly or in groups, are recommended in the literature for defining space in residential or institutional situations. **Ornamental** describes species which can be either used singly or in groups as specimens, or as ornamental focal points in the landscape, and which could be chosen solely for their ornamental value.

Key

FORM

LIFE STYLE AND TYPE
- A - Annual
- B/A - Biennial grown as an annual
- P/A - Perennial grown as an annual
- B - Biennial
- P/B - Perennial grown as a biennial
- HP - Herbaceous perennial
- D - Deciduous
- E - Evergreen
- S - Shrub
- V - Vine
- T - Tree

SHAPE (refers to ground layer plants and vines only)
- cb - Climbing
- tr - Trailing
- cp - Clumping
- sp - Spreading
- mt - Mat
- up - Upright

ADAPTATIONS/TOLERANCES

CLIMATIC ZONE
- 1-7 - See Sunset Western Garden Book or Sunset New Western Garden Book climatic zone maps and designations (Number in parenthesis indicates marginal adaptation).

SHADE TOLERANCE
- O - Prefers/tolerates full sun
- ◑ - Prefers/tolerates partial shade
- ● - Prefers/tolerates full shade

HABITAT
- l - Lowland - Habitats at low elevations, or in lowland positions relative to the surrounding topography.
- u - Upland - Habitats at high elevations, or in upland positions relative to the surrounding topography.
- w - Wet - Habitats which are humid, have high precipitation, or are adjacent to watercourses, lakes, swamps or ponds.
- m - Moist - Habitats which have moderate rainfall or humid conditions. Includes moist forest habitats.
- d - Dry - Arid, exposed habitats of low rainfall, or habitats with seasonally arid conditions.
- fa - Freshwater aquatic - Areas that are usually underwater.
- s - Saltwater margins - Habitats that are subject to salt spray or occasional inundation. Includes tidal marshes and seashores.

DRAINAGE TOLERANCE
- √ - Tolerates poor drainage

TEXTURAL TOLERANCE
- G - Will grow in gravelly soil.
- H - Will grow in heavy clay soil.
- R - Will grow in rocky soil.
- S - Will grow in sandy soil.

pH TOLERANCE
- (-) = Acidic soil < 5.6
- (+) = Alkaline soil > 7.5

N - NATIVE to the Pacific Northwest

USES

SOIL IMPROVEMENT/SITE PROTECTION
- C - Cover crop and/or green manure
- E - Erosion control
- G - Ground cover and/or living mulch
- H - Hedgerow; will create a barrier to stock passage
- +N - Nitrogen fixing
- W - Windbreak component

EDIBLE
- f - Food
- s - Seasoning/tea/coffee/wine/beer
- m - Medicinal (ingested)

DOMESTIC ANIMAL FORAGE
- B - Bees
- F - Fowl
- H - Hoofed stock
- P - Pigs
- O - Other

WILDLIFE
- √ - Habitat and/or forage value
- ★ - High habitat and/or forage value
- I - Insectary plant

COPPICING (refers to shrubs and trees only)
- c - Can be coppiced for fuel
- C - Can be coppiced for fuel, stakes, poles and fenceposts

BUILDING MATERIAL (refers to shrubs and trees only)
- B - Building construction
- F - Furniture and/or tool construction
- P - Poles, fenceposts
- S - Specialized uses

OTHER USES
- b - Biocontrol
- c - Chemicals, oils, cosmetics
- d - Dye
- e - Ethanol (high feasibility)
- f - Fiber, including weaving materials
- h - Medicinal for hoofed stock (ingested or external)
- l - Landscaping
- m - Medicinal (external, for humans)
- o - Ornamental
- r - Rootstock
- s - Fragrant flowers or foliage
- x - Other (see Comments)

An abbreviated running key can be found at the bottom of each spread. The plants are listed by scientific and common name in the index in the back of the book.

TREES: Over 50'

		FORM					ADAPTATIONS/TOLERANCES								USES			
		Life Style and Type	Height at Maturity	Climatic Zone	Shade Tolerance	Habitat	Tolerates Poor Drainage	Textural Tolerance	pH Tolerance	Native	Soil Improvement/ Site Protection	Edible	Domestic Animal Forage	Wildlife	Coppicing	Building Material	Other Uses	COMMENTS
Abies amabilis	Pacific silver fir	ET	245'	1-7	○ ◑	lwm-uwm				N				√		BP	fl	Slow-growing, long-lived. Like *A. concolor*, rough construction, but not heavy loads.
Abies concolor	White fir	ET	130-230'	1-7	○ ●	uwm				N				√		BP	flo	Initially fast-growing. Xmas tree. Odorless wood, used for butter tubs, boxes.
Abies grandis	Grand fir	ET	115-300'	1-7	○	lm-um				N		m		√		S	flm o	Fast-growing, long-lived, deep roots. Paper pulp, boxes, crates. Stately.
Abies procera	Noble fir	ET	140-275'	1-7	○	uwm				N				√		B	lo	Slow-growing initially, then fast. Long-lived. Excellent structural timber.
Acer macrophyllum	Bigleaf maple	DT	100'	4-7	○ ◑	lwm-uwm				N	E	f	B	★		FS	fl	Rapid-growing for 50 years, 275 year life span. Maple sugar. Good carving wood.
Acer saccharum	Sugar maple	DT	100'	1-7	○ ◑	lwm			-			f	B	★		FS	lo	Rapid-growing for 35 years, long-lived. Musical instruments, furniture. Fall color.
Ailanthus altissima	Tree of heaven	DT	80'	1-7	○	lmd-umd					EW		B	√ I	c	F	bl	Invasive. Successful in compacted, polluted, infertile urban sites. High fuel production.
Alnus glutinosa	Black alder	DT	70-100'	1-7	○	lwm-uw	√		-		+NW		B	√		F	cdlx	Fast-growing screen, dense from ground up. Inexpensive cabinetry. Shoe soles.
Alnus rhombifolia	White alder	DT	50-90'	1-7	◑ ●	lwm-uw	√	GR		N	+NW		B	√		F		Fast-growing, thicket-forming, drought-tolerant. Rots quickly in soil.
Alnus rubra	Red alder	DT	80'	1-7	○ ◑	lwm-uw	√	GR		N	+N EW	m	BH	√	C	BF PS	d	Fast-growing pioneer tree. 60-80 year life span, good nurse tree. Shallow roots.
Arbutus menziesii	Pacific madrone	ES, T	40-130'	3-7	○ ◑	lwmd-uwmd				N		fm	BH	★		F	clo	Slow-growing to 225 years. Fuel, cabinet-work, woodturning, burlwood, shuttles.
Calocedrus decurrens	Incense cedar	ET	65-225'	1-7	○ ●	umd				N	W			√		BP	lo	Slow-growing to 1,000 years. Compact root system, but windfirm. Durable timber.
Carya laciniosa	Shellbark hickory	DT	60-130'	2-7	○ ◑	lwm	√					f	FP	√		F	fl	Long-lived, slow-growing, bears in 5-15 years. Well developed taproot, as *C. ovata*.
Carya ovata	Shagbark hickory	DT	65-150'	2-7	○	um		H			EW	f	FP	√	C	BFP	fl	As *C. laciniosa*, promising nut tree, top quality tool, fuel wood. Storm tolerant.
Castanea mollissima	Chinese chestnut	DT	40-60'	2-7	○	lmd-umd		S	-		E	f	BH P	√	C	BP	cflo	Staple food tree, resistant to chestnut blight. Coppices for rot resistant posts.
Castanea sativa	Spanish chestnut	DT	40-100'	2-7	○ ◑	lwmd-uwmd			-		E	f	HP	√	C	BP	cflo	Like *C. mollissima*, high quality carbohydrate. Susceptible to chestnut blight.
Castanopsis chrysophylla	Golden chinkapin	ES, T	20-130'	4-7	○ ◑	lm-umd		R		N		f	B	√	C	P	os	20-25' thicket-forming shrub or slow-growing tree. Edible nuts. Low maintenance.
Cedrus atlantica	Atlas cedar	ET	40-130'	2-7	○ ◑	lwm-uwm										B	o	Grows 20' in 15 years, then slow-growing. Deep-rooted, drought tolerant once estab.
Cedrus deodara	Deodar cedar	ET	80-150'	2-7	○ ◑	lmd-umd										B	o	Grows 70' in 40 years. Like *C. atlantica*, needs wind protection. Drought tolerant.
Cedrus libani	Cedar of Lebanon	ET	80-140'	2-7	○	lmd-umd										B	o	Slow-growing. Majestic, picturesque sky-line tree. Pollution intolerant, drought tol.
Chamaecyparis lawsoniana	Port Orford cedar	ET	100-215'	4-6 (7)	○ ●	lwm-uwm				N	W					BF PS	blox	Slow to moderately fast-growing, to 600 years. Durable wood smells like ginger.
Chamaecyparis nootkatensis	Alaska cedar	ET	65-175'	4-6	○ ◑	lwm-uwm		GR		N						BF PS	lo	Slow-growing to 3,000 years. Pest, disease free. Shallow roots. Can be kept small.
Corylus colurna	Turkish hazel	DT	40-70'	2-7	○ ●	lwmd-uwm			-		E	f		√			lo	Full sun for nut production. Nuts inferior to *C. avellana*. Tolerates cities, poor soil.

Cryptomeria japonica	Japanese cryptomeria	ET	50-180'	4-7	○ ◑	lwm-uwm										B	lo	Primary Japanese timber tree. Slow to med. growth, initially fast. Handsome, graceful.
Ginkgo biloba	Ginkgo	DT	40-100'	1-7	○	lwm-uwm						f					dlo	Slow-growing to 3,000 years. Edible nuts, much used in the Orient. Tolerates cities.
Gleditsia triacanthos	Honey locust	DT	30-140'	1-7	○	lwmd-uwm			+		EH	f	BF HP	√		FP	elo	Pods from improved cultivars high in sugar, protein. Suckers from roots.
Juglans nigra	Black walnut	DT	100-150'	1-7	○	lwm		H	+		W	fsm	B	√		BFS	bcd hlmr	Fast-growing, long-lived, wind-firm. Inhibits understory w/root exudate.
Juglans regia	English walnut	DT	50-100'	1-7	○	lwm-uwm			− +			fm	BF HP	√		BF	bcd lmr	Fast-growing when young. Important nut tree. Inhibits understory w/root exudate.
Metasequoia glyptostroboides	Dawn redwood	DT	115-150'	1-7	○ ●	lwm-uwm		H	−			f	H			BP	flo	Fast-growing. Rare, botanically significant, ancient monotypic genus. Edible leaves.
Morus alba	White mulberry	DT	20-80'	1-7	○ ●	lwmd-uwmd		S	− +		W	f	FP O	★		PS	del	Fast-growing, long fruiting season. Silk-worm food, attracts birds. As w/*M. rubra*, wood resists decay. Staple Near East food.
Morus rubra	Red mulberry	DT	40-65'	1-7	○ ●	lwm			− +		W	fsm	FP	★		FPS	efl mo	Fast-growing, short-lived. Superior fruit. Caution—sap and unripe fruit are toxic.
Picea brewerana	Weeping spruce	ET	50-170'	4-7	○ ●	uwmd		R		N				★			lo	Slow-growing, long-lived. Like other *Picea* spp., winter cover for birds.
Picea engelmannii	Engelmann spruce	ET	80-235'	1-6	○ ●	uwmd		R		N				★		BS	cfl	Moderate to fast-growing, long-lived, shallow roots. Used to build gliders.
Picea sitchensis	Sitka spruce	ET	130-215'	4-6	○	lwm-uw	√			N	W	fsm		★		BPS	cfl	Fast-growing, long-lived. Important timber tree, musical instruments, basketry.
Pinus contorta var latifolia	Lodgepole pine	ET	50-150'	1-7	○	lwmd-uwmd	√	GR S		N	EW	m		★		BFP	hmo	Slow-growing to 600 years. Colonizes cut, burned land. Will furnish 9-17' poles.
Pinus lambertiana	Sugar pine	ET	160-270'	1-7	○	ld-um				N	W	fsm		★		B	lo	Slow, then fast growth, long-lived. Deep taproot. Can get white pine blister rust.
Pinus monticola	Western white pine	ET	100-250'	1-7	○ ●	lwmd-uwmd		S		N	E	m		★		B	hlm	Initially fast-growing. Numbers reduced by white pine blister rust. Large taproot.
Pinus ponderosa	Ponderosa pine	ET	100-245'	1-7	○	ld-umd			− +	N	EW	fm		★		BF	hlo	Very fast to moderate growth rate. Dryland windbreak. Free of insect, disease problems.
Pinus radiata	Monterey pine	ET	100-140'	5	○ ◑	lwm-s		S						★		B	c	Grows 50' in 12 years, 150 year life. Very pest-prone, subject to blow-down.
Populus nigra var italica	Lombardy poplar	DT	120'	1-7	○	lwm-uwm	√	H			W		BH	√	c		d	100' in 30 years, short-lived. Invasive, wide-spreading, suckering, thirsty roots.
Populus trichocarpa	Black cottonwood	DT	180'	1-7	○	lwm-uwm	√	GS		N		m	BH	√		PS	cfh ms	40' in 15 years. Colonizes river silt. Easily storm-damaged, not wind-firm, wood brittle.
Prunus avium	Sweet cherry	DT	50-100'	2, 6, 7	○	lwm-uwm			− +			f	BF	★		FS	hor	Longer lived than other *P.* spp. Caution—leaves poisonous to hoofed stock, people.
Prunus serotina	Black cherry	DT	65-130'	1-7	○	lm-um		GS				fsm	B	★		F	o	Climatic zones theoretical. Leaves poisonous. First-rate furniture, tool wood.
Pseudotsuga menziesii	Douglas-fir	ET	160-380'	1-7	○ ◑	lwmd-uwm				N	W	sm		★		BF	dfl mos	Medium growth rate. Wood doesn't warp, takes beautiful finish. Basketry from roots.
Quercus garryana	Oregon white oak	DT	30-90'	4-6	○	lmd-umd				N		fm	HP	★		BF PS	lo	Slow-growing to 500 years. Leaves almost equal to alfalfa as stock feed. Cyclic, low-tannin acorns abundant. Deep roots.
Quercus ilex	Holly oak	ET	40-70'	4-7	○	lwm-s					W	f		★			l	Shrubby in constant ocean wind, moderately fast growth inland. Almost pest-free.
Quercus lobata	Valley oak	DT	65-150'	1-3, 6,7	○ ●	lwm						f	P	★			dlo	Slow to moderately fast-growing to 400 yrs. Abundant, cyclic, low-tannin acorns.

LIFE STYLE AND TYPE: A-Annual, B/A-Biennial grown as annual, P/A-Perennial grown as annual, B-Biennial, P/B-Perennial grown as biennial, HP-Herbaceous perennial, D-Deciduous, E-Evergreen, S-Shrub, V-Vine, T-Tree. SHAPE: cb-Climbing, tr-Trailing, cp-Clumping, sp-Spreading, mt-Mat, up-Upright. SHADE TOLERANCE: ○-Full sun, ◑-Partial Shade, ●-Full shade. HABITAT: l-Lowland, u-Upland, w-Wet, m-Moist, d-Dry, fa-Freshwater aquatic, s-Saltwater margins. DRAINAGE TOLERANCE: √-Poor drainage. TEXTURAL TOLERANCE: G-Gravelly soil, H-Heavy clay soil, R-Rocky soil, S-Sandy soil. pH TOLERANCE: (−)=Acidic soil, (+)=Alkaline soil. NATIVE: N-Native.

SOIL IMPROVEMENT/SITE PROTECTION: C-Cover crop/green manure, E-Erosion control, G-Ground cover/living mulch, H-Hedgerow, +N-Nitrogen fixing, W-Windbreak. EDIBLE: f-Food, s-Seasoning/drink, m-Medicinal (ingested). DOMESTIC ANIMAL FORAGE: B-Bees, F-Fowl, H-Hoofed stock, P-Pigs, O-Other. WILDLIFE: √-Habitat/forage value, ★-High habitat/forage value, I-Insectary plant. COPPICING: c-Coppice for fuel, C-Coppice for fuel/materials. BUILDING MATERIAL: B-Building construction, F-Furniture/tools, P-Poles/fenceposts, S-Specialized. OTHER USES: b-Biocontrol, c-Chemicals/oils/cosmetics, d-Dye, e-Ethanol, f-Fiber, h-Stock medicine, l-Landscaping, m-Medicinal (external), o-Ornamental, r-Rootstock, s-Fragrance, x-Other.

TREES: Over 50'

		FORM					ADAPTATIONS/TOLERANCES					USES						
		Life Style and Type	Height at Maturity	Climatic Zone	Shade Tolerance	Habitat	Tolerates Poor Drainage	Textural Tolerance	pH Tolerance	Native	Soil Improvement/ Site Protection	Edible	Domestic Animal Forage	Wildlife	Coppicing	Building Material	Other Uses	COMMENTS
Robinia pseudoacacia	Black locust	DT	75'	1-7	○	lmd-umd		S	-		+NE W	f	B	✓	C	PS	ds	Edible flowers, but caution—rest of tree poisonous. Suckering, aggressive roots.
Salix alba	White willow	DT	80-100'	1-7	○	lw-uw	✓				E	sm	H	✓	C	PS	cfh lmo	Grows 3-4'/year, short-lived. Medicinal bark compounds closely related to aspirin.
Salix babylonica	Weeping willow	DT	30-100'	1-7	○	lw-uw	✓				E		BH	✓	c		lo	Like *S. alba*, rapidly spreading, branching roots hold streambanks, interfere w/drains.
Sequoia sempervirens	Coast redwood	ET	70-365'	4-7	○◑	lwm				N					C	BF PS	ls	Grows 70-90' in 25 years, long-lived. May get heartrot, otherwise pest-free.
Sequoiadendron giganteum	Giant sequoia	ET	200-345'	1-7	○	lwmd-uwmd									C	BPS	l	First fast-growing, then slow, very long-lived. Largest living thing. Gen. pest-free.
Thuja plicata	Western red cedar	ET	100-250'	1-7	○●	lwm-uwm	✓			N	W	m				BPS	bflm	Primary tree of native people. Slow-growing, long-lived. Wood decays slowly.
Tilia americana	American linden	DT	40-100'	1-7	○●	lm-um		GR				fsm	B	✓	C	FPS	flms	Fast-growing, moderately long-lived, pest problems. Musical instruments, raffia, rope. Edible leaf buds, fruit and flowers.
Tsuga heterophylla	Western hemlock	ET	125-250'	1-7	○●	lwm-uw			-	N		fsm		★		BP	cdf lm	Moderate to fast-growing, shallow-rooted, not wind-firm. Best construction hemlock.

TREES AND SHRUBS: 20-50'

		FORM					ADAPTATIONS/TOLERANCES					USES						
		Life Style and Type	Height at Maturity	Climatic Zone	Shade Tolerance	Habitat	Tolerates Poor Drainage	Textural Tolerance	pH Tolerance	Native	Soil Improvement/ Site Protection	Edible	Domestic Animal Forage	Wildlife	Coppicing	Building Material	Other Uses	COMMENTS
Acer circinatum	Vine maple	DS, V, T	35-50'	1-6	○●	lwm-uwm	✓			N	E		BH	★	C	P	cflo	Excellent soil-binding roots in top 4". Basketry, espalier, household items, paint.
Albizia julibrissin	Silk tree	DT	40-50'	(2-6) 7	○◑	lmd-umd			+		+N	f					o	Likes heat, plant near S, W wall. Flowers in summer, gives filtered shade. Ornamental.
Alnus sinuata	Sitka alder	DS, T	35'	1-7	○	lwm-uwm	✓	GR		N	+NE		BH	✓	c		l	Fast-growing, short-lived. Nurse crop for Doug-fir, fair sheep browse. Colonizer.
Alnus tenuifolia	Mountain alder	DS	35'	1-3	○	uw				N	+N		BH	✓	c			Fast-growing, short-lived. Dense thickets w/willows. Early spring bee pollen for brood.
Amelanchier alnifolia	Saskatoon serviceberry	DS, T	15-40'	1-7	○◑	lwmd-uwmd	✓			N	E	f	BH	★		F	elo	Short dense thickets or ornamental tree. Alternate host to cedar apple fungus.
Amelanchier canadensis	Shadblow serviceberry	DS, T	25-35'	1-4, 6, 7	○●	lwm-uwm						fs	B	★		F	elo	Moderately fast growth. Also hosts cedar apple fungus. Important food of Indians.
Asimina triloba	Pawpaw	DS, T	20-35'	6, 7	○●	lwm-uwm						f	B	✓			cdf	Moderately fast-growing, long root system. Fruits may be toxic to some people.
Castanea pumila	Allegheny chinkapin	DS, T	25'	4-7	○●	lmd-umd	✓	HR S			E	fs	P	✓	C	P	o	Will form dense stands on slopes. Starchy nuts can serve as staple or cocoa substitute.

Celtis reticulata	Netleaf hackberry	DT	35'	1-3	○	umd		GR	+	N	EW	f		✓		P	l	Moderate to slow-growing. Deep-rooted, can be used next to buildings, pavement.
Cercis occidentalis	California redbud	DS,T	25'	2-7	○	ld-umd				N	E	fsm					flo	Long lateral roots, profuse magenta spring flowers. Flowers, buds, pods edible.
Cercocarpus ledifolius	Curlleaf mountain mahogany	ES,T	40'	1-7	○	ud		GR		N	+N			★		S	do	Slow-growing. Dense, hard, fine-grained wood for carving or fuel. Red dye.
Corylus avellana	European filbert	DS	10-30'	1-7	○	lwm-uwm			– +		E	f		✓	C	P	l	Fast-growing, 150 + years. Susceptible to Eastern filbert blight. Needs wind shelter.
Corylus maxima	Giant filbert	DT	30'	2-7	○	lwm-uwm			–			f		✓	C	P	lo	Slow to moderate growth, depending on variety. Like *C. avellana*, self-sterile.
Crataegus columbiana	Columbia hawthorn	DS	30'	1-3	○◑	lmd-umd		S	+	N	EH W	fm	BF H	✓	C	PS	er	These species are alternate hosts for cedar apple rust, are subject to fireblight and disease. All are tough, thorny hedge-row plants. *C. douglassii* produces most abundant, best fruits for human consumption. *C. douglassii* and *C. columbiana* have beautiful hard, white wood w/red heart. *C. monogyna* makes a narrow, dense screen.
C. crus-galli	Cockspur hawthorn	DS,T	35'	1-7	○	lwmd-uwmd		S	+		H	f	B	★			elo	
C. douglasii	Black hawthorn	DS,T	40'	1-7	○◑	lwmd-uwmd		S	+	N	EH W	fm	BF H	★	C	PS	er	
C. monogyna	English hawthorn	DT	40'	1-7	○◑	lwmd-uwmd		S	+		EH	fm	BH	★	C	PS	elo	
Diospyros kaki	Japanese persimmon	DT	30-40'	(4-6) 7	○	lm			– +			f	FH P	✓		F	elo	Both species are difficult to transplant, slope tolerant, have hard, dark wood, fall color, persistent fruit. *D. virginiana* does well in cities. *D. kaki* has numerous cultivars.
Diospyros virginiana	American persimmon	DT	30-130'	3-7	○◑	lwm			– +			f	BF HP	★		FS	elo	
Elaeagnus angustifolia	Russian olive	DS,T	25-65'	1-3, 7	○	lmd-umd		S	+		+NE HW	f	B	★			los	Fast-growing, pest-free. Makes a dense hedge when clipped. Tolerates poor soil.
Ficus carica	Fig	DS,T	15-30'	4-7	○	lmd			– +			f	FP				lo	Fast-growing, 3 years to bearing. Likes warm location, reflected heat. Espalier.
Ilex aquifolium	English holly	ES,T	20-80'	4-7	○●	lwm-uwm					HW	m	BH	★		S	hlo	Caution—berries mildly poisonous. Spiny, tolerates much pruning, makes good hedge.
Laburnum alpinum	Scotch laburnum	DS,T	35'	1-7	○◑	lm-um					+N						los	Caution—all parts poisonous, remove pods around children. Med. growth, short-lived.
Morus nigra	Black mulberry	DT	30-65'	4,5, 7	○●	lwmd-uwmd			– +			fm	BF P	★			el	Mod. growth rate, long-lived. Leafs out late. Attractive, rugged. Tastiest mulberry.
Myrica californica	California wax myrtle	ES,T	30'	4-6	○●	lwm-s	✓	S		N	+NE			✓			lsx	Drought, poor soil tolerant. Well adapted as informal hedge. Source of wax.
Pinus edulis	Pinyon pine	ET	15-40'	2-7	○	lmd-umd		R	– +			f	H	★			o	Very slow-growing, long-lived. Tolerates drought, heat, poor soil. Edible nuts.
Populus tremuloides	Quaking aspen	DT	40-100'	1-7	○	uwmd		HR	+	N	E	m	BH	★	C	PS	fm	Fast-colonizing. Filtered shade supports understory. Habitat, food for wide range of wildlife, including pest organisms.
Prunus americana	American plum	DT	15-35'	1-3	○●	lmd-uwmd		RS	+		EH W	fm	B	✓			os	Fast-growing, short-lived. Thicket-forming, spiny. Pollution intolerant.
P. amygdalus	Almond	DT	20-30'	(4-7)	○◑	lmd-umd			+			f	B				co	Hall's Hardy most promising, has bitter nut. Protected spring location, good air drainage.
P. armeniaca	Apricot	DT	12-25'	1-3, 5-7	○	lwmd-uwmd		RS	– +			f	B	✓		F	clor	Danger of rot in humid, foggy areas. Good air drainage, protected location, espalier.
P. cerasus	Sour cherry	DS,T	65'	1-7	○◑	lwm-uwm	✓		– +			f	BF	★			lor	Poisonous leaves. Fast-growing. Hedge. Easier to grow than sweet cherries.
P. domestica	Common plum	DT	25'	2-7	○	lwmd-uwmd	✓		– +		EW	f	BF	✓		FS	lo	Will form dense thickets. Prune tree. High maintenance. Many varieties, pests.
P. emarginata	Bitter cherry	DS,T	25-100'	1-7	○	lwmd-uwmd				N	E	f		★	c	F		Leaves poisonous to humans, stock. High polish wood. Thicket-forming colonizer.

LIFE STYLE AND TYPE: A-Annual, B/A-Biennial grown as annual, P/A-Perennial grown as annual, B-Biennial, P/B-Perennial grown as biennial, HP-Herbaceous perennial, D-Deciduous, E-Evergreen, S-Shrub, V-Vine, T-Tree. SHAPE: cb-Climbing, tr-Trailing, cp-Clumping, sp-Spreading, mt-Mat, up-Upright. SHADE TOLERANCE: ○-Full sun, ◑-Partial Shade, ●-Full shade. HABITAT: l-Lowland, u-Upland, w-Wet, m-Moist, d-Dry, fa-Freshwater aquatic, s-Saltwater margins. DRAINAGE TOLERANCE: ✓-Poor drainage. TEXTURAL TOLERANCE: G-Gravelly soil, H-Heavy clay soil, R-Rocky soil, S-Sandy soil. pH TOLERANCE: (-)=Acidic soil, (+)=Alkaline soil. NATIVE: N-Native.

SOIL IMPROVEMENT/SITE PROTECTION: C-Cover crop/green manure, E-Erosion control, G-Ground cover/living mulch, H-Hedgerow, +N-Nitrogen fixing, W-Windbreak. EDIBLE: f-Food, s-Seasoning/drink, m-Medicinal (ingested). DOMESTIC ANIMAL FORAGE: B-Bees, F-Fowl, H-Hoofed stock, P-Pigs, O-Other. WILDLIFE: ✓-Habitat/forage value, ★-High habitat/forage value, I-Insectary plant. COPPICING: c-Coppice for fuel, C-Coppice for fuel/materials. BUILDING MATERIAL: B-Building construction, F-Furniture/tools, P-Poles/fenceposts, S-Specialized. OTHER USES: b-Biocontrol, c-Chemicals/oils/cosmetics, d-Dye, e-Ethanol, f-Fiber, h-Stock medicine, l-Landscaping, m-Medicinal (external), o-Ornamental, r-Rootstock, s-Fragrance, x-Other.

TREES AND SHRUBS: 20-50'

		FORM					ADAPTATIONS/TOLERANCES						USES					
		Life Style and Type	Height at Maturity	Climatic Zone	Shade Tolerance	Habitat	Tolerates Poor Drainage	Textural Tolerance	pH Tolerance	Native	Soil Improvement/ Site Protection	Edible	Domestic Animal Forage	Wildlife	Coppicing	Building Material	Other Uses	COMMENTS
Prunus persica	Peach	DS, T	5–55'	1–7	○	lwmd-uwmd		S	– +			f	B	✓		F	lo	Needs good air drainage, tolerates smog. Genetic dwarfs can be grown as hedge to 6'.
P. subcordata	Klamath plum	DS, T	30'	1–3, 6, 7	○	lmd-umd		GH RS		N	EW	f	H	✓				Thicket-forming. Sheep browse, but killed by overgrazing. Tolerates poor soil, slopes.
P. virginiana	Chokecherry	DS, T	15–40'	1–3	○●	lmd-uwmd		S		N	EW	fm	BF P	★			los	Caution—can be fatal to stock if browsed. Unripe fruit toxic to humans. Colonizer.
Pyrus baccata	Siberian crabapple	DT	20–50'	1–7	○	lwm-uwm		H	–			f	BH	★		FS	dors	Several varieties, variable resistance to fireblight, rust. Medium growth, smog tol.
P. communis	Pear	DT	12–65'	1–7	○	lwm-uwm	✓	H	– +			f	BH	★	C	FPS	lor	Long-lived. Quince grafted, dwarfed varieties especially tolerant of wet conditions.
P. fusca	Oregon crabapple	DT	40'	1–7	○	lwm				N		fm	BH	★		F	o	Slow-growing, thicket-forming. Hard, fine-grained wood, excellent fuelwood.
P. pumila	Apple	DT	6–50'	1–7	○	lwm-uwm		H	– +			fs	BH	★		FS	or	3,000+ varieties. Excellent fuelwood, hard, close-grained, red-brown wood for tools.
P. pyrifolia var culta	Japanese sand pear	DT	30–40'	1–7	○	lmd-umd			– +			f	BH				or	Fast-growing. Fireblight resistant. Can be used as rootstock for *P. communis.*
Quercus chrysolepis	Canyon live oak	ES, T	15–100'	5–7	○	uwm		S		N			P	★		BFS	l	Slow-growing. Hard, shock-resistant wood. Dense groundcover when shrubby.
Rhamnus carthartica	European buckthorn	DS, T	20–60'	1–3	○	lmd-umd			– +		HW	m	BH	✓			hl	Both species are alternate hosts to oat rust. Caution—berries, leaves, bark are medicinally cathartic, can be poisonous in excessive doses. Poisonous to stock.
Rhamnus purshiana	Cascara sagrada	DS, T	20–60'	1–7	○●	lwmd-uwmd		GS		N		m	B	★	c		dlmo	
Rhus glabra	Smooth sumac	DS, T	10–25'	1–7	○	lwmd-uwmd			–	N	E	fsm	B	★			lmo	Fast-growing, short-lived pioneer in open areas. Tolerates steep slopes, poor soil.
Salix laseolepis	Arroyo willow	DS, T	16–40'	1–7	○	lwm-uwm				N	E	f	H	✓			f	Fast-growing, short-lived. Roots hold soil along streambanks. Material for basketry.
Sorbus aucuparia	European mountain-ash	DT	20–60'	1–7	○◐	lwm-uwm		GS			W		FH	★	C	FPS	o	Tough, resilient wood readily bent. Susceptible to many pests. Pollarded for stock.
Taxus brevifolia	Western yew	ET	30–80'	1–6	○●	lwm-uwm				N		m				FS	l	Caution—leaves, fruit, twigs poisonous to people, stock. Slow-growing, long-lived.

SHRUBS, TREES AND PERENNIALS: 2-20'

		FORM					ADAPTATIONS/TOLERANCES						USES					
		Life Style and Type	Height at Maturity	Climatic Zone	Shade Tolerance	Habitat	Tolerates Poor Drainage	Textural Tolerance	pH Tolerance	Native	Soil Improvement/ Site Protection	Edible	Domestic Animal Forage	Wildlife	Coppicing	Building Material	Other Uses	COMMENTS
Arbutus unedo	Strawberry tree	ES	8–20'	4–7	○◐	lmd-umd-s						fs					clo	Fruits edible but mealy, available fall and winter. Warmest microclimates. Deer-proof.
Artemesia tridentata	Big sagebrush	ES	15'	1–7	○	ld-ud				N	W	fsm	H	★			f	Fall and winter range plant, possibly toxic to horses, sheep. Bark for weaving.

Asparagus officinalis	Garden asparagus	HP	3-6'	1-7	○	lwmd-uwmd-s		S	- +			fm	B				lo	2-3 years to production, then productive for 10-15 years. Tolerates wasteland soil.
Atriplex canescens	Four-wing saltbush	ES	3-6'	2-7	○	lmd-ud-s			+	N	EW		H	√ I			bl	Nutritious, palatable stock forage, but has caused sheep poisoning, scours in cattle.
Baccharis pilularis	Coyote brush	ES	9"-3½'	5-7	○	lwmd-uwmd-s	√	S		N	E		B	√ I			bl	Attracts wide range of beneficial insects, Sept.-Oct. Dependable, low-maintenance.
Baptisia australis	Blue false indigo	DS	3-6'	1-7	○	lmd-umd					+N		B				o	Drought-resistant, has tap root. Indigo-blue flowers early summer.
Berberis aquifolium	Tall Oregon grape	ES	6"-11'	1-6 (7)	○ ◑	umd				N	W	fsm	B	√			dl	Hedge, screen. Has thick foliage. Edible berries, wine. Yellow dye, tea from roots.
Caragana arborescens	Siberian peashrub	DS, T	20'	1-7	○ ◑	lwmd-uwmd		S			+NH W	f	BF	√			lrs	Fast-growing, heat, cold, drought, salt tolerant. Insect, disease-free.
Castanopsis sempervirens	Bush chinkapin	ES	1-8'	1-7	○	ud		R		N		f					lo	Slow-growing, thicket-forming. Edible nuts. Usually ignored by hoofed stock—concentrated feeding results in poisoning.
Ceanothus americanus	New Jersey tea	DS	3'	4-7	○ ●	lmd-umd					+N	sm	B	√			l	Tastes like Oriental tea, but has no caffeine. Large root system, with taproot.
C. cordulatus	Snowbush	ES	2-5'	1,7	○ ●	umd				N	+NE H	s	H	★			lo	Thicket-forming, can spread to 12'. Thorny. Sheep browse, but grazing-resistant.
C. integirrimus	Deer brush	ES	3-10'	1-7	○	ld-ud				N	+NE	s	BH	★			l	Fast-growing. Important ceanothus browse species, but some consider it poisonous.
C. thyrsiflorus	Blueblossom	ES, T	15'	5	○ ●	lwmd-uwmd				N	+NE	s	B	√			lo	Pioneer, invasive on cut-over sites, burns. Lives 15-20 years. Not wind-firm.
C. velutinus	Snowbush ceanothus	ES	15'	1-7	○	ud		G		N	+NE W		BH	√			lo	Pioneer, thicket-forming in cut, burned areas. Grazed lightly. Butterfly favorite.
Cercocarpus montanus	Plume tree	ES	4-9'	1-7	○ ◑	ld-ud				N	+N		BH	√			l	Palatability good for all stock, but suspected of causing poisoning. Killed by overgrazing.
Cornus mas	Cornelian cherry	DS, T	15-45'	1-6	◑	lm-um			+			fs		√		F	lo	Slow-growing, long-lived. If from seed, may produce only male flowers for 15 years.
Corylus americana	American hazelnut	DS	15'	1-7	○ ◑	lm-um					H	f	H	√	c		fl	Medium to fast growth, thicket-forming. Small, high quality nut with thick shell.
C. cornuta var californica	Western hazel	DS	15-40'	1-7	○ ●	lwm-uwm				N	EH	f	H	★	C	P	fl	Resistant to Eastern filbert blight. All *C.* spp. need full sun for best nut production.
C. heterophylla	Siberian hazel	DS	12-20'	1-7	○ ◑	lwm-uwm			-		EH	f		√	C	P	f	These 3 *C.* spp. are hedgerow plants, but likely to need protection from high winds.
Cydonia oblonga	Quince	DS, T	25'	1-7	○	lmd-um			- +			f	B				ro	Long-lived, will produce for 40 years. Fire-blight susceptible. Doesn't like wind.
Cynara scolymus	Globe artichoke	HP	4-5'	5, 6	○	lwmd			- +			fm					o	Needs rodent protection, is likely to host aphids. Silvery-green foliage.
Cytisus praecox	Warminster broom	ES	3-5'	2-7	○	ld-ud-s		GS			+NE W						lo	Covers hot, dry banks. Tolerates poor soil. Deep-rooted—doesn't transplant.
Elaeagnus commutata	Silverberry	DS	12'	1-3	○ ●	lwmd-uwmd-s		G	+	N	+NE W	f	H	√			los	Drought, heat tolerant, thicket-forming. Edible fruits. Slender, open form.
E. multiflora	Cherry elaeagnus	DS	10'	2-7	○ ◑	lwmd-uwmd-s			+		+NW	f		√			los	Like *E. commutata*, used as a screen, has fragrant flowers. Also heat, drought tol.
E. umbellata	Autumn olive	DS	15'	1-7	○	lwmd-uwmd-s			+		+NE	f		√			los	Fast-growing, makes a high screen. Thicket-forming. Used as N-fixing nurse tree.
Gaultheria shallon	Salal	ES	10'	1-7	○ ●	lmd-umd		S	-	N	E	f m	B	★			lm	Vigorous but not invasive. Ground cover in wide range of conditions. Low maintenance.
Helianthus tuberosus	Jerusalem artichoke	HP	4-15'	1-7	○ ◑	lwmd-uwm	√	S	- +		W	f					el	Fast, temporary windbreak. Inulin carbohydrate edible by diabetics. May be invasive.

LIFE STYLE AND TYPE: A-Annual, B/A-Biennial grown as annual, P/A-Perennial grown as annual, B-Biennial, P/B-Perennial grown as biennial, HP-Herbaceous perennial, D-Deciduous, E-Evergreen, S-Shrub, V-Vine, T-Tree. SHAPE: cb-Climbing, tr-Trailing, cp-Clumping, sp-Spreading, mt-Mat, up-Upright. SHADE TOLERANCE: ○ Full sun, ◑-Partial Shade, ●-Full shade. HABITAT: l-Lowland, u-Upland, w-Wet, m-Moist, d-Dry, fa-Freshwater aquatic, s-Saltwater margins. DRAINAGE TOLERANCE: √-Poor drainage. TEXTURAL TOLERANCE: G-Gravelly soil, H-Heavy clay soil, R-Rocky soil, S-Sandy soil. pH TOLERANCE: (-)=Acidic soil, (+)=Alkaline soil. NATIVE: N-Native.

SOIL IMPROVEMENT/SITE PROTECTION: C-Cover crop/green manure, E-Erosion control, G-Ground cover/living mulch, H-Hedgerow, +N-Nitrogen fixing, W-Windbreak. EDIBLE: f-Food, s-Seasoning/drink, m-Medicinal (ingested). DOMESTIC ANIMAL FORAGE: B-Bees, F-Fowl, H-Hoofed stock, P-Pigs, O-Other. WILDLIFE: √-Habitat/forage value, ★-High habitat/forage value, I-Insectary plant. COPPICING: c-Coppice for fuel, C-Coppice for fuel/materials. BUILDING MATERIAL: B-Building construction, F-Furniture/tools, P-Poles/fenceposts, S-Specialized. OTHER USES: b-Biocontrol, c-Chemicals/oils/cosmetics, d-Dye, e-Ethanol, f-Fiber, h-Stock medicine, l-Landscaping, m-Medicinal (external), o-Ornamental, r-Rootstock, s-Fragrance, x-Other.

SHRUBS, TREES AND PERENNIALS: 2-20'

		FORM					ADAPTATIONS/TOLERANCES						USES					
		Life Style and Type	Height at Maturity	Climatic Zone	Shade Tolerance	Habitat	Tolerates Poor Drainage	Textural Tolerance	pH Tolerance	Native	Soil Improvement/ Site Protection	Edible	Domestic Animal Forage	Wildlife	Coppicing	Building Material	Other Uses	COMMENTS
Hemerocallis fulva	Tawny daylily	HP	3-6'	1-7	○ ◑	lwmd-uwmd		GH RS			E	f					lo	Edible shoots, flowerbuds, flowers, tubers. Low maintenance, relatively pest-free.
Hippophae rhamnoides	Sea buckthorn	DS,T	20'	1-6	○	lwmd-uwmd-s	√	S			+NH	fs				F	lo	Spiny, tough, drought-resistant. Fruits marginally edible raw. Gray-green.
Holodiscus discolor	Ocean spray	DS	3-20'	1-7	○ ◑	lwmd-uwmd-s				N	EW	fm	H	√		F	lmo	Pioneer. Short where dry, taller where moist. Wood hard, straight, slow to burn.
Lavandula angustifolia	English lavender	ES	3-4'	4-7	○	lmd-umd		S	+		E	fsm	BH				chl os	Edible as potherb, tea from flowers, medicinal oil. Relished by sheep, goats.
Lupinus albifrons	Silver lupine	ES	4'	1-3, 6,7	○	lm-um				N	+N			√			o	May be poisonous to stock, particularly sheep. Yellow and violet flowers, Mar.-July.
Lupinus polyphyllus	Big-leaved lupine	HP	3-4'	1-7	○ ◑	lwm-uwm		H		N	+N			√			o	Caution—poisonous to stock, particularly sheep, esp. in summer when seeds are borne.
Myrica pennsylvanica	Northern bayberry	E/DS	9'	4-7	○ ●	lwm		HS	–		+NE	s		√			lx	Deciduous or ptly. evergreen. Berries yield scented wax. Leaves, berries seasoning.
Oemleria cerasiformis	Indian plum	DS	15'	1, 3-7	◑ ●	lwm-uwm				N		f		★			o	Fruits not tasty raw, better cooked, dried. Relished by birds, small mammals.
Phytolacca americana	Pokeweed	HP	4-10'	1-7	○ ●	lwm-uwm			– +			fm		√			cdlm	Shoots a good edible boiled potherb, but caution—roots, stems, seeds poisonous.
Prunus besseyi	Western sand cherry	DS	3-6'	1-3	○	lmd-umd		S		N	EW	f	B	★			lor	Leaves poisonous to stock. Hedge. Dwarfing rootstock for other *Prunus* spp.
P. institia	Damson plum	DS,T	20'	2-7	○	lwm-uwm	√		–		H	f	BF	√			lor	Dense, spiny, can be used as hedge. Fruitful for 30-40 years. Fruits ripen by summer.
P. salicina	Japanese plum	DT	20'	2-7	○	um		H	–			f	B				o	Generally requires more maintenance than European plums. Flowers early.
P. tomentosa	Nanking cherry	DS	6-10'	1-3	○	lmd-umd					W	fs	B	★			lor	Dwarfing rootstock. Alternate host to Western X cherry disease. Leaves poisonous.
Rhamnus californica	California coffeeberry	ES,T	20'	4-7	○ ●	lwmd-uwmd				N		m	BH	★ I			bl	Poor to fair sheep browse. Forage for beneficial insects. Hosts aphids, aphid-eaters.
Ribes aureum	Golden currant	DS	3-6'	1-7	○ ◑	lwm-uwm				N		fs	BH	★			los	Caution—*Ribes* spp. are alternate hosts of white pine blister rust, and are illegal to own in some counties of OR, WA. Will infect white pine if w/in ½ mile. Fruit availability: *R. aureum*—high quality, June/July; *R. cereum*—best as jelly, wine, June—August; *R. divaricatum, sanguineum*—July/August; *R. nigrum* vars., July-Sept.; *R. lacustre*—August. *R. sanguineum* has ornamental pink to crimson flowers in April/May, attracts hummingbirds. *R. aureum* has ornamental fragrant yellow flowers, spring.
R. cereum	Wax currant	DS	1-5'	1-3	○	lmd-umd				N		fsm	BH	★				
R. divaricatum	Common gooseberry	DS	3-10'	1-7	○ ◑	lwm-uwm		H		N		fm	BH	★			hm	
R. lacustre	Prickly currant	DS	2½-5'	1-7	○ ●	lwm-uwm				N		fm	B	★			m	
R. nigrum	Black currant	DS	7'	1-7	○ ●	lwm-uwm	√		– +			fsm	B	★			h	
R. sanguineum	Red-flowering currant	DS	4-12'	4-7	○ ●	lwmd				N		f	BH	★			lo	
Robinia hispida	Rose acacia	DS	6-8'	1-7	○ ●	lmd-umd		S			+NE		B				lo	Spreads by suckers—may be invasive. Best in full sun. Espalier. Tolerates salt, poor soil.

Romneya coulteri	Matilija poppy	HP	4-8'	5-7	○	lwmd-uwmd		GH			E						los	Excellent erosion control. Caution-invasive, spreads by underground roots.
Rosa canina	Dog rose	DS	4-10'	1-7	○	lwmd-uwmd					EH	fsm	BH	★		F	hors	These 3 spp. all have edible flower petals, hips. *R. rugosa* bears largest hips with
R. nutkana	Wild rose	DS	2-5'	1-7	○	lwm-uwm				N	E	fsm	BH	★			hm os	greatest amount of Vit. C per unit weight, bears most profusely, makes an excellent
R. rugosa	Ramanas rose	DS	3-8'	1-7	○	lwmd-uwmd-s		HS			EH W	fs	BH	★			hlos	hedge. *R. canina* is a hardy rootstock for other *R.* spp. *R. nutkana* best stock forage.
Rubus procerus	Himalayan blackberry	DS	15'	4-6	○◑	lwmd-uwmd			− +		EH	f	BH	★			e	Covers ground fast with 20-30' long canes. Shoots edible. Caution—vigorous, invasive.
R. idaeus var strigosus	Red raspberry	DS	4-6'	4-6	○	lwm-uwm		HR	− +	N	E	fsm	BH	√			eh	As with all *R.* species, secondary palatability to stock. *Verticillium* wilt susceptible.
R. laciniatus	Evergreen blackberry	ES	4-8'	4-6	○	lwmd-uwmd			− +		EH	fs	BH	★			el	Easier to contain than other blackberries, most pest-free. Trellis cover. Shoots edible.
R. occidentalis	Black raspberry	DS	3-6'	4-6	○◑	lwm-uwm			− +	N	E	f	BH	√			e	*Verticillium* wilt susceptible. As other *R.* spp., used as erosion control groundcover.
R. spectabilis	Salmonberry	DS	3-9'	1-7	○◑	lwm-uwm			− +	N	E	fsm	BH	★			be	Shoots edible in February. Caution—invasive. Inhibits growth of nearby plants.
Sambucus canadensis	American elder	DS	5-12'	1-7	○●	lwm-uwm	√		− +			fsm	BH	★			elo	Caution—roots, stems, leaves, unripe fruit poisonous to humans, possibly stock.
Sambucus caerulea	Blue elderberry	DS, T	6-50'	1-7	○◑	lwm-uwm				N	W	fsm	BH	★			l	Caution—parts of this plant may be poisonous to humans, stock. Fruit, flowers edible.
Scirpus acutus	Viscid bulrush	HP	3-9'	1-7	○	lw-uw fa-s	√		+	N		f		★			cfl	Prefers water 1-4" deep. Mat-weaving, thatch, baskets, shipping nursery stock.
Sheperdia canadensis	Canadian buffaloberry	DS	3-12'	1-7	○●	lwm-uwm				N	+N	fs	H	√				Fruits contain saponin, a frothing agent. Edible cooked, whipped. Ferments quickly.
Symphoricarpos albus	Common snowberry	DS	2-6'	1-7	○●	lwmd-uwmd		H	+	N	E	m	BH	★ I			lo	Fast-growing. Thicket-forming, ground cover. Spreads by suckers, may be invasive.
Syringa vulgaris	Common lilac	DS	8-30'	1-7	○◑	lwmd-uwmd					W				c	F	clos	Slow to med. growth rate, long-lived. Root crown sprouts, but spreads slowly. Dense.
Typha latifolia	Broad-leaved cattail	HP	3-12'	1-7	○●	lw-uw fa	√			N	E	f		★			flox	Parts edible year-round. Weaving material. Down used as tinder. Caution—poisonous to stock if eaten in quantity.
Urtica dioica	Stinging nettle	HP	2-9'	1-7	○●	lwm-uwm	√		− +	N		fsm	FH				cdf hmx	Leaves high in protein, iron. Fiber, rennet source. Phosphorus accumulator. Caution—painful sting on contact until cooked.
Vaccinium corymbosum	Highbush blueberry	DS	4-12'	4-6 2,3,7	○◑	lwm-uwm	√	S	−			f	B	★			elo	Peat, slope, smog tolerant. Disease, insect susceptible. Med. textured hedge, fall color.
V. membranaceum	Thin leaf huckleberry	DS	1-3'	1-4	○◑	uwm				N		f	BH	★				As with *V. corymbosum*, full sun best for berry production. Fruit available July-Sept.
V. ovatum	Evergreen huckleberry	ES	2-12'	4-7	○●	lwm-uwm		S		N		f	B	★			lox	Tasty berries, produced best in partial shade. Branches used by florists.
V. parvifolium	Red huckleberry	DS	3-12'	2-7	◑●	lwm-uwm			− +	N	E	fsm	BH	★			lo	Grows in disturbed areas on rotting wood, especially cedar.
Viburnum trilobum	Cranberry bush	DS	8-12'	1-7	○●	lwm-uwm						fs	B				elo	Medium growth rate. Fruit edible raw, in jelly, syrup, wine, September-October.

LIFE STYLE AND TYPE: A-Annual, B/A-Biennial grown as annual, P/A-Perennial grown as annual, B-Biennial, P/B-Perennial grown as biennial, HP-Herbaceous perennial, D-Deciduous, E-Evergreen, S-Shrub, V-Vine, T-Tree. SHAPE: cb-Climbing, tr-Trailing, cp-Clumping, sp-Spreading, mt-Mat, up-Upright. SHADE TOLERANCE: ○-Full sun, ◑-Partial Shade, ●-Full shade. HABITAT: l-Lowland, u-Upland, w-Wet, m-Moist, d-Dry, fa-Freshwater aquatic, s-Saltwater margins. DRAINAGE TOLERANCE: √-Poor drainage. TEXTURAL TOLERANCE: G-Gravelly soil, H-Heavy clay soil, R-Rocky soil, S-Sandy soil. pH TOLERANCE: (-)=Acidic soil, (+)=Alkaline soil. NATIVE: N-Native.

SOIL IMPROVEMENT/SITE PROTECTION: C-Cover crop/green manure, E-Erosion control, G-Ground cover/living mulch, H-Hedgerow, +N-Nitrogen fixing, W-Windbreak. EDIBLE: f-Food, s-Seasoning/drink, m-Medicinal (ingested). DOMESTIC ANIMAL FORAGE: B-Bees, F-Fowl, H-Hoofed stock, P-Pigs, O-Other. WILDLIFE: √-Habitat/forage value, ★-High habitat/forage value, I-Insectary plant. COPPICING: c-Coppice for fuel, C-Coppice for fuel/materials. BUILDING MATERIAL: B-Building construction, F-Furniture/tools, P-Poles/fenceposts, S-Specialized. OTHER USES: b-Biocontrol, c-Chemicals/oils/cosmetics, d-Dye, e-Ethanol, f-Fiber, h-Stock medicine, l-Landscaping, m-Medicinal (external), o-Ornamental, r-Rootstock, s-Fragrance, x-Other.

VINES

		FORM						ADAPTATIONS/TOLERANCES				USES					COMMENTS
		Life Style and Type	Shape	Length at Maturity	Climatic Zone	Shade Tolerance	Habitat	Tolerates Poor Drainage	Textural Tolerance	pH Tolerance	Native	Soil Improvement	Edible	Domestic Animal Forage	Wildlife	Other Uses	
Actinidia chinensis	Kiwi	DV, S	cb	30'	4-7	○ ◑	lmd-um						f	B		l	Needs sunny, hot, protected location for fruiting. Moist soil. 4-5 years to crop. Vigorous.
Akebia quinata	Fiveleaf akebia	E, DV	cb	15-40'	1-7	○ ●	lwm-uwm						f			l	Slow-growing. Needs warm, humid site against S, E wall, partial shade for fruiting.
Humulus lupulus	Common hop	HP	cb	15-25'	1-7	○	lwm-uwm			− +			fsm	H		dfh lsx	Very fast-growing shade for greenhouses, buildings. Potherb, edible flowers, seeds. Basketry, textiles, pulp.
Passiflora incarnata	Maypop	HP	cb, tr	20-30'	1-7	○	lmd-umd					EG	fm	B		lo	Roots need protection in cold winters. Arbor cover, holds slope soil. 3" flowers.
Phaseolus coccineus	Scarlet runner bean	P/A	cb	10'	1-7	○	lwmd-uwmd		S	− +		+N	f			lo	Young pods, or pink and purple streaked beans edible. Covers fences. Scarlet bloom.
Polygonum aubertii	Silver lace vine	DV	cb, tr	20-60'	1-7	○ ◑	lm-um-s			+		EG	f	B	I	bl	White flowers July-Sept. attract beneficial insects. Can cover 100 sq.'/season. Arbors.
Tropaeolum majus	Garden nasturtium	P/A	cb, tr	6'	1-7	○ ◑	lwm-uwm		S			E	fsm			lo	Beautiful leaves, flame flowers, cut flowers. Also dwarf clumping form, see Ground Layer.
Vitis labrusca	Fox grape	DV, S	cb	3-12'	1-7	○	lwmd-uwmd			− +			fs	B	★	lr	Human and wildlife uses incompatible. Deer browse can damage, kill plants. Ripening fruit will attract bird flocks to site. *V. lab-*
Vitis vinifera	European grape	DV, S	cb	3-12'	3, 4, 6, 7	○	lmd-umd		GR	− +			fs	B	★	l	*rusca* is phylloxera-resistant, used as rootstock for *vinifera* vars. *V. vinifera* leaves edible cooked. Espalier, lives 50+ yrs.
Wisteria floribunda	Japanese wisteria	DV, S, T	cb	30-500'	1-7	○ ◑	lwm-uwm							B		los	Long-lived. Shape is adaptable. Landscaping, shade. Beautiful, very fragrant flowers.

BAMBOO

Except where noted, all the following species are running bamboos, tolerate full sun and partial shade, are used for erosion control, windbreaks, food plants, fowl forage, tools, stakes and poles. All are perennial grasses.

		FORM		ADAPTATIONS/ TOLERANCES			USES		COMMENTS
		Maximum Stem Diameter	Maximum Height	Climatic Zone	Habitat	pH Tolerance	Fuel	Structural Timber	
Arundinaria amabilis	Tonkin cane	$2\frac{1}{2}$"	40'	5	lwm-uwm	−		✓	Strong, straight, long internodes. Commercially valuable for split fishing rods.
Arundinaria simonii	Madake Simon bamboo	$1\frac{1}{4}$"	25'	5-7	lwm-uwm	−			Spreads at a moderate rate.
Bambusa textilis		2"	50'	5	lwm-uwm	−		✓	Clumping species. Split stems used in handicrafts.
Phyllostachys aurea	Golden bamboo	$1\frac{3}{4}$"	20'	4-7	lwm-uwm	− +			Withstands some drought. Dense, not too invasive. Strong, used for fishpoles, stakes. Most abundant bamboo in PNW.
P. aureosulcata	Yellow-groove bamboo	$1\frac{1}{2}$"	30'	3-7	uwm	− +			Very hardy. Invasive. Stems useful but of inferior quality.
P. bambusoides	Madake, Giant timber bamboo	$5\frac{3}{4}$"	70'	4-7	uwm	− +	✓	✓	Largest found in the Pacific Northwest. Important construction material in Japan, China.
P. dulcis	Sweetshoot bamboo	$2\frac{1}{2}$"	40'	3-7	uwm	− +			Invasive. Most important edible bamboo of Central China. Used in artisanry, but stem strength inferior.

P. henonis	*P. nigra*, cv. *henon* Henon bamboo	3"	50'	4-7	uwm	− +	√	√	Utensils, construction.
P. meyeri	Meyer bamboo	2"	35'	4-7	uwm	− +			Invasive. Strong stems.
P. nigra	Black bamboo	$1\frac{1}{4}$"	25'	4-7	uwm	− +			Very strong, hard. Artisanry, cabinetry. Widely used as an ornamental.
P. nuda		$1\frac{1}{2}$"	35'	3-7	lwm-uwm	− +			Inferior quality stems.
P. pubescens	Moso	5"	70'	5-7	lwm-uwm	− +	√	√	Major edible shoot source in Japan. Hard to establish, propagate. Good structural material. Presently unavailable in PNW.
P. rubromarginata		1"	30'	3-7	lwm-uwm	− +			
P. viridi glaucescens		2"	35'	5-7	lwm-uwm	− +		√	
P. viridis		$3\frac{1}{4}$"	50'	3-7	lwm-uwm	− +	√	√	Invasive. Artisanry, light construction. Good technical quality.
P. vivax	Smoothsheath	$3\frac{3}{4}$"	45'	3-7	lwm-uwm	− +	√	√	Rare. Artisanry.
Sasa palmata		$\frac{1}{2}$"	12'	4-7	uwm	− +			Aggressive spreader. Good erosion control plant. Big leaves, provides stock forage. Not used for edible shoots. Not a windbreak plant.
Semiarundinaria fastuosa	Narihira bamboo	$1\frac{1}{4}$"	25'	4-7	lwm-uwm	− +			Spreads slowly, makes a dense hedge, windbreak. Stately. Artisanry.

GROUND LAYER PLANTS: Annuals & Low Perennials

		FORM: Life Style and Type	Shape	Height at Maturity	ADAPTATIONS/TOLERANCES: Climatic Zone	Shade Tolerance	Habitat	Tolerates Poor Drainage	Textural Tolerance	pH Tolerance	Native	USES: Soil Improvement	Edible	Domestic Animal Forage	Wildlife	Other Uses	COMMENTS
Abutilon theophrasti	Indian mallow	A	up	6'	1-7	○										f	Stems yield fiber for jute, fabric, carpets. Paper fiber. Weedy, invasive.
Achillea millefolium	Common yarrow	HP	cp, up	3'	1-7	○	lmd-umd						sm	H	I	hlsx	Medicinal forage for stock, esp. sheep. Traditional I Ching stalks. Drought resistant.
Agastache foeniculum	Anise hyssop	HP	up	3'	1-7	○	lmd-umd				N		fs	B			Short-lived perennial, 3 years or so. Bees prefer this plant over others, June-Sept.
Allium acuminatum	Rock onion	HP, P/A	up	9"	1-5	○ ◑	ld-ud		R		N		fsm				Food of native peoples-eaten with Camas. Attractive rose flowers, May-June.
A. ampeloprasum var *porrum*	Leek	B, B/A	up	2-3'	1-7	○ ◑	lwmd-uwmd			− +			fs			b	Slow-growing, extremely winter hardy. Repels insects, may be naturalized.
A. cepa var *viviparum*	Egyptian top set onion	HP	cp, up	2-4'	1-7	○	lwmd-uwmd						fs				Self-sows. Hardy to -25° F. Disease-resistant. Tops, bulblets edible. Perennial onion.
A. fistulosum	Scallions	P/A, P/B, HP	up	1'	1-7	○	lm-um						fs				Non-bulbing, clump-forming green onion. Winter hardy.

LIFE STYLE AND TYPE: A-Annual, B/A-Biennial grown as annual, P/A-Perennial grown as annual, B-Biennial, P/B-Perennial grown as biennial, HP-Herbaceous perennial, D-Deciduous, E-Evergreen, S-Shrub, V-Vine, T-Tree. SHAPE: cb-Climbing, tr-Trailing, cp-Clumping, sp-Spreading, mt-Mat, up-Upright. SHADE TOLERANCE: ○-Full sun, ◑-Partial Shade, ●-Full shade. HABITAT: l-Lowland, u-Upland, w-Wet, m-Moist, d-Dry, fa-Freshwater aquatic, s-Saltwater margins. DRAINAGE TOLERANCE: √-Poor drainage. TEXTURAL TOLERANCE: G-Gravelly soil, H-Heavy clay soil, R-Rocky soil, S-Sandy soil. pH TOLERANCE: (-)=Acidic soil, (+)=Alkaline soil. NATIVE: N-Native.

SOIL IMPROVEMENT/SITE PROTECTION: C-Cover crop/green manure, E-Erosion control, G-Ground cover/living mulch, H-Hedgerow, +N-Nitrogen fixing, W-Windbreak. EDIBLE: f-Food, s-Seasoning/drink, m-Medicinal (ingested). DOMESTIC ANIMAL FORAGE: B-Bees, F-Fowl, H-Hoofed stock, P-Pigs, O-Other. WILDLIFE: √-Habitat/forage value, ★-High habitat/forage value, I-Insectary plant. COPPICING: c-Coppice for fuel, C-Coppice for fuel/materials. BUILDING MATERIAL: B-Building construction, F-Furniture/tools, P-Poles/fenceposts, S-Specialized. OTHER USES: b-Biocontrol, c-Chemicals/oils/cosmetics, d-Dye, e-Ethanol, f-Fiber, h-Stock medicine, l-Landscaping, m-Medicinal (external), o-Ornamental, r-Rootstock, s-Fragrance, x-Other.

GROUND LAYER PLANTS: Annuals & Low perennials

		FORM						ADAPTATIONS/TOLERANCES					USES				
		Life Style and Type	Shape	Height at Maturity	Climatic Zone	Shade Tolerance	Habitat	Tolerates Poor Drainage	Textural	pH Tolerance	Native	Soil Improvement	Edible	Domestic Animal Forage	Wildlife	Other Uses	COMMENTS
A. sativum	Garlic	P/A, P/B, HP	up	2-3'	1-7	○	lmd-umd			− +			fsm	FH		bhm	Fungicide in feed for fowl, medicinal forage component for stock, but will flavor milk.
A. schoenoprasum	Chives	HP	cp	1½'	1-7	○ ◑	lwm-uwm		G	− +	N		s			l	Used as perennial border, evergreen in mild winter areas, can be brought indoors.
A. tuberosum	Garlic chives	HP	cp	2½'	1-7	○	lmd-umd			− +			s			bl	Masks presence of plants to insects w/in 3'. More likely than chives to go winter dormant.
Amaranthus hypochondriacus	Prince's feather	A	up	3-8'	2-7	○ ◑	lwmd-uwmd			− +		C	fm	FHP	✓	o	Needs 90-day growing season. Weed tolerant. High protein grain, silage crop.
A. cruentus	Purple amaranth	A	up	6'	2-7	○ ◑	lwmd-uwmd			−		C	f	HP	✓	o	Like *A. hypochondriacus*, indicates potassium, phosphorus in soil.
A. retroflexus	Pigweed	A	up	1-10'	1-7	○	lwmd-uwm						f	H	★		Nutritious green. Can become toxic in quantity if grown on nitrate-fertilized land.
Arachis hypogaea	Peanut	A	cp	10"-1½'	2, 3, 7	○	lwmd		S	− +		+NC	f	HP	✓		Needs 150-day growing season, irrigation. Breeding being done for northern strains.
Armoracia rusticana	Horseradish	HP	up	2-4'	1-7	○	lwm-uwm	✓		−			fsm			m	Best in cool climates, cool seasons. Besides root, leaves edible raw, cooked.
Asarum caudatum	Wild ginger	HP	tr	7-10"	1-7	◑ ●	lwm-uwm				N	G	sm			lm	Handsome groundcover, forest understory plant. Ginger substitute.
Asclepias cornuti	Common milkweed	HP	up	3-6'	1	○							s	B		cf	Grows on poor soil. Stem fiber mixed in fabric, used in paper. Silky material from pods.
Asclepias incarnata	Swamp milkweed	HP	cp, up	5'	1-7		lw	✓					f	B		cf	Grows on flood-prone bottomlands. Replant every 4-5 years. Strong stem fibers.
Avena sativa	Oats	A	up	3-4'	1-7	○	lwm-uwm			− +		C		H			Will mature in 90-110 days. Effective green manure on land too poor for clover.
Balsamorhiza sagittata	Arrowleaf balsamroot	HP	cp	1-2'	1-3, 7	○	lmd-umd		RS		N		f				All plant parts edible. Roots contain inulin, carbohydrate usable by diabetics.
Berberis nervosa	Oregon grape	ES	mt, up	1-2½'	1-7	○ ◑	lmd-umd		RS		N	G	fs	B	✓	dl	Spreads by suckers, outcompetes weeds, needs no pruning. Fruit edible raw, as wine.
Beta vulgaris var cicla	Swiss chard	B/A, B	cp, up	2-5'	1-7	○	lwmd-uwmd						f			o	Roots, to 6', will open subsoil. Self-seeding. Ruby chard spectacular. Mulch to overwinter.
Borago officinalis	Borage	A	up	1-3'	1-7	○ ●	lwmd-uwmd			− +			fsm	BH		hox	Generally pest, disease-free. Tolerates poor soil. Self-seeding. Flowers in potpourri.
Brassica napus	Rape	A, B	cp, up	3'	1-7	○	lwm-uwm		S	− +		C	f	H		cf	Cool growing season, matures 8-10 weeks. Late season forage. Lubricant, cooking oil.
B. nigra	Black mustard	A	up	8'	1-7	○	lwmd-uwmd	✓		− +		C	fsm	BH	✓	hm	Edible greens, root and flowers May-Oct. Smog, weed tolerant. Honey thin, ferments.
B. oleracea var acephala	Kale	B/A, B	cp, up	3'	1-7	○	lwmd-umw			− +		C	f	H			Very nutritious fall, winter green. Good cow forage. Susceptible to *Brassica* pests.
B. oleracea var gemnifera	Brussels sprouts	B/A, B	up	3'	1-7	○	lwmd-umw		HS	− +			f				Edible throughout winter, zones 4-7. Smog tolerant. Susceptible to *Brassica* pests.
B. oleracea var italica	Italian broccoli	P/A, P/B	cp, up	3'	1-7	○	lwmd-umw		HS	−			f				Varieties w/different soil drainage adaptations available. Plant to mature in winter.
B. rapa var septiceps	Turnip	B/A	cp	3'	1-7	○	lwmd-umw			− +		C	f	BHP			Matures quickly, then becomes woody. Sow early spring, late summer.
Calendula officinalis	Calendula	A	up	2-3'	1-7	○	lwmd-uwmd		S	− +			fm	H		hlmo	Potherb, edible flowers. Goat, sheep forage. Blooms spring to frost. Flower borders.

Calochortus nitidus	Big-pod mariposa lily	HP	up	6"–1¼'	1–3	○	ld				N		f			o	Edible bulb. Lavender flower. Dwindling in wild—should not be wild collected.
Camassia quamash	Common camas	HP	up	1–2'	1–7	○ ◑	lwm–umw	√	H		N		f	H	√	o	Staple food for native peoples. Caution—very similar to death camas. Make positive identification before collecting, transplant.
Cannabis sativa	Hemp	A	cp, up	3–10'	1–7	○	lwmd–uwmd		S	– +			m	F		cfm	Multiple-use medicinal and fiber plant. Can be grown only with govt. permit in the US.
Carthamus tinctorius	Safflower	A	up	1–3'	1–7	○	lwmd–uwmd	√	S	– +			fsm	B		cdo	Lamp, industrial, cosmetic, edible oil. Saffron substitute. Dried or cut flowers.
Ceanothus prostratus	Mahala mat	ES	sp, mt	1–2"	1–7	○	umd				N	+NG		B	√	l	Attractive evergreen groundcover, flowers white to dark blue, nectar source, May–July.
Chenopodium album	Lamb's Quarters	A	cp, up	2–10'	1–7	○ ◑	lwmd–uwmd			– +			f		★		Self-sowing camp follower on disturbed land. More protein than spinach, calcium-rich.
Chlorogalum pomeridianum	Soaproot	HP	up	2–4'	1–3, 7	○ ◑	ld–umd				N		f			c	Bulb edible, also used for soap, shampoo. Edible shoots, leaves. Stupefies stream fish.
Cichorium intybus	Chicory	HP	up	2–6'	1–7	○	lmd–um		S	– +			fsm	H		mo	Edible leaves. Root makes coffee-like beverage. Mineral-rich herbal ley component with deep taproot. Tolerates poor soil, slopes.
Claytonia lanceolata	Lance-leaved springbeauty	HP	cp	1'	1–3	○	uwmd				N		f		√	o	Bulb, stem, leaves, flowers edible raw or cooked. White to red flowers May–July.
Claytonia siberica	Siberian montia	A, HP	cp	1¼'	1–7	◑ ●	lwm–uwm						fm		√	m	Edible bulb. Used for internal and external native medicine. Lowland to mid-mountain.
Cornus canadensis	Bunchberry dogwood	HP	tr, sp	4–9"	1–6	◑ ●	lwm–uwm				N	G	fm		√	lox	Needs humus, woodland soil. Groundcover—creeping roots. Feathered bark toothbrush.
Dryas octopetala	Mountain avens	ES	mt	3–4"	1–6	○ ◑	uwmd				N	+NG	s			lo	Pest-free ground cover. Transplants poorly once established. White flowers June–July.
Epilobium angustifolium	Fireweed	HP	up	3–8'	1–7	○	lwm–uwm				N	E	fs	BH	★	f	Invasive colonizer on disturbed land. Some people allergic—eat small amount first time.
Equisetum arvense	Common horsetail	HP	up	3'	1–7	○ ◑	lwmd–uwm		GS		N		fm		√	hmx	Caution—both spp. poisonous to stock, esp. young horses, though *E. arvense* is used
Equisetum hyemale	Common scouring rush	EHP	up	4'	1–7	○ ◑	lw–uw				N		f		√	mx	medicinally. Both used as an abrasive tool. Invasive on wetlands.
Eriogonum spp.	Wild buckwheats	A, B, HP	cl, sp mt	2'	1–3, 5	○	lmd–ud		GR S		N	E	f	B	√ I	bo	15+ spp. of wild buckwheat grow in the PNW. One—*E. latifolium*—grows wild in zone 5. The rest are generally mountain & dryland plants.
Eurotia lanata	Winterfat	DS	cl	2'	1–3, 7	○	ld–ud			+	N			H	★		Valuable fall, winter forage for cattle and sheep. Rapid-growing, deep root system.
Fagopyrum esculentum	Buckwheat	A	up, sp	3–4'	1–7	○	lwmd–uwmd		HS	– +		C	f	B	√ I	b	Fast-growing, matures in 60 days, outcompetes weeds, grasses. Deep, extensive roots.
Foeniculum vulgare	Fennel	B, B/A, HP	up	3–8'	1–7	○ ●	lwmd–uwmd			– +			fsm	BH	√ I	bch ms	Tolerates drought, poor soil, weeds. Tonic forage, but cows may overeat. Flavoring.
Fragaria chiloensis	Chilean strawberry	HP	cl	1'	4–7	○ ◑	lwm–s		S	–	N	EG	fs	B	√	l	Both *F.* spp. are ancestors of cultivated strawberry. Both will grow on sand, have
Fragaria virginiana	Virginia strawberry	HP	cl	1'	1–7	○ ●	lwmd–uwmd		S	–	N	EG	fs	B	√	l	4–5 yr life. *F. virginiana* spreads rapidly, is weed tolerant.
Fritillaria lanceolata	Checker lily	HP	up	2½–3'	1–7	○ ◑	lwm–uwm				N		f			o	Slow-growing. Edible bulb. Becoming rare in the wild, purchase from bulb dealer.
Fritillaria pudica	Yellow fritillary	HP	up	6"–1'	1–7	○ ◑	lmd–umd				N		f			o	Bulb edible. Disappearing from former range. Yellow-red flower, April–June.

LIFE STYLE AND TYPE: A-Annual, B/A-Biennial grown as annual, P/A-Perennial grown as annual, B-Biennial, P/B-Perennial grown as biennial, HP-Herbaceous perennial, D-Deciduous, E-Evergreen, S-Shrub, V-Vine, T-Tree. SHAPE: cb-Climbing, tr-Trailing, cp-Clumping, sp-Spreading, mt-Mat, up-Upright. SHADE TOLERANCE: ○-Full sun, ◑-Partial Shade, ●-Full shade. HABITAT: l-Lowland, u-Upland, w-Wet, m-Moist, d-Dry, fa-Freshwater aquatic, s-Saltwater margins. DRAINAGE TOLERANCE: √-Poor drainage. TEXTURAL TOLERANCE: G-Gravelly soil, H-Heavy clay soil, R-Rocky soil, S-Sandy soil. pH TOLERANCE: (-)=Acidic soil, (+)=Alkaline soil. NATIVE: N-Native.

SOIL IMPROVEMENT/SITE PROTECTION: C-Cover crop/green manure, E-Erosion control, G-Ground cover/living mulch, H-Hedgerow, +N-Nitrogen fixing, W-Windbreak. EDIBLE: f-Food, s-Seasoning/drink, m-Medicinal (ingested). DOMESTIC ANIMAL FORAGE: B-Bees, F-Fowl, H-Hoofed stock, P-Pigs, O-Other. WILDLIFE: √-Habitat/forage value, ★-High habitat/forage value, I-Insectary plant. COPPICING: c-Coppice for fuel, C-Coppice for fuel/materials. BUILDING MATERIAL: B-Building construction, F-Furniture/tools, P-Poles/fenceposts, S-Specialized. OTHER USES: b-Biocontrol, c-Chemicals/oils/cosmetics, d-Dye, e-Ethanol, f-Fiber, h-Stock medicine, l-Landscaping, m-Medicinal (external), o-Ornamental, r-Rootstock, s-Fragrance, x-Other.

GROUND LAYER PLANTS: Annuals & Low Perennials

		FORM						ADAPTATIONS/TOLERENCES					USES				COMMENTS
		Life Style and Type	Shape	Height at Maturity	Climatic Zone	Shade Tolerance	Habitat	Tolerates Poor Drainage	Textural Tolerance	pH Tolerance	Native	Soil Improvement	Edible	Domestic Animal Forage	Wildlife	Other Uses	
Galium aparine	Cleavers	A	cb,tr	2-6'	1-7	○ ◑	lwmd-uwmd				N		fsm	F		fhm	Weedy. Seeds are coffee substitute. Goose forage. Pleasant-smelling mattress stuffing.
Gaultheria procumbens	Wintergreen	ES	mt	6"	2-7	○ ●	lmd-umd		S	-		G	sm			clm	Slope, smog tolerant evergreen ground-cover. Wintergreen oil. Medicinal.
Genista pilosa	Broom	D,ES	sp	1½'	2-7	○	lwmd-uwmd-s		R			+NE G				lo	Fairly fast-growing but not invasive. Deciduous, but acts as evergreen. Grey-green.
Gossypium hirsutum	Upland cotton	A	cp	5'	2-7	○	lwm-uwm		S	- +			m	B		f	Important fiber plant. Bark of root will induce uterine contractions.
Helianthus annuus	Common sunflower	A	up	3-12'	1-7	○	lwmd-uwmd	✓	S	- +	N		f	FH	★	cdlo	Quality protein seed. Deep taproot, long laterals. Tolerates poor soil. Yellow dye, edible and fuel oil, varnish, paint.
Hordeum vulgare	Barley	A	up	1½-3'	1-7	○	lwmd-uwmd	✓		- +		C	fsm	FH	★	m	Adapted to northern, high-altitude conditions. Winter cover crop in mild areas.
Lemna minor	Common duckweed	HP	mt	<1"	1-7	○	fa				N			FHP	✓	o	In quiet water. Experimental chicken, pig, stock feed. Duck, goose, fish forage.
Linum usitatissimum	Flax	A	up	2-4'	1-7	○	lwmd-uwmd			- +		C	fm	BHP		bcf hm	Pre-WWII Willamette Valley fiber industry. Repels potato beetle, carrot fly. Conditions soil. May be poisonous to stock.
Lotus corniculatus	Bird's-foot trefoil	HP	tr	3'	1-7	○	lwmd-uwmd	✓	HS	- +		+NC EG		BH	✓	l	Will grow in marginal soils, but can't tolerate initial competition. Non-bloating stock forage. Lawn substitute.
Lupinus albus	White lupine	A	up	2-3'	1-7	○	lwmd	✓	HS	- +		+NC	f	H	✓		Fast-growing, opens clay soils, makes sandy soils productive. Generally toxic to stock, but non-toxic strains being tested.
L. littoralis	Seashore lupine	HP	mt	2½'	4,5	○	s		S		N	+N GE	f				Roots edible in early spring, roasted. Caution—poisonous raw. Dune colonizer.
L. luteus var sativus	Yellow lupine	A	cp,up	1-2'	1-7	○	lwmd-uwmd		S	- +		+NC	f	H	✓	os	Reclaims poor soils. Caution—somewhat toxic, but used as silage, hay, pasture.
Medicago lupulina	Black medic	A,B,HP	tr,sp mt	6"-1'	1-7	○	lwmd-uwm			- +		+NC	f	H	✓		Edible seeds, flowers, leaves. Understory in corn, pastures. Prefers moist soils.
Medicago sativa	Alfalfa	HP	cp	1-3'	1-7	○	lwmd-uwmd	✓		- +		+NC	fsm	BH	✓		Varieties adapted throughout region. Prefers well-drained loam. Taproot opens subsoil but not hardpan.
Melissa officinalis	Lemon balm	HP	up	2½'	1-7	○ ●	lm-um			- +			fsm	B		chsx	Culinary herb. Good bee plant. Oil used in perfume, sachets. Used in arrangements.
Mentha piperita	Peppermint	HP	up	3'	1-7	◑ ●	lwm-uwm	✓		- +			fsm	BH		chm	Widely used flavoring, tea. As milk treatment, retards curdling. Will dry off stock.
Mentha spicata	Spearmint	HP	up	2'	1-7	◑ ●	lwm-uwm	✓		-			fsm	BH		chm	Same uses as *M. piperita*. Tolerates muck, peat, lower pH. Is less alkali tolerant.
Mimulus guttatus	Monkey flower	A,HP	up	1-2'	1-7	◑	lwm-uwm				N		f			o	Edible raw in salads—somewhat bitter. Yellow and reddish-brown flowers.
Montia perfoliata	Miner's lettuce	A	cp	1'	1-7	◑ ●	lwm-uwm		GS		N		f	H	★		Edible leaves, stems, flowers, roots. Good stock forage.
Myrrhis odorata	Sweet cicely	HP	cp	2-3'	1-7	◑ ●	lwm-uwm			-			fs			o	Sugar substitute, anise-like flavor. Edible leaves, roots. Leaves, seeds are seasoning.
Nasturtium officinale	Watercress	A,P/A HP	tr,mt	3"-1¼'	1-7	○ ●	lw-uw-fa	✓		- +			fsm	FH	✓	l	Caution — do not eat if grown in polluted water; resembles poisonous water hemlock.

Nelumbo lutea	American lotus	HP	cp	3-6'	1-7	○	fa	√					f		★	o	Both spp. grow in water 8-12" deep. Excellent wild and domestic fish habitat.
Nelumbo nucifera	East Indian lotus	HP	cp	3-6'	1-7	○	fa	√		–			f		★	o	Roots of both are edible. Indians ate seeds of *N. lutea*. Both have insect pests.
Nuphar polysepalum	Spatterdock	HP	cp,mt	1'	1-7	○◐	fa	√			N		f		√	mo	Edible roots, seeds. Tolerates muddy water, fluctuating level. Ponds, slow streams.
Onobrychis viciifolia	Sainfoin	HP	up	6"-4'	1-3,7	○	lmd-umd			– +		+NC EG		BFH	√		Good quality non-bloating forage, hay. Does well with bunchgrasses. Deep taproot.
Opuntia polyacantha	Plains prickly pear	HP	cp, sp	1'	1-5	○	ld-ud		R		N	E	fs	BH	√	o	Fruits, stem-pads edible raw, cooked, dried, pickled, as syrup, preserves. Seed ground.
Oryza sativa	Rice	A	up	3-6'	3(4) 6,7	○	lwm	√		– +			f		★	f	Staple crop in Asia. Northern varieties need testing in the Pacific Northwest.
Panax quinquefolius	Ginseng	HP	up	1-2'	4-7	◐●	lwm-uwm			–			sm				Primary use is as medicinal root. Understory in moist, cool, rich woods.
Panicum miliaceum	Millet	A	up	3-4'	1-3,7	○	lmd-umd	√		– +		C	f	FHP	√		Warm, short growing season. Catch crop. High protein grain for people, animals.
Papaver somniferum	Opium poppy	A	cp,up	3-4'	1-7	○	lwmd-uwmd			– +			fsm	FH		cho	Edible seeds, oil. Stock, fowl eat seed-cake. Poisoning from milky sap rare, has occurred.
Perideridia gairdneri	Yampah	HP	up	4'	1-7	○◐	lwm-uwm				N		fs				Roots have pleasant nutty taste, edible raw, cooked, dried and ground. Seeds parched.
Physalis pruinosa	Ground cherry	P/A	tr cp,sp	1½'	2-7	○	lm-um						fs		√	o	Long, hot growing season to fully ripen fruits. Edible raw, in preserves, sauces.
Pisum arvense	Field pea	A	cb,tr	1-6'	1-7	○	lm-um			+		+NC		H	√		Main use is green manure. Prefers mod. rainfall, mod. temperatures. Austrian winter pea is most frost-hardy variety.
Plantago major	Common plantain	HP	cp	6"-2'	1-7	○◐	lwmd-uwmd						fsm	FH	√	h	Camp-follower. Leaves, seeds edible, leaves make tea. Sheep, goat, chicken forage.
Portulaca oleracea	Purslane	A	sp,mt	4"	1-7	○◐	lwm		S	– +			f		√		Nutritious stems, leaves, shoots, and seeds. High in iron, calcium. Tolerates poor soil.
Potentilla anserina	Silverweed	HP	tr,cp	1-3'	1-7	○	lwm-s	√	S	+	N		fsm		√	m	Roots edible, leaves make antispasmodic tea. Caution—may be invasive on low pastures.
Raphanus sativus	Radish	A, B	cp,up	1-3'	1-7	○◐	lwm-uwm			– +		C	fm				Many varieties, much variation. Root, leaves edible, seed pods pickled, pressed for oil.
Rheum rhabarbarum	Rhubarb	HP	cp	6'	1-7	○●	lwm-uwm			– +			f			bd hl	Stalks edible, used for dye. Caution—leaves poisonous to people, stock, yield insecticide.
Rubus ursinus	Pacific blackberry	HP	tr, cb	2'	1-7	○◐	lwm-uwm			– +	N	EG	f	BH	★	l	Will cover a trellis, colonize logged, burned land. Gave rise to boysenberry, loganberry.
Rumex crispus	Yellow dock	HP	up	1-6'	1-7	○	lwm-uwm	√					fsm			o	Leaves edible, nutritious. Caution—can irritate skin, mucus membranes in stock.
Sagittaria latifolia	Wapato	HP	cp	3-4'	1-7	○	fa-lw	√			N		f		√	o	Grows in shallow, slow water or wet ground. Tuberous, starchy root edible cooked.
Salicornia virginica	Virginia glasswort	HP	up		4,5	○	s	√	S		N	EG	fs		√	c	Saltmarsh, beaches. Edible stems, growing tips. Ashes high in soda, make soap, glass.
Salvia sclarea	Clary	B/A B, P/B	cp	1½-4'	1-7	○	lwmd-uwmd			–			s	B		clm os	Decorative border, generally pest-free. As seasoning, in wine, cosmetics, soap.
Santolina virens	Santolina	ES,HP	cp,sp	2'	1-7	○	lmd-umd					EG	m			bcl ox	Fast-growing, fire-retardant. Insecticide against moths. Low hedge, bank cover.
Satureja douglasii	Yerba buena	ES	tr	1'	4-7 (1-3)	○◐	lwm-uwm				N		sm			los	Prostrate, spreading to 3'. Tea, medicinal herb, aromatic foliage, flowers April-Sept.
Satureja montana	Winter savory	DS	sp	6"-2'	1-7	○	lwmd-uwmd		S				sm	B		los	Tolerates drought, poor soil. Fire-resistant. Culinary, medicinal herb, edging plant.

LIFE STYLE AND TYPE: A-Annual, B/A-Biennial grown as annual, P/A-Perennial grown as annual, B-Biennial, P/B-Perennial grown as biennial, HP-Herbaceous perennial, D-Deciduous, E-Evergreen, S-Shrub, V-Vine, T-Tree. SHAPE: cb-Climbing, tr-Trailing, cp-Clumping, sp-Spreading, mt-Mat, up-Upright. SHADE TOLERANCE: ○-Full sun, ◐-Partial Shade, ●-Full shade. HABITAT: l-Lowland, u-Upland, w-Wet, m-Moist, d-Dry, fa-Freshwater aquatic, s-Saltwater margins. DRAINAGE TOLERANCE: √-Poor drainage. TEXTURAL TOLERANCE: G-Gravelly soil, H-Heavy clay soil, R-Rocky soil, S-Sandy soil. pH TOLERANCE: (-)=Acidic soil, (+)=Alkaline soil. NATIVE: N-Native.

SOIL IMPROVEMENT/SITE PROTECTION: C-Cover crop/green manure, E-Erosion control, G-Ground cover/living mulch, H-Hedgerow, +N-Nitrogen fixing, W-Windbreak. EDIBLE: f-Food, s-Seasoning/drink, m-Medicinal (ingested). DOMESTIC ANIMAL FORAGE: B-Bees, F-Fowl, H-Hoofed stock, P-Pigs, O-Other. WILDLIFE: √-Habitat/forage value, ★-High habitat/forage value, I-Insectary plant. COPPICING: c-Coppice for fuel, C-Coppice for fuel/materials. BUILDING MATERIAL: B-Building construction, F-Furniture/tools, P-Poles/fenceposts, S-Specialized. OTHER USES: b-Biocontrol, c-Chemicals/oils/cosmetics, d-Dye, e-Ethanol, f-Fiber, h-Stock medicine, l-Landscaping, m-Medicinal (external), o-Ornamental, r-Rootstock, s-Fragrance, x-Other.

GROUND LAYER PLANTS: Annuals & Low Perennials

		FORM				ADAPTATIONS/TOLERENCES						USES					
		Life Style and Type	Shape	Height at Maturity	Climatic Zone	Shade Tolerance	Habitat	Tolerates Poor Drainage	Textural Tolerance	pH Tolerance	Native	Soil Improvement	Edible	Domestic Animal Forage	Wildlife	Other Uses	COMMENTS
Scorzonera hispanica	Black salsify	HP, P/A, P/B	cp	2–3'	1–7	○ ●	lwm–uwm						f				Black-skinned roots edible cooked, taste like oysters. Leaves edible raw or cooked.
Secale cereale	Rye	A	up	3–6'	1–7	○	lwmd–uwmd	✓	HS	– +		C	f	FH	✓		Germinates readily, grows rapidly, makes complete cover, produces much organic matter. Caution—host to ergot fungus.
Stellaria media	Common chickweed	A	cp,mt	3"–1½'	1–7	○ ●	lwm–uwm	✓	HS			C	fm	BFH PO	★	hm	Great winter salad green. Fast life cycle on disturbed land. Sheep, lambs will overeat.
Symphytum officinale	Comfrey	HP	cp	1–5'	1–7	○ ●	lwm–uwm	✓	H	– +			fsm	BF HPO		bhm	Superior protein source for all animals. Alkaloids shown not to be dangerous using traditional methods & quantities. Invasive.
Tagetes erecta	African marigold	A	up	6"–4'	1–7	○	lwm–uwm									bd os	Root exudate may be nematode repellent. Flowers early summer to frost—dye for wool.
Taraxacum officinale	Common dandelion	HP	cp	3"–3'	1–7	○ ●	lwm–uwm		HS	– +			fsm	BH		h	Leaves, roots, flowers edible. Also, flowers make wine, leaves tea. Roots—coffee subst.
Thymus serpyllum	Creeping thyme	ES	tr,mt	2–6"	1–7	○ ●	lwm–uwm		HS	– +		G	sm	BH		hl ms	Groundcover where traffic is light. Tonic forage for sheep, goats. Prefers sun.
Thymus vulgaris	Common thyme	ES	cp,sp	6"–1½'	1–7	○ ◑	lwmd–uwmd		HS	– +		G	sm	B		bcl ms	Used in insecticides, perfumes, embalming fluid, deodorants, antiseptics, flavorings.
Tragopogon porrifolius	Salsify	B/A	cp	2–4'	1–7	○	lwmd–uwmd		S	– +			fm				Edible leaves, flowers in spring, summer. Oyster flavor roots in fall, winter, spring.
Trifolium fragiferum	Strawberry clover	HP	tr	2"–1'	1–7	○	lwm–uwm –s	✓		– +		+NC		H			Excellent pasture. Disease susceptible in cool, moist conditions. Drought tolerant.
T. hybridum	Alsike clover	B,HP	cp	1–3'	1–7	○	lwm–uwm	✓	H	– +		+NC		BH	✓		Pasture, hay but may be somewhat toxic to individual sensitive animals.
T. incarnatum	Crimson clover	A	cp	1–3'	4–6	○ ●	lwm–uwm		HS	– +		+NC		BH			Winter annual. Excellent forage, but may form hair balls in animal stomachs.
T. pratense	Red clover	B,HP	cp	6"–2'	4–6	○	lwm–uwm	✓		– +		+N CE	fsm	BH	✓I	bhm	Varieties differ in characteristics. Used in many rotations. Deep tap root.
T. repens	White clover	HP	mt,tr	2–9"	1–7	○ ●	lwm–uwm		H	– +		+NC EG	f	BH	✓I	bl	Best ground cover for natural farming rotations. New Zealand variety slug resistant, contains hydrocyanic acid. Lawn substitute.
T. subterraneum	Subterranean clover	A	tr	2–6"	4–7	○	lwm–uwm			+		+NC		H			Self-seeding winter annual. Foundation of southern Oregon sheep industry.
Trigonella foenum-graecum	Fenugreek	A	up	1–2'	2–3,7	○	lwmd		S	– +		+NC	fsm	H		dhm	Matures in 75–90 days. Long taproot. Pot-herb. Medicinal seeds also condiment. Rain at maturity ruins seed crop.
Triticum aestivum	Winter wheat	A	up	4'	1–7	○	lwmd–uwmd			– +		C	f	FH	★		Drought tolerant. Inferior to oats as hay. Extremely high wildlife value.
Tropaeolum majus	Garden nasturtium	P/A	cp	1'	1–7	○ ◑	lwm–uwm		S				fsm			lo	Fast-growing, easy culture. Edible leaves, flowers, seed pods.
Vaccinium angustifolium	Lowbush blueberry	DS	sp,mt	3–8"	1–7	○ ◑	lm–uwmd		S	–		G	f	B	★	l	Avoid frost pockets. Rebounds from fire or mowing in one year with a good crop.
V. macrocarpon	Cranberry	EV	tr,mt	10"	1–7	○	lw–uw		S	–			f	B	✓		Competes best at pH 4.5–5.0. Fruit must ripen before hard frost. Peat bogs, sand.
V. oxycoccos	Swamp cranberry	ES	tr	4"	1–6	○	lw–uw	✓		–	N		f		✓		Peat, sphagnum bogs. Small edible fruits have good flavor, August-October.

V. vitis-idaea	Cranberry	ES	tr sp,mt	1'	1-7	○	lwm-uwm	√	S	-	N		fm			lo	Wetlands, marshes, woodlands. Sour fruit, used in syrup, preserves. Groundcover.
Verbascum thapsus	Common mullein	B	cp,up	2-6'	1-7	○	lwmd-uwmd						m	F		hlm os	Tall, stately. Borders. Medicinal for people, stock. Powdered root fed to fowl.
Veronica anagallis	Water speedwell	B,HP	up	6-10"		○	lw-uw -fa	√					fm				Moist to wet ground, rivulets, swamps. Short-lived perennial. High in Vit C.
Vicia faba	Broad bean	A	up	2-4'	1-6	○	lwm-uwm			- +		+NC	f	H			Leaves, beans, pods edible. Caution—some people get toxic reaction from beans, pollen. Can fix 100 lbs N/acre in 6 weeks.
V. sativa	Common vetch	A	cb,tr	1½-3'	1-7	○	lwm-uwm		GS	- +		+NC	f	H	√		Young stems edible as potherb. Caution—seeds may poison stock, humans. *Sativa* is
V. villosa	Hairy vetch	A,B	cb,tr	3-4'	1-7	○	lwmd-uwmd		S	- +		+NC	f	BH	√		heat-intolerant. *Villosa* is the best vetch for fall sowing in the north.
Viola cornuta	Viola	B,HP	cp	6-8"	1-7	○ ◑	lwm-uwm					G	fs			lo	Both *V.* spp. have edible leaves raw, in soup, flowers raw in salad, desserts, candied.
Viola odorata	Sweet violet	HP	cp,mt	2-10"	1-7	○ ◑	lwm-uwm					G	fsm			hl os	Dried leaves make tea. Both used as ground-covers, esp. w/ bulbs, and for borders.
Zizania aquatica	Northern wild rice	A	up	2-9'	1-7	○	fa	√		+			f	FH	★	l	Grows in clear, pure, slow-moving water 3"-6' deep fluctuating less than 6"/year.

LIFE STYLE AND TYPE: A-Annual, B/A-Biennial grown as annual, P/A-Perennial grown as annual, B-Biennial, P/B-Perennial grown as biennial, HP-Herbaceous perennial, D-Deciduous, E-Evergreen, S-Shrub, V-Vine, T-Tree. SHAPE: cb-Climbing, tr-Trailing, cp-Clumping, sp-Spreading, mt-Mat, up-Upright. SHADE TOLERANCE: ○-Full sun, ◑-Partial Shade, ●-Full shade. HABITAT: l-Lowland, u-Upland, w-Wet, m-Moist, d-Dry, fa-Freshwater aquatic, s-Saltwater margins. DRAINAGE TOLERANCE: √-Poor drainage. TEXTURAL TOLERANCE: G-Gravelly soil, H-Heavy clay soil, R-Rocky soil, S-Sandy soil. pH TOLERANCE: (-)=Acidic soil, (+)=Alkaline soil. NATIVE: N-Native.

SOIL IMPROVEMENT/SITE PROTECTION: C-Cover crop/green manure, E-Erosion control, G-Ground cover/living mulch, H-Hedgerow, +N-Nitrogen fixing, W-Windbreak. EDIBLE: f-Food, s-Seasoning/drink, m-Medicinal (ingested). DOMESTIC ANIMAL FORAGE: B-Bees, F-Fowl, H-Hoofed stock, P-Pigs, O-Other. WILDLIFE: √-Habitat/forage value, ★-High habitat/forage value, I-Insectary plant. COPPICING: c-Coppice for fuel, C-Coppice for fuel/materials. BUILDING MATERIAL: B-Building construction, F-Furniture/tools, P-Poles/fenceposts, S-Specialized. OTHER USES: b-Biocontrol, c-Chemicals/oils/cosmetics, d-Dye, e-Ethanol, f-Fiber, h-Stock medicine, l-Landscaping, m-Medicinal (external), o-Ornamental, r-Rootstock, s-Fragrance, x-Other.

Bibliography

Major References

The phrases at the end of each major reference refer to the categories of information in the chart to which the reference often contributed.

Benoliel, Doug. *Northwest Foraging.* Seattle: Signpost Books, 1974. Habitat; edibility.

Dirr, Michael A. *Manual of Woody Landscape Plants.* Champaign: Stipes, 1975. Height; shade, drainage, textural and pH tolerance; landscaping and ornamental uses.

Duke, James A. "The Quest for Tolerant Germplasm." In *Crop Tolerance to Suboptimal Land Conditions.* ASA Special Publication Number 32. Madison: American Society of Agronomy, Crop Science Society of America, Soil Science Society of America, 1978. Names; life style and type; climatic zone, shade, habitat, drainage, textural and pH tolerance.

Elias, Thomas S. *Trees of North America.* New York: Van Nostrand Reinhold, 1980. Names; life style and type; height; climatic zone; habitat; origin; uses.

Gunther, Erna. *Ethnobotany of Western Washington.* Seattle: University of Washington Press, 1945. Native uses.

Hafenrichter, A. L.; Schwendiman, John L.; Harris, Harold L.; MacLauchlan, Robert S.; and Miller, Harold W. *Grasses and Legumes for Soil Conservation in the Pacific Northwest and Great Basin States. Agriculture Handbook 339.* Washington, D.C.: U.S. Department of Agriculture, 1968. Tolerances; uses.

Hitchcock, C. Leo, and Cronquist, Arthur. *Flora of the Pacific Northwest.* Seattle: University of Washington Press, 1973. Names; height; habitat.

Kirk, Donald R. *Wild Edible Plants of Western North America.* Happy Camp, CA: Naturegraph, 1970. Habitat; edibility; other uses.

L. H. Bailey Horatorium staff. *Hortus Third.* Revised and expanded. New York: Macmillan, 1976. Names; bamboos.

Lust, John B. *The Herb Book.* New York: Bantam, 1974. Medicinal uses.

Martin, Alexander C.; Zim, Herbert S.; and Nelson, Arnold L. *American Wildlife Plants.* New York: Dover, 1951. Wildlife value.

McLeod, Edwin. *Feed the Soil.* 665 Meadowdale, Ferndale, MI 48220. Forthcoming. A grower's guide to green manures, cover crops, and cultivated legumes.

Peterson, Lee. *A Field Guide to Edible Wild Plants of Eastern and Central North America.* Boston: Houghton Mifflin, 1978. Height; habitat; edibility.

Quinney, John; Barnhart, Earle; and Massion, Maia. *Farming in the Image of the Forest.* The New Alchemy Institute, 237 Hatchville Road, East Falmouth, MA 02536. Forthcoming. Form; tolerances; uses.

Sunset Magazine and Sunset Books, ed. *Sunset New Western Garden Book.* Menlo Park: Lane, 1979. Form; tolerances; climatic zone maps and designations; landscaping and ornamental uses.

Taylor, Roy L., and MacBryde, Bruce. *Vascular Plants of British Columbia.* Vancouver: The University of British Columbia Press, 1977. Names; climatic zone and habitat tolerances; origin; edibility; toxicity; other uses.

Van Dersal, William R. *Native Woody Plants of the United States.* Washington, D.C., USGPO, 1938. Tolerances; erosion control; stock use; wildlife value.

Whyte, R. O.; Nillson-Leissner, G.; and Trumble, H. C. *Legumes in Agriculture.* Food and Agriculture Organization of the United Nations, 1953. Tolerances; uses.

Other References

Angier, Bradford. *Field Guide to Edible Wild Plants*. Harrisburg: Stackpole, 1974.

Bailey, L. H. *The Standard Cyclopedia of Horticulture*. Vols. 2, 4, 5, and 6. New York: Macmillan, 1916.

Baron, Lloyd C., and Stebbins, Robert. *Growing Filberts in Oregon. Extension Bulletin 628.* Oregon State University Extension Service, 1978.

Burroughs, Kate. "Insectary Plant Index." See Page 34.

Clapham, A. R. *The Oxford Book of Trees*. London: Oxford University Press, 1975.

Colebrook, Binda. *Winter Gardening in the Maritime Northwest*. Arlington, WA: Tilth, 1977.

Grant, John A., and Grant, Carol L. *Trees and Shrubs for Pacific Northwest Gardens*. Seattle: Dogwood Press, 1943.

Grieve, Mrs. M. *A Modern Herbal*. Vol 2. New York: Dover, 1971.

Hall, Alan. *The Wild Food Trailguide*. New York: Holt, Rinehart, 1973.

Hedrick, U.P., ed. *Sturvetant's Edible Plants of the World*. NY: Dover, 1974.

Hunn, Eugene. *Native Plants, Native Americans*. Catalog Notes. Seattle: University of Washington Press, 1981.

International Tree Crops Institute. "Perennial Insectary Plants." Winters, CA: ITCI, 1980.

Jacobs, Betty E. M. *Growing and Using Herbs Successfully*. Charlotte, VT: Garden Way, 1976.

Pellet, Frank C. *American Honey Plants*. New York: Orange Judd, 1947.

Phillips Petroleum Company, *Pasture and Range Plants*. Bartlesville, OK: Phillips Petroleum, 1963.

Reader's Digest Association, Ltd., The. *Reader's Digest of Garden Plants and Flowers*. London: Reader's Digest, 1971.

Shoemaker, James S. *Small Fruit Culture*. Westport, CT: AVI, 1978.

Smith, J. Russell. *Tree Crops*. New York: Harper Colophon, 1950.

Spurr, Joy. *Wild Shrubs*. Seattle: Pacific Search Press, 1978.

FRANK AND ERNEST by Bob Thaves

Courtright, Gordon. *Trees and Shrubs for Western Gardens*. Forest Grove: Timber Press, 1979.

Crockett, James U. *Trees*. Edited by Time-Life Books. New York: Time-Life, 1972.

Domico, Terry. *Wild Harvest*. Seattle: Hancock House, 1979.

Fernald, Merritt L., and Kinsey, Alfred C. *Edible Wild Plants of Eastern Northern America*. Revised by Reed C. Rollins. New York: Harper & Row, 1958.

Franklin, Jerry, and Dryness, C. P. *Natural Vegetation of Oregon and Washington. General Technical Report Pacific Northwest 8*. USDA Forest Service, 1973.

Frankton, Clarence. *Weeds of Canada*. Ottawa: Canada Department of Agriculture, 1955.

Furlong, Marjorie, and Pill, Virginia. *Wild Edible Fruits and Berries*. Happy Camp, CA: Natureograph, 1974.

Gibbons, Euell. *Stalking the Wild Asparagus*. New York: David McKay, 1962.

Gilkey, Helen M. *Weeds of the Pacific Northwest*. Oregon State College, 1957.

Gilkey, Helen M., and Dennis, La Rea J. *Livestock-Poisoning Plants of Oregon*. Corvallis: Oregon State University Cooperative Extension Service, 1969.

Jepson, Willis L. *A Manual of the Flowering Plants of California*. Berkeley: University of California Press, 1925.

Johns, Leslie, and Stevenson, Violet. *The Complete Book of Fruit*. London: Angus & Robertson, 1979.

Levy, Juliette de Bairacli. *Herbal Handbook for Farm & Stable*. London: Faber and Faber, 1963.

Li, Hui-Lin. *Metasequoia, A Living Fossil*. American Science 52:93-109.

Lovell, Harvey B. *Honey Plants Manual*. A. I. Root Company, 1966.

Martel, Donald J., and Fredeen, George N. *Plant Materials for Landscaping*. Pacific Northwest Cooperative Extension, 1978.

Mollison, Bill, and Holmgren, David. *Permaculture 1*. Winters, CA: International Tree Crops Institute, 1981.

Muenscher, Walter C. *Poisonous Plants of the United States*. New York: Macmillan, 1939.

Musick, Mark. *Nitrogen-Fixing Trees and Shrubs for the Pacific Northwest*. Arlington, WA: Tilth, 1980.

Pacific Northwest Cooperative Extension Service. *Trees Against the Wind*. Pacific Northwest Cooperative Extension, 1975.

Sunset Magazine and Sunset Books, ed. *Sunset Western Garden Book*, Menlo Park: Lane, 1967.

Taylor, T. M. C. *The Lily Family of British Columbia*. British Columbia: Department of Recreation and Conservation, 1966.

Thompson, Mary, and Thompson, Steven. *Huckleberry Country*. Berkeley: Wilderness Press, 1977.

Turner, F. Newman. *Fertility Pastures and Cover Crops*. Pauma Valley, CA: Bargyla and Gylver Rateaver, 1955.

Uphof, J. C. *Dictionary of Economic Plants*. Monticello, NY: Lubrecht and Cramer, 1968.

Williams, Gregory. "Tree Crops for Energy Production in Appalachia." In *Tree Crops for Energy Co-Production on Farms*. Springfield, VA: National Technical Information Service, 1980.

Winkler, A. J. *General Viticulture*. Berkeley: University of California Press, 1974.

Yepsen, Roger B., Jr., ed. *Trees for the Yard, Orchard and Woodlot*. Emmaus: Rodale, 1976.

Seed & Nursery List

Robert Williamson

MICHAEL PILARSKI

There are thousands of nurseries, greenhouses and seed companies in the United States and Canada. Local nurseries are especially numerous. In preparing this listing I have given special emphasis to Pacific Northwest sources. I have not hesitated, however, to list out-of-the-region companies which offer rare, unique or useful plants. The current price for each catalog is included when available.

This list is not comprehensive. It emphasizes smaller companies serving home gardeners and small farmers. We would appreciate references to seed and nursery sources that we are not aware of, as well as references to new companies as they are formed.

To the right of each entry in the Seed and Nursery List are code numbers for the kinds of plants available. The numbers refer to the Key at the head of the list. They are only a general guide to the offerings of each company, and are not intended to be a complete description.

OTHER LISTINGS FOR PLANT MATERIALS

Forest Tree Seed Orchards, US Forest Service, US Department of Agriculture, Washington, DC 20250.
A directory of industry, state and federal forest tree seed orchards in the U.S.

Nursery Newsletter, Office of the Oregon State Forester, 2600 State Street, Salem, OR 97310.
A comprehensive listing of Oregon and Washington sources of conifer seedlings, including such hard-to-find species as Port Orford cedar, Sitka spruce, and giant sequoia.

Nursery Source Guide, Brooklyn Botanic Garden, 1000 Washington Avenue, Brooklyn, NY 11225. $1.75.
A listing of major retail and wholesale nurseries in the United States. Includes descriptions of over 1200 tree and shrub species and cultivars, with references to which nurseries supply them.

Riker, Tom. *The Guide to Buying Plants*. New York: William Morrow, 1976. 256 pp, $12.50.
Most of the book is taken up by descriptions of horticultural plants. The most valuable part of the book is the International Buyers' Guide chapter, which includes a great deal of information on which plants can be imported into the United States without restriction, which need two-year quarantine, which are totally restricted, and the pests and diseases which are being guarded against. The steps needed to import plant seeds and materials from other countries are clearly explained—useful information which is hard to come by elsewhere. The book also includes a listing of nurseries in foreign countries which will export to the U.S.

Sources of Native Seeds and Plants, Soil Conservation Society of America, 7515 NE Ankeny Road, Ankeny, IA 50021. $2.50.
Directory of seed sources and native plant nurseries in the United States.

CAUTIONARY NOTE ON SPECIES INTRODUCTION

In the process of moving plant materials around the globe, people have also introduced insect and disease pests into new areas. Because of the lack of natural controls, widespread damage has been caused by insects such as the Gypsy moth or diseases such as the chestnut blight, Dutch elm disease, and the white pine blister rust, all of which were introduced along with imported plant material.

Keeping this in mind, the best source of useful seeds and plants is a reputable seed company or nursery. If you are collecting seeds in your travels or are having friends in other parts of the continent send plant materials to you, please examine them carefully for insects or their eggs, and for signs of disease damage. Carefully watch new plantings imported from outside your area. If an unknown disease appears, the plant should be destroyed by burning.

Importing foreign plant materials into the United States is lawful only by permit. Certain plants are under quarantine and may not be imported. A list of restricted plants and import permits can be obtained from:

> Permit Unit, Plant Protection and Quarantine Programs, Animal and Plant Health Inspection Service, Rm 638, Federal Building, Hyattsville, MD 20782.

Canadians wishing to purchase plants or plant materials from the United States must first write to the following address to receive an information packet and an identification number which the customer then sends to the US nursery along with their order:

> The Food Production & Marketing Branch, Plant Quarantine Division, Agriculture Canada, Ottawa, Ontario K0A 0C6

Key

1. Aquatic & Bog Plants
2. Exotic (including Bamboo), Heirloom, Rare, or Native
3. Fruit, Vines
4. Flowers, Wildflowers
5. Grains, Grasses, Forbs
6. Herbs
7. Nitrogen-Fixing, Reclamation Plants
8. Nuts
9. Shrubs
10. Trees
11. Vegetables
12. Wholesale
13. Wildlife, Livestock

ABUNDANT LIFE SEED FOUNDATION 2,4,6,9 10,11
PO Box 772
Port Townsend, WA 98368
Non-profit foundation with an emphasis on seed of native and naturalized plants for the North Pacific Rim. Catalog is $2.

AL BACHARACH, INC 8,12
17490 West View Rd
Lake Oswego, OR 97304
Wide range of grafted nuts. New Butler and Ennis filbert varieties must be ordered two to three years in advance. Wholesale only.

ALEXANDER & BROWN 4,6,11
PO Box 13
Perth, Scotland PH1 5NY
Winter vegetables and herbs for maritime climates.

APPALACHIAN WILDFLOWER NURSERY 2,4 9,10
Route 1, Box 275A
Reedsville, PA 17084
Don Hackenberry sells more than wildflower plants and seeds. His nursery sells rare trees, shrubs and perennial plants, many of which are grown from seed obtained from botanical gardens in Asian locations such as Kirghizstan, Tashkent, Ashkabad, Hokkaido and Kashmir. He also carries North American species. Descriptive catalog $1.00.

ARMSTRONG NURSERIES 2,3
Box 473
Ontario, CA 91764
Wide range of fruit varieties, including exotics such as jujubes, black mulberry, persimmon, figs, and kiwis.

THE BANANA TREE 2,3
715 Northampton Street
Easton, PA 18042
They carry seed of a wide variety of rare and uncommon plants including kiwi fruit, giant bamboo, and bananas, too. Catalog is 25 cents.

BARNEY'S GINSENG PATCH 6
Rt 2-H, Box 43
Montgomery City, MO 63361
Ginseng and goldenseal seeds and plants. Their informative catalog costs $1.

BAUM'S NURSERY 3
RD 2
New Fairfield, CT 06810
More than 150 American and European varieties of apples, plums and pears. A booklet, "Fruit Varieties of America and Europe," provides detailed variety descriptions (cost $1).

BEAR CREEK FARMS 3,7,8,9,10
PO Box 248
Northport, WA 99157
Cold hardy and drought-resistant fruits, nuts, shrubs and grafted trees for northern growers. Retail and wholesale. Catalog 50 cents.

BEAVERLODGE NURSERY 3,9, 10,11
Box 127
Beaverlodge, Alberta T0H 1E0
Hardy, northern-grown fruits, vegetables and ornamentals.

BOB SWEDBERG NURSERY 3,8,9
Battle Lake, MN 56515
Hardy grafted fruits and nuts.

BOUGHEN NURSERIES 2,3,4,9
Box 12
Valley River, Manitoba
Canada R0L 2B0
A 67-year-old family business which sells cold-hardy fruit trees and bush fruits. Their catalog is dotted with planting and growing tips encouraging the organic approach.

BOUNTIFUL RIDGE 3,8,9,12
Princess Anne, MD 21853
Long-established family nursery carrying a wide range of berries, nuts and fruits, including kiwi. Wholesale and retail.

BRIMFIELD GARDENS 2,9,10
109 Main St
Rocky Hill, CT 06067
Wide collection of plants including bamboos, heathers, ivies, heaths, ferns, hollies, witch hazels, and rare shrubs and trees. Catalog $1.

BURNT RIDGE NURSERY 3,8, 10,12
432 Burnt Ridge Rd
Onalaska, WA 98570
Grafted nut trees available: filberts, Persian and black walnuts, Chinese and hybrid chestnuts. Also fig trees, kiwi vines, honeylocust, and numerous varieties of grapes.

BYLAND'S NURSERIES 3,9,10
1600 Byland Road
Kelowna, BC V1Z 1H6
Grape vines, fruit trees, and container and field grown ornamentals.

CALCHICO KIWI CO 2,3
PO Box 978
Chico, CA 95926
Specialists in kiwi plants.

CARTER SEED CO 2,4,9,10
PO Box 4006
Sylmar, CA 91342
Wide variety of tree and shrub seed, along with seed for cactus, succulents, flowers and ferns.

CASA YERBA 6
Star Rt 2, Box 21
Days Creek, OR 97429
Casa Yerba carries one of the largest selections of medicinal and culinary herb seeds in the country. Many of the herbs are also sold as plants. "We grow all our herbs naturally, organically, without chemical fertilizers or poisonous sprays. Except for a few of the tropicals, herb seeds are hand gathered from our own plants at the peak of ripeness." Catalog $1.

Shelterbelts, Hedges & Windbreaks

It is our view that the first and perhaps most important step in planting food trees is to provide a sheltered and secure environment for optimum growth.

The use of dense, semi-permeable hedges of shrubs and trees in rows surrounding and sometimes interspersed within an orchard has several important benefits, including diversion of winds to reduce evaporation, blossom damage, fruit fall, and winter freeze damage. Frost control can be facilitated by blocking and diversionary windbreaks. Hedges can provide feeding and refuge sites for beneficial insects and wildlife, create privacy, provide wood products during thinning, and, in time, can become effective fences. They will moderate heating loads around buildings, control drifting snow and erosion, and protect livestock. Orchardists might well consider interplanting nitrogen producing trees and shrubs in the orchard to provide on-site nitrogen production.

—Bear Creek Farms

Collecting Seeds

Seed collecting can offer a pleasant hobby or part-time employment. There is a large demand by the nursery trade for seeds of almost all native woody plants and for many non-native woody plants which have been introduced to our region. Hourly wages vary widely depending on the quality of the stand and the experience of the collector.

Observe the trees and plants you pass during your daily travels. Take careful note of the varying stages of seed ripening of the best trees you come across. Some seeds must be collected within several days or several weeks of ripening; others can be collected over a period of several months. Keeping a journal of ripening dates will be helpful in future years, since plant species ripen their seeds in a definite sequence each year. Where possible, collect seed from plants which are in their prime habitat. Avoid collecting in stands containing many poorly-formed, off-color, diseased, or otherwise abnormal plants. Seed collected from suppressed or subdominant trees yields poor quality stock.

Although very young and very old trees may yield excellent seed, vigorous trees in middle age are best for seed collection. The best seed grows at the top of the crown and on the ends of the side branches. Due to seasonal variation, trees produce an abundant seed crop in some years and a light crop in others. In general, it is better to obtain seed from trees which are bearing an abundant crop rather than a light one.

It is important that the seeds are fully mature. In most cases you will be able to tell when the seeds are mature by some visible indicator such as color, firmness of fleshy fruits, drying of cone scales or bracts, or the appearance of the seed itself. For many species a common indicator of maturity is the natural falling of nuts, or release of seeds. For heavy fruits, such as walnut, oak, hickory, beech and chestnut, the first fruits to fall are often wormy or otherwise defective and should not be collected for propagation.

HARVESTING METHODS

There are many methods of harvesting seed. Some species can be harvested with specially designed machinery such as vacuums and shakers, but most are still collected by hand. The seed of some conifers is harvested by tree-climbers equipped with spurs, belts and ropes. Cones can also be obtained from logging operations where the trees being cut have cones at the correct stage of ripeness. Elderberries and other terminal-borne fruits can be harvested with pole pruners. Serviceberries and other berries are usually picked with special combs or stripped by hand. Some seeds cling tenaciously to the plant, while others are readily dislodged at the proper stage of ripeness into a bag or onto canvas spread out for the purpose. Various cleaning equipment is needed depending on the type of seed. Some can easily be made out of readily available materials, while others such as fanning mills must be purchased. Small lots of seed of almost all species can be properly cleaned with simple home methods, but the economical cleaning of large lots of some species calls for more sophisticated machinery.

STORAGE

Seeds breathe, and the slower we can keep their respiration the longer they will maintain viability. The best conditions for most seeds are low humidity and low temperatures (30° to 40° F.). Some seeds lose viability quickly even under the best of conditions. Some of these are ash, oak and aspen. Seeds which ripen in the spring or early summer should be planted in that season, since their natural cycle is to germinate the same year they ripen. As long as they are kept dry, cardboard and cloth make good containers for seeds. The finer the seed, the closer should be the weave of the cloth or the more air-tight the container.

—Michael Pilarski

CD SCHWARTZ NURSERY 3,10
2302 Tacoma Road
Puyallup, WA 98371
A long-established nursery specializing in fruit trees adapted to the maritime Northwest.

CHESTNUT HILL NURSERY 8
Rt 3, Box 477
Alachua, FL 32615
Limited quantity of the blight-resistant Dunstan hybrid chestnut.

CLIFTY VIEW NURSERY 3,8,9,10,12
Rt l, Box 509
Bonners Ferry, ID 83805
Hardy ornamentals, fruits and nuts.

CLOUD MOUNTAIN NURSERY 3,8,10
6906 Goodwin Road
Everson, WA 98247
Specialists in quality, disease resistant trees for the maritime Northwest. Apples, pears, plums, cherries, figs, walnuts, grapes, currants, gooseberries, hardwoods.

CLYDE ROBIN SEED CO 4,7,9,10,12
PO Box 2855
Castro Valley, CA 94546
Seeds of native trees, shrubs and wildflowers for arid lands. Wide selection of manzanitas and ceanothus. Also wildflower mixes and foreign tree seeds. Packet or bulk amounts. Wholesale and retail. Catalog $2.

CONVERSE NURSERY 2,3
Amherst, NH 03031
Large selection of grafted antique apple varieties.

CORWIN DAVIS NURSERY 2,3
20865 Junction Road
Bellevue, MI 49021
Pawpaw seed and grafted plants. A pamphlet on pawpaws is available for $1.

DAVE WILSON NURSERY 3,8,9,10
Box 4G
Hughson, CA 95326
Source for wide range of fruits and nuts for home and commercial plantings. They have excellent climate zone maps for every region in the Western states, including charts of cultivar evaluations. Their "Planting Guide Supplement for Chestnuts, Filberts, Citrus, Kiwi Fruit and Roses" contains very good information on growing the kiwi vine. Cost is $1.

DE GIORGI COMPANY 4,11
PO Box 413
Council Bluffs, IA 51502
Wide selection of traditonal, non-hybrid vegetable varieties, with detailed cultural information for home and commercial growers. Numerous flowers, too. Catalog $1.00.

DUTCH MOUNTAIN NURSERY 3,9,10,13
Augusta, MI 49012
Specialists in plants that attract birds and other wildlife. Over 150 varieties, mainly trees and shrubs. Also available is a card file of five hundred ornamental plants, describing native habitat, form, color, soil preference, culture, growth rate, flower, and fruiting season. The price for the set is $8.75.

EARL DOUGLAS CHESTNUT SEEDS 8
Red Creek, NY 13143
A source for blight-resistant American chestnut seed.

ENVIRONMENTAL SEED PRODUCERS
PO Box 5904 4,7,9,10,12
El Monte, CA 91734
Large selection of wildflower mixes. Each mix contains between eleven and thirty-five species. Seed is sold in pound lots. Also trees and shrubs for land reclamation. Wholesale and retail.

EUGENE WHOLESALE NURSERY 3,10,11,12
55 Delay Drive
Eugene, Or 97404
Hardwoods and ornamentals for landscaping, plus a wide variety of fruit trees.

EXOTICA SEED COMPANY 2,3,8,12
8033 Sunset Blvd, Suite 125
West Hollywood, CA 90046
Seeds of rare fruit and nut trees adapted to Mediterranean or subtropical climates, including many species almost impossible to find from other sources. Wholesale and retail.

FAIRACRE NURSERY 3
Rt 1, Box 1068
Prosser, WA 99350
Specialists in certified virus-free stock of a wide variety of grapes, gooseberries and currants.

FALL CITY HERB FARM 2,6,9
32804 Issaquah-Fall City Road
Fall City, WA 98024
Wide range of common and unusual plants for cooking, teas and landscaping.

FORESTFARM 2,3,7,9,10,12,13
990 Tethertil Rd
Williams, OR 97544
Plants for multiple use. Large selection of native plants and unusual ornamentals: plants for wild fruit, birds and wildlife, bees, natural dyeing, revegetation, windbreaks, solar landscaping, nitrogen fixation. Containerized starter-plants at affordable prices. Informative catalog $1.00. See box at right.

FOX HILL FARM 4,6
444 W Michigan Ave, Box 7
Parma, MI 49269
Wide selection of potted herbs and scented geraniums.

FRANK T SESOCK - SEEDSMAN 2,10
6045 Foley Lane
Central Point, OR 97052
Among the most comprehensive listings of seeds from trees native to the Western U.S.

FW SCHUMACHER CO 9,10,12
36 Spring Hill Rd
Sandwich, MA 02563
One of the largest selections of tree and shrub seeds. Over 7000 species and varieties. Wholesale and retail.

G SEED COMPANY 2,4,6,11
PO Box 702
Tonasket, Washington 98855
Vegetables, herbs, flowers, wildflowers and cover crops, plus compost inoculants and books. Specialties: Mexican peppers, Native American corns, beans, and squashes, heirloom vegetables and wildflower mixtures custom blended to your microclimate. Send 40 cents in stamps for catalog.

GARDEN CITY SEEDS 4,6,11
625 Phillips
Missoula, MT 59802
Self-reliant living is the goal of this new company. They are contracting with local organic farmers and gardeners to grow open pollinated varieties of seeds. They carry vegetable, flower and herb seed, and herb plants. Their "Catalog & Planting Guide" is 50 cents.

GARDENS OF THE BLUE RIDGE 1,4,6,9,10
PO Box 10
Pineola, NC 28662
Hardy native plants of the Blue Ridge Mountains. Herbaceous perennials, bog plants, hardy vines, ferns, climbers, creepers and ground covers, deciduous trees and shrubs.

Mike Karpinski

Forestfarm, a nursery in Williams, Oregon, offers one of the best examples of the nursery renaissance underway in the Pacific Northwest. In 1973, Ray and Peg Prag started Forestfarm on 30 acres in the Siskiyou Mountains of southern Oregon. By 1981, six to twelve part-time employees were needed to care for the more than 300 species of trees, shrubs and vines they offer.

Forestfarm's specialty is native Western plants for rehabilitating damaged lands and for low-maintenance landscapes. Their second specialty is plants for self-sufficiency, such as native fruits, nitrogen-fixing plants, dye plants, plants supplying bee forage, and plants beneficial to wildlife. The third specialty is native and uncommon ornamentals. The Forestfarm Nursery catalog contains detailed information on the uses, hardiness zones, and origins of all plants offered.

Ray and Peg point out that "nursery work in general is demanding and time-consuming. Plants have to be watered every day in the summer, greenhouses have to be kept warm on cold nights, plants must be carefully tended. We work long hours almost every day of the year." They enjoy what they are doing, and the nursery helps them achieve their goal of "working together as partners in a rural environment, and living closely with nature, while making a creative contribution at least to our small part of the world."

GLECKLER'S SEEDSMEN 2,11
Metamora, OH 43540
Outstanding selection of heirloom and unusual vegetable varieties. Catalog 50 cents.

GOVERNMENT TREE SALES 10
Local Farm Advisors, County Extension offices, Soil Conservation Service (SCS), or Agricultural Stabilization and Conservation Service (ASCS) often have spring tree sales for conservation or reforestation projects. Often at very low prices.

GREENMANTLE 3,12
3010 Ettersburg Rd
Garberville, CA 95440
Small nursery specializing in old homestead apple varieties, plus curlless peaches and Moyer prunes. Retail and wholesale.

GRIMO NUT NURSERY 3,8,10
RR #3 Lakeshore Road
Niagara-on-the-Lake
Ontario, Canada L0S 1J0
Ernie Grimo specializes in nut cultivars, sweet-pit apricots and mulberries.

GROOTENDORST NURSERIES 3,12
Lakeside, MI 49116
One of the widest selections of rootstocks for apples, pears, plums, peaches, and cherries, plus custom grafting. Serving both commercial and home orchardists.

GURNEY'S 3,8,11
Yankton, SD 57079
Northern-grown vegetables and fruits, with lots of unusual varieties for adventurous gardeners.

HARRIS SEEDS 4,11,12
Moreton Farm
Rochester, NY 14624
-or-
HARRIS SEEDS WEST
PO Box 432
Gresham, OR 97030
Good source of seed for home and commercial growers, with a separate commercial catalog for wholesale orders. Oregon branch also has special commercial varieties for the Northwest not listed in main catalog.

HEISE WAUSAU FARMS 6
Route 3
Wausau, WI 54401
Source of ginseng and goldenseal roots.

HENDERSON'S BOTANICAL GARDENS
Greensburg, IN 47240 1,4,6
Henderson specializes in bog and aquatic plants, hardy wildflowers, native ferns, lilies, and medicinal plants.

HENRY LEUTHARDT NURSERIES 3
Montauk Hwy
East Moriches, NY 11940
Specialists in espaliered fruit trees. Their catalog (50 cents) includes useful information on espalier techniques.

HERBS & HONEY NURSERY 6,11
Rt 2, Box 205
Monmouth, OR 97361
More than 100 varieties of herb plants, organically grown. They also carry a variety of vegetable and herb seeds, books, herbal salves and other items.

INTERNATIONAL TREE CROPS INSTITUTE 2,3,8,9,10,13
PO Box 666
Winters, CA 95694
Wide variety of fruit, nut, forage and windbreak species, including four-wing saltbush, persimmon, fig, honey locust, and many others.

ISLAND SEED CO LTD 4,11
PO Box 4278, Station A
Victoria, BC V8X 3X8
Vegetable and flower seeds for British Columbia gardeners.

ISRAEL BROTHERS NURSERY 3,8,10
707 S Elm
Toppenish, WA 98948
Specialists in fruit, nut and ornamental trees adapted to Eastern Washington conditions. Grape plants also available for wine, juice, raisin, or table use. Will custom grow trees and vines for large orders. Stock is grown with love and care and is guaranteed to grow.

JACKLIN SEED CO 1,5,7,9,12
West 5300 Jacklin Ave
Post Falls, ID 83854
Reclamation plants for difficult sites. Wide variety of legumes, native shrubs, wildflowers, bunchgrasses and aquatic plants including wild rice and tuberous food plants native to the Northwest. Wholesale only.

JAMIESON VALLEY GARDEN 2,3,4,9,10
Jamieson Road, Rt 3
Spokane, WA 99203
They carry many rare and unusual wildflowers, both Eastern and Western natives; also many native bulbs, orchids, alpines, perennials, berries, shrubs and trees. Their catalog reflects a strong conservationist orientation.

JERSEY CHESTNUT FARM 3,8
58 Van Duyne Ave
Wayne, NJ 07470
Persimmons and the Orrin Chestnut.

JL HUDSON, SEEDSMAN 4,6,9,10,11,12
PO Box 1058
Redwood City, CA 94064
"A World Seed Service." Wide variety of native and non-native trees, shrubs, herbs, wildflowers and vegetables, as well as wild ancestors of common vegetables. Retail and wholesale. Catalog $1.00.

JOHNNY'S SELECTED SEEDS 5,11
Albion, ME 04910
Specialists in hardy, disease-resistant vegetables for home and commercial organic growers. They also carry field grains, including amaranth, wheat, oats, buckwheat, and winter rye, plus rape, oilseed sunflowers, northern soybeans, and 'Stampede,' a large, hardy Jerusalem artichoke. The Johnny's catalog is filled with cultural information and is an utstanding model for regional seed companies.

Hybrid vs. Open Pollinated

What do "Hybrid" and "Open-pollinated" mean? Most seed catalogs list many "F-1 hybrids." This seed is produced by crossing two different strains, usually inbreds. Seed saved from hybrid plants will normally be unable to produce plants with the uniformity and similarity expressed in the hybrid parent. Segregation occurs, or in other words the plants didn't have the ability to stay the way they were. "Open-pollinated" or "Standard" varieties are true breeding. Seed saved from such plants will produce a crop resembling the previous generation. One might accuse us of stretching science and becoming too philosophical here. But, since personal health and well-being are certainly influenced by the quality of foods, this matter must receive consideration. Hybrids do routinely occur in Nature and often express "vigor" just as do the man-made hybrids. So, to totally reject hybrid plants and their use as food would be to deprive oneself of one of Nature's most exciting events. And we certainly don't feel that life should be too awfully solemn. However, in the long run it seems more valuable to enjoy the "constant warmth" of well maintained varieties than to depend on the short-lived "fire" of the hybrids.

—Johnny's Selected Seeds

KEITH SOMMERS TREES 8,9,10,13
10 Tillson Ave
Tillsonburg, Ontario N4G 2Z6
This small, family-run business had its 37th year of selling trees in 1982. Their specialties are nut trees and shrubs, particularly shrubs which attract wildlife. They carry two "buartnut" cultivars (cross between heartnut and butternut). They also sell woodland plants and tree seeds. They ship coast to coast in Canada, but will not ship to the US.

KELLY BROS NURSERIES 3,8
Dansville, NY 14437
They carry a wide variety of strawberries, blueberry bushes, nut trees and grape vines. Will ship to Canada.

KELOWNA NURSERIES, LTD 3,9
PO Box 178
Kelowna, BC V1Y 7N5
Fruit trees, prairie hardy shrubs.

KESTER'S WILD GAME FOOD NURSERIES 1,5,7,13
PO Box V
Omro, WI 54963
Kester's carries many aquatic plants including wapato, wild rice, cattails, many rushes, and specialized seed mixes for game birds and wildlife. They also carry a wide variety of grains, including millet and a perennial grain. Their catalog contains valuable information on establishing permanent wildlife plantings. (See feature on page 127.)

KITIZAWA SEED CO 11
356 W Taylor St
San Jose, CA 95110
Specialists in Oriental vegetables.

LAMB NURSERIES 2,9
E 101 Sharp Ave
Spokane, WA 99202
A wide selection of hardy shrubs, perennials, alpines and rock plants. Native plants and exotics.

LAWYER'S NURSERY 9,10,12
Rt 2, Box 95
Plains, MT 59859
One of the largest suppliers of hardy tree and shrub seed. Sell in bulk quantities only.

LESLIE WILMOTH NURSERY 8
Rt 2, Box 469
Elizabethtown, KY 42701
Butternuts, hickories, chestnuts and walnuts. Custom grafting.

LOUIS GERARDI NURSERY 2,3,8
RR 1, Box 143
O'Fallon, IL 62269
Northern nuts and fruits, including chinquapin, grafted persimmons, grafted pawpaws, and mulberries.

Animals Sowing Seeds

Nature has devised many ways to spread seeds. The forces of wind and gravity help plants on slopes to seed the area below them. Moving water carries seeds downstream in search of suitable habitats for colonization. Animals are especially active plant propagators. Many rodents make caches of seeds, some of which are forgotten, not needed, or simply not used. A number of birds also carry seeds and hoard them, thus lending their help to nature's planting scheme.
These hoarding rodents and birds are one of the main reasons that forests can climb mountains in relatively short periods of time. The propagation of some timberline tree species, such as the whitebark pine, are almost totally dependent on animal plantings.

Some birds take a positive pleasure, it seems, in carrying things about simply for the sake of carrying them about. This propensity is especially strong among the crows, ravens, jays and magpies, and nuts are one of their most treasured items. In fact jays help themselves to part of the Northwest's domestic filbert harvest every year. They are also one of the chief sowers of wild hazelnuts and are helpful in spreading other plants as well.

The seeds of fruits such as raspberry, blackberry, and mulberry need to pass through the digestive system of a bird or a mammal before it will germinate. In the process the pulp is removed, and the seed is given an acid bath and mechanical scarification before "planting." Many other plants also bear a fruit that tempts animals to eat the seed and carry it away from the parent plant to be deposited with a fresh supply of fertilizer.

MAKIELSKI BERRY FARM AND NURSERY 3
7130 Platt Road
Ypsilanti, MI 48197
Their specialty is bramble fruits, blackberries, purple and red raspberries and an especially wide assortment of black raspberries. Also currants, gooseberries, and rhubarb.

McLAUGHLIN'S WILDFLOWER SEED 2,4
PO Box 550
Mead, WA 99021
They offer one of the most complete selections of native wildflowers of the Pacific Northwest and Rocky Mountains. Packet lots, bulk prices on request.

MEADOWBROOK HERB GARDEN 6
Whispering Pines Road
Wyoming, RI 02898
Wide selection of biodynamically grown herbs. Catalog 50¢.

MELLINGER'S 2,3,4,8,10,11,12
2310 W South Range Rd
North Lima, OH 44452
Surprising array of seeds and plants. Retail and wholesale.

MILES W FRY & SON NURSERY
RD 3 10,13
Ephrata, PA 17522
1982 is their 27th year of growing the Hybrid Poplars. They sell a wide variety of clones and have special mixtures of clones for differing conditions such as alkaline soil and acid soil.

They have also started the Wood Power Energy Corporation to develop and test aspects of energy programs based on fuelwood from short-rotation woodlots. Their catalog contains excellent information on wood gasification, a progress report on wood ethanol, and hybrid poplars for municipal sewage sludge composting, cattle feed, and reclamation of coal spoil banks.

MISTLETOE SALES 2,7,9,10,12
247 West Alamar Ave
Santa Barbara, CA 93105
Wide selection of trees and shrubs, including eight acacia, two alder, seventeen eucalyptus, one ginkgo, twelve pine, and four sumac species. Wholesale only.

MOORE WATER GARDENS 1
Port Stanley, Ontario, Canada N0L 2A0
Canada's most extensive offering of bog plants and water lilies. Their catalog offers much useful information on planting and care of plants.

MOSES LAKE CONSERVATION DISTRICT NURSERY 9,10,13
Rt 3, Box 415
Moses Lake, WA 98837
Trees and shrubs for shelterbelts, windbreaks and wildlife.

MUSSER FORESTS 9,10,12
Box 340-M
Indiana, PA 15701
They claim to be the largest nursery in the country for evergreen trees and shrubs, ground covers and hedges.

The Wonder of a Tree

Trees are a magical, wonderful balance of grace and quietness, power and strength, guardians of the earth. The oldest and largest living beings on the earth, some, like the bristlecone pine, grow to be over 4000 years old. Some, like the giant sequoia, grow to be over 350' tall. They are made of primarily sunlight, water, and air. One chestnut tree in Sicily is large enough for 100 people to stand under it out of the rain and so big that 30 people could not reach around it.

Trees are responsible for creating over half the world's oxygen supply, and one acre of trees creates enough oxygen to support 18 people. A large elm tree can have over 6 million leaves, while the leaves from a mature maple cover over a half acre of ground. Many times more strong than the most efficient suction pump, an acre of 40 apple trees will collect from the ground and discharge in the air over 450 tons of water in a month, all carried through channels the size of a hair. One tree can have as much as 50,000 cubic feet of wood, enough to build a community of six room houses. But still with all the magic that trees provide, they are diminishing gradually from the land. The Tillamook Burn in Oregon burned over 350,000 acres of trees, and one issue of the New York Sunday paper takes over 250 acres to make.

from *Treesong*, Children of the Green Earth

NATIVE PLANTS INC 4,5,7,9,10,12
360 Wakara Way
Salt Lake City, UT 84108
Offers a wide variety of seed from native trees, shrubs, wildflowers, grasses and legumes for semiarid regions. Wholesale only.

NATURAL WOODLAND NURSERY 2,8,9,10
544 Conestogo Road West
Waterloo, Ontario N2L 4E2
Canadian native flowers, trees and shrubs. Canadian customers only.

THE NATURALIST SEED CO 6,13
PO Box 435
Yorktown Heights, NY 10598
Seeds of herbs, native useful plants, medicinals, natural pesticides, and plants which attract wildlife and songbirds. Catalog $1.00.

NEW ALCHEMY INSTITUTE 2
c/o Earle Barnhart
237 Hatchville Rd
E Falmouth, MA 02536
Eighteen hardy bamboo species available. Will propagate on order.

NEW YORK STATE FRUIT TESTING ASSOCIATION 2,3
PO Box 462
Geneva, NY 14456
One of the oldest and most prestigious fruit testing stations in the United States. Their yearly catalog lists many new and antique fruit varieties, scionwood, and dwarfing rootstocks. Custom propagation. "A donation to help defray the cost of the catalog would be appreciated."

NICHOLS GARDEN NURSERY 2,4,6,11
1190 North Pacific Hwy
Albany, OR 97321
Herb, vegetable and flower seeds. Sprouting seeds. Herb plants, saffron crocus bulbs, Oriental vegetables, peanuts.

NORTHPLAN SEED PRODUCERS 4,5,7,9,10
PO Box 9107
Moscow, ID 83843
Specialize in seeds for native trees, shrubs, forbs, grasses and wildflowers of the Western states, especially for semiarid and arid regions.

NORTHWOODS NURSERY 3,7,8,10,13
2525 Kingsley Road
Hood River, OR 97031
Trees for food crops, wildlife enhancement, timber, energy production, and beauty. Among those offered are several varieties of nut trees, woodlot and soil conservation species such as black locust and Russian olive, and ornamental trees including paper birch and Norway and sugar maple.

OKANOGAN SEED SUPPLIERS 2,4,6,9,10
PO Box 1064
Tonasket, WA 98855
A newly-formed cooperative seed collecting service. Native tree, shrub and wildflower seeds for Idaho, Montana and Eastern Washington.

OREGON ORGANICALLY GROWN SEEDS 6,11
46525 McKenzie Hwy
Vida, OR 97488
Small selection of quality home-grown seed, adapted to the Willamette Valley. A good example of small farm seed production. Catalog is 50 cents.

OTTO RICHTER & SONS 6,11
Goodwood, Ontario L0C 1A0
Large selection of herb seeds (over 300 varieties), alpine flower seeds, and gourmet vegetable seeds. They also carry one of the most extensive listings of books available in the fields of herbal medicine, herb dyes and culinary uses. Catalog $1.00.

PALOUSE SEED CO 2,5,7,9,10,13
PO Box 291
Fairfield, WA 99012
Specialists in native grasses, legumes, herbs and shrubs for re-establishing forage and ground cover. Over 40 species of containerized native trees, shrubs and forbs, including Saskatoon berries, winterfat, and perennial flax. Will contract to grow species not in catalog.

PEACEFUL VALLEY FARM SUPPLY 3,4,11
11173 Peaceful Valley Rd
Nevada City, CA 95959
Services and supplies for gardeners, homesteaders and orchardists, including a transplanting service for vegetables, herbs, flowers and grapes. Nursery stock selected from California, Washington and Oregon. Catalog $2.00.

PECOFF BROTHERS 4,5,7,9,10,12
Rt 5, Box 215-R
Escondido, CA 92025
Seed for many trees, shrubs, forbs and wildflower mixtures, especially for California climates. Specialists in plants for land rehabilitation. Wholesale only.

PELLETT GARDENS 2,4,7,9,10
Atlantic, IA 50022
Specialists in honey plants. Excellent selection of nectar sources.

PHOENIX SEEDS 2,3,6,8,10
PO Box 661
Burnie, Tasmania, Australia 7320
Serving the Australian permaculture movement, the Phoenix Seeds catalog is based on Bill Mollison's design concepts. Vegetables, herbs, fruits and nuts arranged by permaculture zones. Many exotic species listed. Catalog $3.00.

PLANTS OF THE SOUTHWEST 2,4,5,7,9,10,11
1570 Pacheco St
Santa Fe, NM 87501
Drought-resistant plants for arid climates, including reclamation grasses and Southwestern vegetables. Catalog $1.00.

PULVERMACHERS GINSENG 6
Richland, WI 53581
Source of ginseng seed and roots.

RAINTREE NURSERY 2,3,8,10
265 Butts Road
Morton, WA 98356
Fruit and nut trees for the maritime Northwest. 25 varieties of grafted apples, Japanese pears, figs, persimmons, kiwi, berries, grapes, walnuts, Chinese chestnuts, almonds, woodlot trees, bamboo, and giant rose hips.

REDWOOD CITY SEED CO 2,4,6,11,12
PO Box 361
Redwood City, CA 94064
Carries many hard-to-find tree and shrub seeds, plus vegetable, flower and herb seeds. Catalog is 50 cents. Their catalog supplement on tree crops is also worth obtaining (50 cents). Retail and wholesale.

RH SHUMWAY 3,11,12,13
PO Box 777
Rockford, IL 61105
Good selection of traditional vegetable varieties, plus some fruits and forage plants. Will sell vegetables in wholesale lots.

ROBINSON'S OAKVILLE NURSERY 3
29730 Harvest Drive
Albany, OR 97321
Highbush blueberries.

RUSSELL GRAHAM 2,4,9
4030 Eagle Crest Rd NW
Salem, OR 97304
Large selection of native plants, including four species of camas, avalanche lilies, fritillaria, and Oregon grape.

S & S SEEDS 2,4,7,12
910 Alphonse
Santa Barbara, CA 93103
Collectors and suppliers of a wide variety of plants suited to reclamation projects, including several nitrogen-fixing species. Wholesale and retail.

SANCTUARY SEEDS 4,6,11
1913 Yew Street
Vancouver, BC V6K 3G3
Non-hybrid vegetable seed, grown organically as much as possible. Also a variety of herb and wildflower seed.

SEED SAVERS EXCHANGE 6,9,10,11
c/o Kent Whealy
Rural Route 2
Princeton, MO 64673
Active network of gardeners saving and sharing heirloom varieties. Membership $6/yr US, $8/yr Canada.

SELF-SUFFICIENT SEEDS 6,11
Barr Hollow Road
Woodward, PA 16882
Devoted to serving homesteaders, this new company specializes in herb and vegetable varieties best suited for home production and preservation. Catalog includes detailed cultural, harvesting and storage information.

SHARP BROTHERS SEED CO 5,9
Healy, KS 67850
Specialists in native grass seed. Also seeds for a few forbs and shrubs.

SILVASEED CO 10,12
PO Box 118
Roy, WA 98580
Seed for major Pacific Northwest conifers. Bulk quantities, wholesale and retail.

SHEPPARD'S BULB FARM 4
6707 Bradner Rd, RR #1
Mount Lehnan, BC V0X 1V0
Flower bulb specialists.

SKINNER'S NURSERY 3,9,13
Roblin, Manitoba
Grafted fruits, berries, hedge plants, vines, and hardy perennials.

Reaping What We Sow

seeds and the crisis in agriculture

CARY FOWLER

All major food crops grown in North America originated elsewhere. Most come from poor, Third World countries—generally via Europe. On his second voyage to America in 1493, Columbus brought wheat, barley, sugarcane and grapes. The Spanish introduced alfalfa, oats, lemons, oranges, walnuts, cabbage, lettuce and peas. American Indians brought tomatoes and corn from Central and South America. Benjamin Franklin brought rhubarb and kale from Europe. And Thomas Jefferson smuggled upland rice out of Italy in his coat to introduce it to South Carolina. Each wave of immigrants came with seeds from their homeland, enriching and diversifying American agriculture.

Nearly all contemporary food crops are of ancient origin. Our Stone Age ancestors began to identify and domesticate plants for food thousands of years ago, saving their best seeds each year for replanting the following year. Our early ancestors developed and safeguarded certain varieties not only for their nutritional value and taste, but also for their resistance to local diseases and pests.

Human efforts and natural selection processes resulted in different varieties of food crops becoming adapted to "different niches in the ecosystem." The result was thousands of varieties of wheat, rice, corn and other crops as genetically distinct as beagles and Great Danes. We now know that this diversity was nature's way of protecting itself. As pests and diseases changed or mounted more powerful attacks, plants evolved different or better defenses. These defenses were represented by the genetically diverse varieties of each crop. Modern agriculture changed all that.

By the early 1950's, major efforts were underway at research centers supported by private and government sources to breed grains which would produce high yields when pumped full of fertilizers and water. Food crops, especially vegetables, were also bred to fit the demands of brutal harvesting machinery and the rigors of long-distance transportation.

Modern agriculture needs predictability; therefore, plant breeders strive for uniformity. Plants are bred and inbred to develop the desired characteristics. The result has been the creation of new varieties that are extremely genetically limited.

In 1970, a corn blight struck the U.S. crop. Old, open-pollinated varieties were not affected, but most farmers were growing the new, hybrid models. Nearly 15% of the nation's crop was

(Continued on next page)

SOUTHMEADOW FRUIT GARDENS 2,3,9,13
2363 Tilbury Place
Birmingham, MI 48009
Extensive collection of rare culinary apples, plus medlar and a wide variety of other deciduous fruits, gooseberries, grapes, quince, and native fruits for wildlife hedges. Dwarfing rootstock also available. No charge for their basic catalog. In addition, Southmeadow publishes a 112 page illustrated catalog with complete descriptions available of each variety's history, characteristics and culture ($8.00).

ST LAWRENCE NURSERIES 3,8,10
RD 2, State Rt 56A
Potsdam, NY 13676
American chestnut (seedlings and scionwood, exceptionally blight-resistant), beechnut, walnut, butternut, linden and a wide variety of grafted apples and pears, specializing in fireblight and scab resistant varieties and rootstocks.
Bill MacKently is now carrying on the work of the late Fred Ashworth, an internationally known plant breeder. They offer more apple varieties than any nursery in the US, all cold hardy (temperatures at the nursery routinely reach -30° F every winter).

STEVE RAY'S BAMBOO GARDENS 2
909 79th Place South
Birmingham, AL 35206
Bamboo growers for over 25 years. Their catalog describes the species they sell and gives the temperature they are hardy to. They have 50 acres planted in over 50 varieties of bamboo.

STOCK SEED FARMS 2,4,5,7
PO Box 112
Murdock, NE 68407
Specialists in native prairie grasses and wildflowers.

STRONGHOLD INC 10
Rt 1
Dickerson, MD 20753
A source of blight-resistant chestnut seed and seedlings. They have been involved in the cause of reestablishing the American chestnut for many years.

SUTTER NURSERY 2,3
3220 Silverado Trail North
St Helena, CA 94574
Offers scionwood of over 70 old time apple varieties. Send SASE for list.

SUTTONS SEEDS LTD 4,11
Hele Rd, Torquay
Devon, England TQ2 7QJ
Good selection of hardy vegetable varieties suited to year-round gardening.

TALBOTT NURSERY 3,8,10
RR 3, Box 212
Linton, IN 47441
Persimmons, hardy pecans, Chinese chestnuts, Carpathian walnuts, pawpaw, and mulberries.

TERRITORIAL SEED CO 11
PO Box 27
Lorane, OR 97451
Serving only gardeners west of the Cascades, Territorial offers quality, locally-adapted vegetable seed for year-round gardens.

TILLINGHAST SEED CO 3,4,5,11
La Conner, WA 98257
Catalog includes several varieties suited for the Maritime, including early tomatoes, "Skagit Gold" melons, and winter cabbages.

TREE CROPS NURSERY 2,3,8,10,13
76500 Short Creek Rd
Covelo, CA 95428
Specialists in deciduous fruit and nut trees, including many older and hard-to-find varieties. Also shrubs and trees for hedgerows, windbreaks, environmental control, forage, and woodlots.

(Continued from previous page)
destroyed. In some southern states the losses topped 50%. It was the disaster many scientists told us could never happen.

Could it happen again? Listen to what is being said:

"The array of diseases that pose threats to wheat, rice, maize, and sorghum is formidable. Except for the case of the Irish potato when neither plant pathology nor genetics had been born, research teams have been able to move fast enough to salvage our crops from complete devastation. But, we are becoming more and more vulnerable and there is no assurance that we can always react in time."

—Dr. Jack Harlan
Professor of Plant Genetics
University of Illinois

"The key lesson of 1970 (the year of the corn blight) is that genetic uniformity is the basis of vulnerability to epidemics. The major question the Committee on Genetic Vulnerability of Major Crops asked was, 'How uniform genetically are other crops upon which the nation depends, and how vulnerable, therefore, are they to epidemics?" The answer is that most crops are impressively uniform genetically and impressively vulnerable."

—National Academy of Sciences

"Genetic wipe-out... might well be tomorrow's greatest single problem—the problem of feeding the human species in the midst of the technological wealth it has created."

—Dr. Erna Bennett, FAO

Bringing diversity back to our food crops, stabilizing world food supplies and insuring the future of agriculture are goals we should all work towards. There is not an individual in the world that cannot do something. Everyone's contribution is important.

What can we do individually and collectively?

1. Support increased funding for collection and storage of our plant genetic resources, before those resources disappear forever. Storing seeds is only a partial solution, however. We need as many varieties as possible out in the real world, growing in their own diverse environments so they can continue to change and adapt. In addition all seeds eventually lose the life they hold within themselves. Periodically, all stored seeds must be taken out of storage and grown into plants whose seeds must be collected and stored afresh. This would be a monumental task; even the Fort Collins inadequate collection contains over 1,300 species.

2. Help promote "plant preserves." Here, wild ancestors of our major food crops could be allowed to live in safety much the same way that lions and elephants are protected in African game preserves. If traditional varieties are to be preserved, their environment must likewise be preserved. At present, "plant preserves" are more concept that reality. Public awareness of the need for plant preserves could help make them a reality.

3. Multinational corporate involvement in the seed industry should be closely monitored. Manufacturers of pesticides and fertilizers should not be allowed to own seed companies. Companies that export seeds to Third World areas should be

TSANG & MA INTERNATIONAL 11
1306 Old Country Road
Belmont, CA 94002
Over a dozen Chinese vegetable varieties.

UPPER BANK NURSERIES 2,3
PO Box 486
Media, PA 19063
Specialists in ornamental plants, including bamboos and jujubes.

VALLEY NURSERY 3,7,9,10,13
PO Box 4845
Helena, MT 59601
Plants are grown from seeds obtained from wild, northern sources or from highly competent, northern plant breeders. Wide selection of trees, shrubs and vines.

VAN NESS WATER GARDENS 1
2460 North Euclid Ave
Upland, CA 91786
Plants, supplies and books for aquatic habitats. Catalog contains information on ecological pond management.

VERMONT BEAN SEED CO. 11
Garden Lane
Bomoseen, VT 05732
Wide selection of garden vegetables for short-season gardens, plus an extensive collection of beans.

WELL-SWEEP HERB FARM 6
317 Mt Bethel Rd.
Port Murray, NJ 07865
Over 300 varieties of herb plants and scented geraniums. Many herb seeds and dried flowers are carried also.

WESTERN ONTARIO FRUIT TESTING ASSOCIATION 3,10
Agriculture Canada Research Station
Harrow, Ontario N0R 1G0
Seeds, trees and budwood for new varieties of apricot, nectarine, peach and pear.

WILDLIFE NURSERIES 1,13
PO Box 2724
Oshkosh, WI 54903
Specialists in plants and seeds for increasing sources of wildlife foods, in particular waterfowl and upland game birds. Their selection includes giant wild rice, wapato, sweet flag, waterlilies, sego pond lily, crownvetch, cicer milkvetch, and many more.

WILLIAM DAM 11,13
PO Box 8400, Dundas
Ontario, Canada L9H 6M1
Untreated seeds for cold-hardy vegetables and livestock crops.

WONDROUS LION NURSERY 3,8,9,10,13
Rt 1, Box 212-A
Sagle, ID 83860
Specialists in hardy, disease and drought resistant varieties of grafted and seedling fruit and nut trees, shrubs and vines, and seedling timber, conservation and rootstock trees adapted to Interior Northwest climates. Custom propagation of cultivars selected for adaptability to short growing season and cold winters. Emphasis on plants suitable for ecological farming.

WORCESTER COUNTY HORTICULTURAL SOCIETY 2,3
30 Elm Street
Worcester, MA 01608
Preservation orchard offering scion and budwood for over 50 varieties of antique apples.

ZILKE BROTHERS NURSERY 3,8,10
Baroda, MI 49101
Large selection of fruit and nut varieties, including American persimmons, mulberry, bush fruits and pawpaw. Catalog has very complete plant descriptions.

---

required to file statements documenting the environment impact of those seed exports. If old varieties will be replaced, the company should be responsible for seeing that they do not become extinct. If a company will not make this guarantee, it should be prohibited from marketing in a given area.

4. Oppose plant patenting legislation. Most governments (like Canada's) do not expect much public awareness over this issue; therefore, a few letters expressing concern would have a big impact.

5. Governments at all levels should be encouraged to offer marketing incentives to small farmers who grow the older varieties, for example, price supports for traditional varieties. Other government farm benefits could be offered to farmers willing to devote a small amount of acreage to endangered varieties.

6. Farmers should consider banding together to buy bulk quantities of traditional seeds and to market the produce in bulk or through farmers' markets. When selling the old varieties, farmers should help educate the public by labeling their produce with the name of the variety. Consumers can then learn to tell the difference and begin to put pressure on supermarkets to carry good produce.

7. Individuals, churches, community groups, colleges and town governments can begin to plant and safeguard the old varieties. In some areas, groups have established small "preservation" orchards devoted to traditional varieties of fruit trees. These efforts bring people together, offer good, healthy recreation and excellent food. They also serve to promote awareness of the problem, while contributing to its solution.

In the end, the future of agriculture can be insured only by healthy, vibrant small farms. The old varieties are threatened today, not because they taste bad or are nutritionally deficient, but because they do not suit the requirements of the factory farmers and the food processing industry. The California plantation owner who grows tomatoes to be shipped all over the country cannot grow the old, tasty varieties. Their skins are not tough enough. Their insides are not hard. If the old varieties are to flourish they must be, as they always have been, grown by small farmers and sold to a local market. This system of agriculture has provided sustenance to people for well over 10,000 years. It is an enduring agriculture that we tamper with only at great risk.

Seeds are a unique product of the efforts of people and nature. In seeds, culture and agriculture are linked. This bond dissolved, both are threatened. Our ancestors knew this and lived accordingly. Thomas Jefferson once professed his belief that "the greatest service which can be rendered to any country is to add a useful plant to its culture." For our generation, the challenge will be to preserve the useful plants we already have.

This excerpt was reprinted by permission from *The Graham Center Seed Directory*, available for $1.00 from the Frank Porter Graham Center, Rt 3, Box 95, Wadesboro, NC 28170.

Glossary

Acid rain: sulfur dioxide, released primarily by the burning of fossil fuels, combines with water in the atmosphere to create sulfuric acid. This acidic rain then falls on forests, soils, streams and lakes, making them less capable of supporting life.

Acid soil: soil with a pH value of 6.6 or below.

Actinomycetes: a group of soil microorganisms intermediate between bacteria and fungi that secrete digestive enzymes which break down organic matter in the soil.

Adaptation: an inherited structural, functional or behavioral characteristic that improves an organism's chances for survival in a particular habitat.

Aeration: the circulation of air in the soil. Decomposition of organic matter is fueled by oxygen, so the amount of air contributes to the well-being of the soil. Aeration is increased when soil structure is improved.

Afforestation: the process of introducing trees into an ecosystem where they did not previously grow.

Algae bloom: occurs when an algae population grows and covers the surface of water.

Alkaline soil: a soil with a pH value of 7 or above.

Allelopathy: the process whereby certain plants release toxins via roots, leaves or microbial action which inhibit the growth of other plants. Plants with allelopathic toxins include barley, black walnut, Canadian thistle, Jerusalem artichoke, nutsedge and sunflowers.

Annual: a plant that completes its life cycle from seedling to mature seed-bearing plant during a single growing season and then dies.

Apiary: a place where bees are kept.

Aquaculture: the intentional raising of fish and other water organisms for consumption, rather than harvesting from the wild.

Azotobacter: a group of soil bacteria which fixes nitrogen from the air.

Bare root plants: dormant plants dug from the nursery, with roots freed of soil and then protected against drying out prior to planting.

Bast fiber: fibrous material from certain plants, such as flax, hemp, yucca and nettle, which can be made into rope, mats, textiles or other products.

Beescape: landscape designed with flowering plants especially to meet the needs of bees.

Biennial: a plant that lives for two growing seasons, producing leaves during the first season, then flowering and setting seeds during the second.

Biodynamics: a science of agriculture inspired by Rudolf Steiner which recognizes the role of cosmological influences in agriculture and utilizes special composting methods.

Biomass: the weight of living material, usually expressed as dry weight, in all or part of an organism, population, or community. It is commonly expressed as weight per unit area, or biomass density.

Bioregion: an area with unique interlocking webs of life that are distinguishable from those of neighboring regions. The geographical area of a bioregion is often defined by watersheds.

Biosphere: the totality of the earth's ecological systems.

Brassica: plants in the *Cruciferae* family, include cabbage, broccoli, kale, cauliflower, radish and turnip.

Canopy: the layer formed by the leaves and branches of the forest's tallest trees.

Carbohydrates: compounds of oxygen, hydrogen and carbon that are formed and stored by plants.

Carrying capacity: the upper limit on the number of individuals of a particular kind of animal that can be supported within a given unit of habitat, such as an area of a stream.

Cation exchange capacity: a measure of a soil's ability to make nutrients available to plants.

Cellulose: the main ingredient in the material of plant cell walls. Cellulose rots relatively slowly and can contribute to soil quality by increasing the amount of organic matter. This improves the soil's water-holding capacity and its store of nutrients.

Chaparral: densely woven thickets of shrubs with stiff branches and thick, hard, evergreen leaves.

Chlorophyll: a group of green pigment compounds in plants. The site of photosynthesis.

Climax: a stable state reached by a plant community in a particular set of climatic and soil conditions.

Cloche: a movable translucent covering placed directly over plants in the garden. Cloches are used in cold weather, during the early stages of growth, and to aid in summer ripening.

Community: all the plants and animals in a particular habitat that are bound together by food chains and other interrelationships.

Compost: carbonaceous material that is formed by the slow decomposition of plant and animal materials. It is rich in nutrients essential for plants and microorganisms in a readily usable form.

Coppicing: the cutting of trees or shrubs which regenerate by sprouts or root suckers. Species suited to coppicing include filberts, alder, maple, lilac, dogwood, chestnut, and black locust.

Cover crops: after a main crop has been harvested, cover crops are sown to protect the soil from weather, fix nitrogen and, when incorporated in the soil, increase its organic matter content. Cover crops include annual rye, buckwheat, oats, barley, lupine, vetch, winter peas, and crimson clover.

Crop rotation: a crop system designed to grow successive crops with differing nutrient requirements on the same piece of ground over a number of growing seasons. Crop rotation helps prevent soil depletion, releases nutrients, controls weeds, and breaks disease cycles.

Decomposers: plants and animals, chiefly fungi and bacteria, that live by extracting energy from the decaying tissues of dead plants and animals. In the process they release simple chemical compounds, making them available to plants.

Degree-days: a measure of heat requirements. Degree-days are an indication of the extent of departure from a standard of mean daily temperature. Each crop has a different standard, and the number of degree-days indicates where plants will get enough heat to mature and ripen.

Desertification: the process of land becoming increasingly arid as a result of groundwater depletion, salinization, soil erosion and the destruction of vegetation.

Ecology: the scientific study of the relations of living things to one another and to their environment.

Economic botany: the study of plants with economic value, such as foods, fibers or medicines.

Ecosystem: all living things and their environment in an area of any size. All are linked by energy and nutrient flows.

Entropy: the influence of the second law of thermodynamics which states that, while the energy in the cosmos is always in constant supply and cannot be created or destroyed, it is always moving from an available state (eg. coal) to an unavailable state (eg. heat). Entropy measures the rate of change from available to unavailable energy.

Espalier: a tree or shrub trained to grow against a wall, trellis or other support.

Ethanol: alcohol derived from grain or other vegetable matter.

Ethnobotany: the study of the uses of plants by native peoples.

Exotic species: plants, fish or animals introduced into an ecosystem from elsewhere.

Exudates: substances passed into the soil by living plant roots.

Flax: a fibrous plant that is treated to separate its fibers, which can be spun and woven into linen. Flax seeds yield linseed oil.

Forb: any herbaceous plant other than grass which grows in a pasture or rangeland.

Gaia: the Greek earth goddess. Also a name for the earth regarded as one living organism.

Geomancy: the art of divining and channeling subtle energy currents within the earth. Also the study of earth as a living system.

Grafting: the union of a shoot or bud from one plant with another, so that the new tissue will be accepted by the host and continue to live, grow and bear fruit.

Green manure: plants grown specifically to turn back into the soil while still green, as a store of nutrients to increase fertility.

Habitat: a place where an organism normally lives, often characterized by a dominant plant form or physical characteristic, such as stream or forest habitat.

Herbaceous: a plant with soft tissue that dies back to the ground every year.

Herbal ley: temporary pasture composed of grasses, legumes and herbs (including burnet, chicory and plantain) grown as a sod crop for 2 or more years in rotation with cultivated crops to feed livestock and enrich the soil.

Host: a living plant or organism which provides food or living space for another organism.

Humus: the stable remains of decomposed organic matter. Humus retains moisture, and maintains soil structure and fertility.

Inoculation: the introduction of bacteria or other microorganisms into root cells or other parts of an organism or ecosystem.

Insectary plant: a plant which attracts insects.

Insolation: the amount of solar radiation falling on a given area.

Integrated pest management: an array of strategies for controlling pests by encouraging beneficial insects, manipulating vegetation and, if necessary, using pesticides.

Intercropping: growing two or more crops together on the same piece of ground.

Interior: in the Pacific Northwest, the Interior is the region between the Cascades and the crest of the Rockies.

Keyline: system developed by P.A. Yeomans in Australia for conserving water using underground channels and ponds on the contours of hillsides.

Land trust: a legal agreement that removes land from the speculative market, protecting it by covenants that determine its use.

Legumes: plants of the *Leguminosae* family which produce seeds in pods, such as peas, beans, lentils, clover, alfalfa, acacia and black locust. Many (but not all) legumes establish a symbiotic relationship with specific bacteria which have the capacity to fix atmospheric nitrogen in the soil.

Maritime: a geographical area where the climate is modified by proximity to an ocean or sea. In the Pacific Northwest the Maritime is the region west of the Cascade mountains.

Microclimate: the climate of a specific area within a larger region.

Microhabitat: a small habitat within a larger one in which environmental conditions differ from those in the surrounding area. A hole in a tree or an animal carcass are microhabitats within a forest.

Microorganism: an animal or plant of microscopic size, especially bacteria, fungi, actinomycetes and protozoa.

Monoculture: the raising of only one crop on a given piece of land.

Morant system: the housing of small animals, such as rabbits, in a fold during the day and enclosing them in a movable hutch at night. The fold and hutch are moved onto an ungrazed spot every day.

Mulch: a layer of organic matter put around growing plants to conserve moisture, control weeds, provide insulation from heat and cold, and provide nutrients as it decomposes.
Multiple story: a diverse mixture of trees, shrubs and ground layer plants.
Mutualism: the relationship between two or more species that benefits all parties to the association. Sometimes survival is not possible outside the association.
Mycorrhizae: bacteria and fungi living in symbiotic association with roots of higher plants.
Mycorrhizal association: relationship between host plants and mycorrhizae.
Natural selection: elimination or enhancement of genetic traits in a population through survivability and reproduction of the individuals bearing the traits.
Naturalize: the ability of some plants, when introduced into a new habitat, to reproduce and become an established part of the environment.
Niche: role or "profession" of an organism in the environment; its activities and relationships in the community. Also, the area occupied by a plant or animal.
Organic matter: material derived from once-living organisms. Also any substance containing carbon compounds.
Paleobotany: the study of the composition and dynamics of vegetation in past geologic ages.
Parasite: a plant or animal that lives in or on another living thing (its host) and obtains part or all of its food from the host's body.
Perennial: a plant whose life expectancy is longer than one or two growing seasons, and which produces fruit or flowers every year.
Permaculture: a word coined by Bill Mollison and the Tagari Group in Australia to describe the conscious design of permanent, self-sustaining agricultural ecosystems.
Pest: any creature in the garden or farm whose presence or activities is detrimental to the plants and animals being raised for consumption.
pH: symbol for potential of hydrogen ions. pH is the measure of acidity or alkalinity in soil.
Photosynthesis: the chemical process in plants in which energy from sunlight is used in the presence of chlorophyll to make complex carbohydrates and proteins in plant tissue.
Pleaching: a method of interweaving branches to form hedges or arbors.
Polyculture: the culture of multiple species in the same field or area.
Pome fruit: a fleshy fruit having seeds but no stone, such as an apple, pear or quince.
Propagation: the activity of producing a number of plants by a variety of means such as seeding, cutting, grafting and layering.
Pulse: any member of the legume family which produces edible seeds such as beans, peas and lentils.
Retting: soaking fiber plants, like flax, under water for varying periods of time to release individual fibers.
Rhizobia: bacteria which colonize the roots of legumes and form nodules in which atmospheric nitrogen is fixed, making it available to the plant.
Rhizome: a thickened, modified stem that grows horizontally under the soil surface.
Riparian: the habitat on the banks of rivers, streams and lakes.
Saline soil: soil with mineral salts concentrated in the root zone or on the surface, rendering it incapable of supporting the growth of most plants.
Savanna: a flat, treeless grassland.
Scarification: the treatment of the outer coat of seeds to insure germination. Procedures include slitting or soaking the seeds in hot water or acid.
Scutching: a process of beating the stems of fiber plants like flax to separate individual fibers from the stem structure and to soften the fibers.
Shelterbelt: a row of trees and shrubs which break the force of the wind and create protected areas behind them. They also provide food and shelter for wildlife.
Short rotation forestry: the management of forests to maximize biomass production in the short term, for fuel, fodder or soil building rather than for timber.
Shrub: a woody plant less than 12 feet tall, usually with more than one stem rising from the ground.
Soil fauna: the abundant animal populations of the soil, both visible and microscopic. They feed on each other or on organic matter in the soil as part of nutrient cycles which foster the continuous generation of life.
Stability: the capacity of any system to adapt to or incorporate change.
Stone fruit: a fruit in which a hard cover encloses the kernel, such as in a cherry or plum.
Steppe: 1. Grass steppe: plains with grass, herbs and wildflowers. 2. Meadow steppe: plains with meadow adjacent to mountain zone with tall, thick grasses. 3. Shrub steppe: plains with shrubs.
Stratification: exposure of certain kinds of seed to frost or cold temperatures enable them to germinate.
Subalpine forest: forest growing in mountainous regions just below the timberline.
Succession: replacement of one kind of community by another kind; the progressive changes in vegetation and animal population.
Sucker: a secondary shoot, rising from the roots of certain tree and shrub species, that can grow into a mature individual.
Sustainable: continuous, enduring, permanent.
Symbiosis: the relationship of 2 or more different organisms in a close association that may be, but is not necessarily, of benefit to each.
Tilth: the structure and quality of soil resulting from the combined effects of parent material, biological activity and management practices. Soil in good tilth is well aerated, retains moisture and is rich is humus.
Transpiration: the process by which water evaporates from the leaves of plants.
Trophic level: functional classification of organisms in a community according to feeding relationships and energy flows.
Water table: the upper surface of ground water or the level below which the soil is saturated with water.
Windbreak: a row of trees or bushes that protects the area on the leeward side from wind and heavy weather.
Woodlot: land planted with trees that are being grown for timber, fuelwood and craft materials.

Index

This Index cross-references common and Latin names used throughout the text and in the Plant Species Index (PSI) on pages 151-170. In most cases, the primary references for a plant named in the text are given under the plant's common name. Seed and Nursery Companies in the list on pages 171-180 are not referenced below. They are indexed with the Key on page 172.

Abies amabilis (Pacific silver fir), PSI 154; *A. concolor* (White fir), PSI 154; *A. grandis* (Grand fir), PSI 154; *A. procera* (Noble fir), PSI 154
Abutilon theophrasti (Indian mallow), PSI 163
Access to land, in cities, 72; in forests, 119; land trusts, 142-144
Acer circinatum (Vine maple), PSI 156; *A. macrophyllum* (Bigleaf maple), PSI 154; *A. saccharum* (Sugar maple), PSI 154
Achillea millefolium (Yarrow), PSI 163
Acorns, as chicken feed, 41; as swine forage 112
Actinidia arguta (Kiwi), 94; *A. chinensis*, 94, PSI 162; *A. kolomikta*, 94
Actinomycetes, definition, 181; actinorhizal plants, 36-38
Aesthetics, importance of, 26
African marigold *(Tagetes erecta)*, PSI 168
Agastache foeniculum (Anise hyssop), PSI 163
Agricultural ecology, bibliography, 28-29
Agricultural land preservation, bibliography, 144; see also Belden, 145
Agricultural organizations, 30
Agriculture, community, 72-75; origins, 70; urban 72-75
Agrisilviculture, 121
Agroforestry, in cities, 109; in third world, 120-121; with bamboo, 87; with nut trees, 99
Agrotriticum, as wildlife plant, 127; perennial grain 50-51
Ailanthus altissima (Tree of heaven), PSI 154
Akebia quinata (Fiveleaf akebia), PSI 162
Alaska cedar *(Chamaecyparis nootkatensis)*, PSI 154, description, 113; native habitat, 110,111
Albizia julibrissin (Silk tree), PSI 156
Alcohol fuel, bibliography, 65; experimental still, 48; from trees, 122; Small Farm Energy, 47; Tilth Producers' Coop, 150
Alder *(Alnus spp.)*, bibliography, 124; as green manure, 116; as nitrogen-fixing plant, 37; for shiitake production, 120; hardwood, 124; in pastures, 61; with bamboo, 88; see also Red alder
Alfalfa *(Medicago sativa)*, PSI 166; as bee forage, 130; as green manure, 30; as mulch 55; in grassland agriculture, 60; in wild garden, 82
Algae, in ponds, 136; red tide, 138; seaweed, 138
Allegheny chinkapin *(Castanea pumila)*, PSI 156; description, 101
Allium acuminatum (Rock onion), PSI 163; *A. ampeloprasum var. porrum* (Leek), PSI 163; *A. cepa var. viviparum* (Egyptian top set onion), PSI 163; *A. fistulosum* (Scallions), PSI 163; *A. sativum* (Garlic), PSI 164; *A. schoenoprasum* (Chives), PSI 164; *A. tuberosum* (Garlic chives), PSI 164
Almond *(Prunus amygdalus)*, description, 100; wild, 102; PSI 157
Alnus glutinosa (Black alder), PSI 154; *A. rhombifolia* (White alder), PSI 154; *A. rubra* (Red alder), PSI 154; *A. sinuata* (Sitka alder), PSI 156; *A. tenuifolia* (Mountain alder), PSI 156
Alpine blueberry, see Blueleaf huckleberry
Alsike clover *(Trifolium hybridum)*, PSI 168
Amaranth, as food crop, 65; in wild garden, 82-83; *see also* Rodale Research Center, 44
Amaranthus spp., as insectary plants 33; *A. cruentus* (Purple amaranth), PSI 164; *A. hypochondriacus* (Prince's feather), PSI 164; *A. retroflexus* (Pigweed), PSI 164
Amelanchier alnifolia (Saskatoon serviceberry), 91, PSI 156; *A. canadensis* (Shadblow serviceberry), PSI 156; *A. utahensis*, 91
American chestnut *(Castanea dentata)*, 101
American elder *(Sambucus canadensis)*, PSI 161
American hazelnut *(Corylus americana)*, PSI 159
Amercian linden *(Tilia americana)*, PSI 156
American lotus *(Nelumbo lutea)*, PSI 167
American persimmon *(Diospyros virginiana)*, PSI 157
American plum *(Prunus americana)*, PSI 157
Anabaena, nitrogen-fixing algae, 38
Angelica *(Angelica spp.)*, as insectary plant, 35
Animals, sowing seeds, 176; *see also* Livestock; Wildlife
Anise hyssop *(Agastache foeniculum)*, PSI 163; as bee forage, 130
Aphelinid parasitic wasps, insectary plants for, 34-35
Aphids, control by predators, 32-33
Apple *(Pyrus pumila)*, PSI 158; trees as wind shelter, 59; for pastures, 61; preferred habitats, 95
Apple, wild, 102
Appropriate technology, bamboo, 88; bibliography, 45-47; in sustainable agriculture, 26-52; organizations, 48, periodicals, 47
Apricot *(Prunus armeniaca)*, PSI 157; preferred habitats, 95; trees as wind shelters, 59; wild, 102; *see also* Sweetpit apricots, 100
Aprovecho, and Bamboo 88; Institute, 48
Aquaculture, definition, 181; bibliography, 136-137; in cities, 72, 73; *see also* Rodale Research Center, 44; Merrill, 45
Arachis hypogaea (Peanut), PSI 164
Arboreta, 22
Arbutus menziesii (Pacific madrone), PSI 154; *A. unedo* (Strawberry tree), PSI 158
Arid lands, bibliography, 63-64; desertification, 63; rangeland, 62
Armoracia rusticana (Horseradish), PSI 164
Arrowleaf balsamroot *(Balsamorhiza sagittata)*, PSI 164
Arroyo willow *(Salix laseolepis)*, PSI 158
Arundinaria amabilis (Tonkin cane), PSI 162; *A. simonii* (Madake, Simon bamboo), PSI 162
Asarum caudatum (Wild ginger), PSI 164
Asclepias cornuti (Milkweed), PSI 164; *A. incarnata* (Swamp milkweed), PSI 164
Ash, as hedgerow plant, 59
Asimina triloba (Pawpaw), PSI 156
Asparagus officinalis (Garden asparagus), in garden, 105; PSI 159
Aspen, native habitat, 112, 113
Asters, for bee forage, 130
Atlas cedar *(Cedrus atlantica)*, 113; PSI 154
Atlases, 19
Atriplex canescens (Four-wing saltbush), PSI 159
Autumn olive *(Elaegnus umbellata)*, for wildlife, 127; PSI 159
Avalanche lily *(Erythronium grandiflorum)*, as food, 20
Avena sativa (Oats), PSI 164
Azolla, as nitrogen-fixing plant, 37; see also FAO, 49
Azotobacter, definition, 181; nitrogen-fixing bacteria, 38

Baccharis pilularis (Coyote brush), PSI 159
Baker, Richard St. Barbe, 26, 140, 147
Balsamorhiza sagittata (Arrowleaf balsamroot), PSI 164
Bamboo, history and use, 86-87; references, 88
Bamboo Species Index, 162-163
Bambusa textilis, PSI 162
Baptisia australus (Blue false indigo), PSI 159
Barley *(Hordeum vulgare)*, in no-till farming, 55; PSI 166
Bats, shelters for, 128
Beechnuts, 103
Beekeeping, 129-130; in forests, 123; supplies, 130
Bees, beebliography, 129-130; for kiwi pollination, 104; in dump heap, 71; in orchard, 106; insectary plants for, 34-35; wind shelter for, 58-59
Berberis aquifolium (Tall Oregon grape), PSI 159; *B. nervosa* (Oregon grape), PSI 164
Berries, 90-93; as market crop, 29; bibliography, 91, 93
Berry, Wendell, 26; on abundance, 9; on agricultural responsibilities, 53; on planting trees, 25
Beta vulgaris var. *cicla* (Swiss chard), PSI 164
Big huckleberry *(Vaccinium membranaceum)*, 91
Big pod mariposa lily *(Calochortus nitidus)*, PSI 165
Bigleaf maple *(Acer macrophyllum)*, PSI 154; native habitat, 110
Big-leaved lupine *(Lupinus polyphyllus)*, PSI 160
Bio-dynamics, definition, 181; Northwest Bio-Dynamic Agricultural Society, 30; Soil tests, see Woods End Laboratory, 53
Bioregion, definition, 15, 181; models, 17
Birch, native habitats, 112
Bird's-foot trefoil *(Lotus corniculatus)*, for wildlife, 127; in grassland farming, 60; PSI 166
Bitter cherry *(Prunus emarginata)*, PSI 157
Bitterbrush *(Purshia tridentata)*, native habitats, 111; *see also* Vallentine, 62
Black alder *(Ainus glutinosa)*, PSI 154
Black bamboo *(P. nigra)*, PSI 163
Black cherry *(Prunus serotina)*, PSI 155
Black cottonwood *(Populus trichocarpa)*, native habitats, 113; PSI 155
Black currant *(P. nigrum)*, PSI 160
Black hawthorn *(Crataegus douglasii)*, PSI 157
Black locust *(Robinia pseudoacacia)*, as

bee forage, 130; wind shelter, 59; PSI 156
Black mulberry *(Morus nigra)*, PSI 157
Black mustard *(B. nigra)*, PSI 164
Black raspberry *(R. occidentalis)*, PSI 161
Black salsify *(Scorzonera hispanica)*, PSI 168
Black walnut *(Juglans nigra)*, as nut crop, 102; as wind shelter, 59; PSI 155
Blackberries, description of species, 92; native habitats, 111
Blackcap raspberry *(Rubus leucodermis)*, 92
Blue elderberry *(Sambucus caerulea)*, description, 92; PSI 161
Blue false indigo *(Baptisia australis)*, PSI 159
Blue mussel *(Mytilus edulis)*, 138
Blueblossom *(Ceanothus thyrsiflorus)*, PSI 159
Blue-green algae, as nitrogen source, 38; see also FAO, 49; R & A Plant/Soil, 68
Blueleaf huckleberry *(Vaccinium deliciosum)*, 91
Borage *(Borago officinalis)*, as bee forage, 130; edible flower, 76; in wild garden, 82; PSI 164
Borago officinalis (Borage), PSI 164
Botanical gardens, 22
Braconid parasitic wasps, insectary plants for 34-35
Brassica, definition, 181
Brassica napus (Rape), PSI 164; *B. nigra* (Black mustard), PSI 164; *B. oleracea* var. *acephala* (Kale), PSI 164; *B. oleracea* var. *gemnifera* (Brussels sprouts), PSI 164; *B. oleracea* var. *italica* (Italian broccoli), PSI 164; *B. rapa* var. *septiceps* (Turnip), PSI 164
Broad bean *(Vicia faba)*, PSI 169; *see also* Fava bean
Broom *(Genista pilosa)*, PSI 166
Brussels sprouts *(Brassica oleracea* var. *gemnifera)*, PSI 164
Buckthorn, native habitat, 110
Buckwheat *(Fagopyrum esculentum)*, as bee forage, 130; as green manure, 30; as insectary plant, 35; PSI 165
Bullfrogs, 137; toads, 127
Bunchberry dogwood *(Cornus canadensis)*, native habitats, 111; PSI 165
Burnet, for pastures, 61; see also University of Idaho Research and Extension Center, 42
Burr oak (Quercus macrocarpa), 103
Bush chinkapin *(Castanopsis sempervirens)*, PSI 159
Butterfly farming, 127
Butternut *(Juglans cinera)*, 102

Cabbage, protection from flea beetles, 31; root fly barrier, 77; *see also Brassica*
Cabbage aphid *(Brevicoryne brassicae)*, 31
Cabbage butterfly *(Pieris rapae)*, 31
Cabbage root fly *(Erioischia brassicae)*, barrier against, 77; using inter-cropping against, 31
Calendula *(Calendula officinalis)*, edible flower, 76; PSI 164
California coffeeberry *(Rhamnus californica)*,PSI 160
California lilac *(Ceanothus* spp.*)*, native habitat, 111
California redbud *(Cercis occidentalis)*, PSI 157
California wax myrtle *(Myrica californica)*, PSI 157
Calochortus nitidus (Big-pod mariposa lily), conservation of, 21; PSI 165
Calocedrus decurrens (Incense cedar), description, 113; PSI 154
Camas *(Camassia quamash)*, PSI 165
Camas lilies *(Camassia* spp.*)*, as food source, 83; conservation of, 21, 83; history, 112; native habitat, 111
Camassia quamash (Camas), PSI 165
Canadian buffaloberry *(Sheperdia canadensis)*, PSI 161
Cannabis sativa (Hemp), PSI 165
Canyon live oak *(Quercus chrysolepis)*, PSI 158
Caragana arborescens (Siberian peashrub), PSI 159
Carpathian walnut, 102
Carrots, as fodder crop, 61
Carthamus tinctorius (Safflower), PSI 165
Carya spp., 100; see also Hickories
Carya laciniosa (Shellbark hickory), PSI 154; *C. ovata* (Shagbark hickory), PSI 154
Cascade Range, climate, 16; vegetation, 111
Cascara sagrada *(Rhamnus purshiana)*, native habitat, 110; PSI 158
Castanea mollissima (Chinese chestnut), PSI 154; *C. pumila* (Allegheny chinkapin), PSI 156; *C. sativa* (Spanish chestnut), PSI 154; *C. sequinii*, 101; *C. vilmoriniana*, 101
Castanopsis chrysophylla (Golden chinkapin), PSI 154; *C. sempervirens* (Bush chinkapin), PSI 159
Catfish, 137
Cattail *(Typha latifolia)*, as food and fuel source, 137; PSI 161
Ceanothus spp., as green manure, 116; in chaparral, 114; nitrogen-fixing shrub, 37; *C. americanus* (New Jersey tea), PSI 159; *C. cordulatus* (Snowbush), PSI 159; *C. integirrimus* (Deer brush), PSI 159; *C. prostratus* (Mahala mat), PSI 165; *C. thyrsiflorus* (blueblossom), PSI 159; *C. velutinus* (Snowbush ceanothus), PSI 159
Cedar of Lebanon *(Cedrus libani)*, 113, PSI 154
Cedrus atlantica (Atlas cedar), PSI 154; *C. deodara* (Deodar cedar), PSI 154; *C. libani* (Cedar of Lebanon), PSI 154
Cellulose, definition, 181
Celtis reticulata (Netleaf hackberry), PSI 157
Cercis occidentalis (California redbud), PSI 157
Cercocarpus ledifolius (Curlleaf mountain mahogany), PSI 157; *C. montanus* (Plume tree), PSI 159
Chalcid parasitic wasps, insectary plants for, 34-35
Chamaecyparis lawsoniana (Port Orford cedar), PSI 154; *C. nootkatensis* (Alaska cedar), PSI 154
Cheat grass *(Bromus tectorum)*, 50, 56-57
Checker lily *(Fritillaria lanceolata)*, PSI 165
Chenopodium album (Lamb's Quarters), PSI 165
Cherry *(Prunus* spp.*)*, for pastures, 61; preferred habitats, 96; PSI 155; see also Blodgett, 98
Cherry elaeagnus *(Elaeagnus multiflora)*, PSI 159
Chestnut blight *(Endothia parasitica)*, 101
Chestnuts *(Castenea* spp.*)*, description, 101; for shiitake, 120; in coppiced woodlot, 122; in mixed planting, 106, 120; in pastures, 61; wild, 102
Chickens, Buff Orpingtons, 133; for eggs, 75; in permaculture design, 40-41
Chickweed *(Stellaria media)*, PSI 168
Chicory *(Cichorium intybus)*, as food source, 83; for pastures, 61; PSI 165
Children's education, 140-141
Children of the Green Earth, 140, 147
Chilean Strawberry *(Fragaria chiloensis)*, PSI 165
China, uses of azolla as fertilizer, 37, 38; use of bamboo, 86-87
Chinese chestnut *(Castanea mollissima)*, 101, PSI 154
Chinese gooseberries, see Kiwi
Chinese hazel *(Corylus chinensis)*, 100
Chinquapin *(Castanopsis* spp.*)*, description, 100; native habitat, 114; PSI 154
Chives *(A. schoenoprasum)*, 76, PSI 164
Chlorogalum pomeridianum (Soaproot), PSI 165
Chlorophyll, definition, 181; and hemoglobin, 36
Chloropoid flies, insectary plants for, 34-35
Chokecherry *(Prunus virginiana)*, PSI 158
Cichorium intybus (Chicory), PSI 165
Cities, agricultural responsibilities of, 53; design for, 40; design series, 17; forests in, 108-109; see also Planet Drum
Clary *(Salvia sclarea)*, PSI 167
Claytonia lanceolata (Lance-leaved springbeauty), PSI 165; *C. siberica (Siberian montia)*, PSI 165
Cleavers, *(Galium aparine)*, PSI 166
Climate, in landscape design, 41; bibliography, 41
Cloches, definition, 181; use, 80-81
Clover, against cabbage root fly, 31; in grassland agriculture, 60-61; in no-till farming, 54-55
The Clover Project, 13
Coast redwood *(Sequoia sempervirens)*, PSI 156
Cockspur hawthorn *(Crataegus crus-galli)*, PSI 157
Coffeeberry *(Rhamnus californica)*, as insectary plant, 35
Cold frames, 80-81
Columbia hawthorn *(Crataegus columbiana)*, PSI 157
Columbia River Valley, climate, 16
Comfrey *(Symphytum officinale)*, as edible flower, 76; as fodder crop, 61; PSI 168
Commercial production, 29; marketing, 66-67
Community, 139-150; agriculture, 72-75; goal of sustainability, 13; importance of, 26; organizations, 146-148; stream rehabilitation, 135; Tilth chapters, 149-150
Community forestry, 119, 120-121; in cities, 108, 109
Community gardening, 74
Companion Planting, against pests, 31; to attract beneficial insects, 32-35
Compost, definition, 181; in cities, 73; in dump heap garden, 71; as energy source, 46; and manure, 131; see also soils bibliography, 53
Composting greenhouses, 46
Computers, in farming, 39, 52
A Continuous Harmony: Essays Cultural and Agricultural, 25
Cooperation, forest workers, 119; urban-rural, 75
Cooperative Extension Services, 68
Coppicing, definition, 182; bibliography, 122-123; as design tactic, 40; for fuel, 122-123; see also Edlin, 45
Coriander *(Coriandrum sativum)*, as insectary plant, 35
Cork oak, *(Quercus suber)*, 103
Corn *(Zea mays)*, blight, 178-179; compared to cattail, 137; for wildlife, 127; in association with alder, 37; in dump heap garden, 71; sod planting, see Rodale Research Center, 44; varietal testing, 84; with blue-green algae, 38
Corn borer *(Ostrinia furnicalis)*, 31
Cornelian cherry dogwood *(Cornus mas)*, as wind shelter, 59; PSI 159
The Cornucopia Project, 10
Cornus canadensis (Bunchberry dogwood), PSI 165; *C. mas* (Cornelian cherry dogwood), PSI 159
Corylus americana (American hazelnut), PSI 159; *C. avellana* (European filbert), as breeding stock, 100; PSI 157; *C. colurna* (Turkish hazel or columnar hazel), PSI 154; *C. cornuta* var. *californica* (Western hazel), native, 100; PSI 159; *C. heterophylla* (Siberian hazel), PSI 159; *C. maxima* (Giant filbert), as breeding stock, 100; PSI 157; see also Filberts
Cottonwood, as bee forage, 130

Coyote brush *(Baccharis pilularis)*, as insectary plant, 35; PSI 159
Coyote thistle *(Eryngium* spp.) as insectary plant, 34
Coyotes, in sheep loss, 134
Crabapple *(Malus diversifolia)*, in hedgerows, 59; native uses, 95; preferred habitats, 95, 110
Cranberries *(Vaccinium* spp.), PSI 168-169; see also Coastal Washington Research and Extension Unit, 43
Crataegus columbiana (Columbia hawthorn), PSI 157; *C. crus-galli* (Cockspur hawthorn), PSI 157; *C. douglasii* (Black hawthorn), PSI 157; *C. monogyna* (English hawthorn), PSI 157
Crayfish, 136, 137
Credit, sources, 68
Creeping thyme *(Thymus serpyllum)*, PSI 168
Crimson clover *(Trifolium incarnatum)*, PSI 168
Crop rotation, definition, 182; importance of grasses and legumes, 60
Crops, for commercial production, 29; new, 65; marketing, 66-67
Cryptomeria japonica (Japanese cryptomeria), PSI 155
Curlleaf mountain mahogany *(Cercocarpus ledifolius)*, PSI 157
Currants, 90
Cydonia oblonga (Quince), PSI 159
Cynara scolymus (Globe artichoke), PSI 159
Cyprus Cedar *(Cedrus brevifolia)*, 113
Cytisus praecox (Warminster broom), PSI 159
Damson Plum *(Prunus institia)*, PSI 160
Dandelion *(Taraxacum officinale)*, in wild garden, 82; PSI 168
Daylily *(Hemerocallis fulva)*, 76; PSI 160
Dawn redwood *(Metasequoia glyptostroboides)*, 105; PSI 155
Deer brush *(Ceanothus integirrimus)*, PSI 159
Degree-days, definition, 182; computers and, see 97
Deodar cedar *(Cedrus deodara)*, 113; PSI 154
Desertification, definition 182; bibliography, 63; planetary, 147
Design, 40-41; gardens, 70-71; of orchards, see Logsdon, 98; principles, 12-13; windbreaks, hedges, and shelterbelts, 58-59; see also Planet Drum, 17
Designing landscapes, 40-41; for livestock, 60-61; for wildlife, 127; see also Robinette, 59
Devil's club, native habitat, 110
Dewberry *(Rubus ursinus)*, 92; PSI 167
Diamondback moth *(Plutella xylostella)*, intercropping to discourage, 33
Digger pine *(Pinus sabiniana)*, as source of pine-nuts, 101
Diospyros kaki (Japanese persimmon), PSI 157; *D. virginiana* (American persimmon), PSI 157; *see also* 96 and 105
Disease resistance, in peas, 43
Diversity, and complexity, 41; importance of, 26; in crops, 178-180; in forestry, 116-117, 122-123
Dog rose *(Rosa canina)*, PSI 161
Dogs, for sheep management, 134
Douglas-fir *(Pseudotsuga menziesii)*, in association with red alder, 124; with bamboo, 88; for grazing livestock, 61; manual release of, 122; native habitat, 110, 111, 112; in Ohme Garden, 23, in reforestation, 114; PSI 155
Downy chess (Cheat grass), 56-57
Draft Animals, 131; Fjord horses, 133; horse logging, 118-119
Dryas octopetala (Mountain avens), PSI 165
Dryland farming, 64; green manures for, 30; in eastern Washington, 49-52; no-till, 54
Ducks, 136, 137
Duckweed *(Lemna minor)*, as feed, 137; PSI 166
Dump heaps, 70-71
Earthcare, Ecological Agriculture in Saskatchewan, 28
Earthworms, as fish food, 137; in gardening, 78; see also Barrett and Darwin, 53; Livingston and Associates, 44
East Indian lotus *(Nelumbo nucifera)*, PSI 167
Eastern filbert blight, 100
Eco-Decentralist Design Series, 17
Ecology, definition, 182; and agriculture, 28-29; in design, 40-41
Economic botany, definition 182; bibliography, 65
Ecosystem, definition, 182; in farming, 28-29
Edible flowers, 76; camas lily, 21, 83, 112; in the garden, 75; yellow avalanche lily, 20; roses, 92; sego lily, 21
Edible Landscaping, see Britz, 40; Creasy, 40; Gessert, 76
Education, 140-141
Egyptian top set onion *(Allium cepa* var. *viviparum*, PSI 163
Elaeagnus angustifolia (Russian olive), PSI 157; *E. commutata* (Silverberry), PSI 159; *E. multiflora* (Cherry elaeagnus), PSI 159; *E. umbellata* (Autumn olive), PSI 159
Elderberry *(Sambucus* spp.), as hedgerow plant 59; in mixed orchard, 105; native fruit, 92
Elm, as hedgerow plant, 59
Employment opportunities, 119, 123; in agriculture, 140
Englemann spruce *(Picea engelmannii)*, native habitat, 111, 112; PSI 155
English hawthorn *(Crataegus monogyna)*, PSI 157
English holly *(Ilex aquifolium)*, PSI 157
English lavender *(Lavandula angustifolia)*, PSI 160
English walnut *(Juglans regia)*, PSI 155
Entropy, definition, 11, 182
Entropy, A New World View, 11
Epilobium angustifolium (Fireweed), PSI 165
Equisetum arvense (Horsetail), PSI 165; *E. hyemale* (Scouring rush), PSI 165
Eriogonum spp. (Wild buckwheat) PSI 165
Ethnobotany, definition, 182; *see also* Native people, 20
Eucalyptus *(Eucalyptus gobulus)*, 105; in China, 120; in urban forestry, 109
European buckthorn *(Rhamnus cathartica)*, PSI 158
European chestnut *(Castanea sativa)*, in Northwest, 101, 109; PSI 154
European filbert *(Corylus avellana)*, PSI 157
European mountain ash *(Sorbus aucuparia)*, PSI 158
Eurotia lanata (Winterfat), PSI 165
Evergreen blackberry *(Rubus laciniatus)*, in Maritime lowlands, 92; PSI 161
Evergreen euonymous *(Euonymous japonica)*, as insectary plant, 35
Evergreen huckleberry *(Vaccinium ovatum)*, PSI 161
Export of crops, balance with green manures, 30; impact on the farm, 29; foreign trade, 9; international politics, 50
Extension Services, 68
Extinction, of rare animal breeds, 133; of seed varieties, 178-180
FAO (Food & Agriculture Organization), 120-121
Fagopyrum esculentum (Buckwheat), PSI 165
Farallones Institute, 48, 73
Farm, credit, 68; organizations, 30; supplies, 68; tools, 45
Farmers Home Administration, 68
Farmers' Markets, 66, 150
Farming, 49-68; bibliography, 49-52; ecological principles in, 28-29; tool sources, 45; with computers, 39
Faust, Robert H., research, 38, 68
Fava bean *(Vicia faba)*, in wild garden, 82; see also Northwest Washington Research and Extension Unit, 44
Fencing, sources, 61; windbreaks, 58-59
Fennel *(Foeniculum vulgare)*, as insectary plant, 34; PSI 165
Fenugreek *(Trigonella foenum-graecum)*, PSI 168
Fertilizer, commercial sources, 68; manure, 131; from fossil fuels 36; from urine, 73
Fiber plants, bibliography, 46, 47
Ficus carica (Fig), PSI 157
Field pea *(Pisum arvense)*, PSI 167
Fig *(Ficus carica)*, 71, 96, 98; in mixed planting, 105, 120; wild, 102; PSI 157
Filberts *(Corylus* spp.), as wind shelters, 59; for pastures, 61; inoculated with truffles, 115; species descriptions, 100
Fire, design considerations, 41; for range improvement, see Eastern Oregon Agricultural Experiment Station, 43; in cheat grass, 57; in juniper and chaparral, 114; see also Lotan, 29; Wright, 62
Fireweed *(Epilobium angustifolium)*, as bee forage, 130; for habitat improvement, 123; PSI 165
Firewood, 122, 123
Fish, aquaculture, 136-137; fisheries enhancement, 135
Fiveleaf akebia *(Akebia quinata)*, PSI 162
Fjord horses, 133
Flax *(Linum usitatissimum)*, definition, 182; as fiber crop, 47, 65, 67; in wild garden, 82; PSI 166
Flea beetles, intercropping against, 31
Flies, insectary plants for, 34-35
Flora and fauna, bibliography, 19
Flowering currant *(Ribes sanguineum)*, description, 90; PSI 160
Flowers, as market crop, 29; wildflowers, 21; see also Edible flowers
Fodder crops, 60-61, 120; coppiced trees, 122; shrubs, 59
Foeniculum vulgare (Fennel), PSI 165
Forb, definition, 182; for habitat improvement, 122-123
Forest ecology, bibliography, 118
Forest management, 116-123; bibliography, 116; see also Research centers, 42-44
Forest organizations, 114, 117, 119; periodicals, 117; workers, 119
Forestry tool suppliers, 45
Forests, bibliography, 110-113; livestock grazing in, 61; mycorrhizae, 115; native Northwest, 110-113; urban, 108-109
Fossil fuels, and entropy, 11; as fertilizer source, 36
Four-wing saltbush *(Atriplex canescens)*, PSI 159; see also Vallentine, 62
Fox grape *(Vitis labrusca)*, PSI 162
Fragaria chiloensis (Chilean strawberry), PSI 165
F. virginiana (Virginia strawberry), PSI 165
Frankia, nitrogen-fixing fungi, 37
Franklin County (WA), history, 49-52
Fritillaria lanceolata (Checker lily), PSI 165; *F. pudica* (Yellow fritillary), PSI 165
Fruit, 89-106; in cities, 108, 109; in forests, 121, 123; as market crop, 29
Fruit and Nut Inventory, 95
Fruit and nut forests, 72, 123; of central Asia, 102; public plantings, 108, 109
Fruit flies *(Oscinella frit)*, 31
Fuel crops, bibliography, 65; for market, 29; in urban forests, 109
Fuelwood, coppiced, 122, 123; shelterbelts for, 58
Fukuoka, Masanobu, development of natural farming, 11-13; interplanting with acacia, 37; grain growing, 54-55; semi-wild vegetable growing, 82
Fungi, 115; see also Mycorrhizae

Galium aparine (Cleavers), PSI 166
Game management, 126, 127

Garden hyssop *(Hyssop officianalis)*, as bee forage, 130
Gardening, 69-88; bibliography, 72-74, 76-77, 80-81
Gardens, medicinal, 79
Garlic *(Allium sativum)*, PSI 164
Garlic chives *(A. tuberosum)*, PSI 164
Garry oak, or Oregon white oak *(Quercus garryana)*, description, 103; PSI 155
Gaultheria procumbens (Wintergreen), PSI 166
Gaultheria shallon (Salal), PSI 159
Geese, for weed control, 57
Genetic preservation, butterflies, 127; Computerized Plant Germplasm Inventory, 39; Fruit and Nut Inventory, 95; Henry Doubleday Research Association, 44; Kusa Research Foundation, 44; Northwest Plant Germplasm Repository, 43; public gardens, 22; rare animals, 133; seeds, 178-180
Genista pilosa (Broom), PSI 166
Giant filbert *(Corylus maxima)*, PSI 157
Giant sequoia *(Sequoiadendron giganteum)*, PSI 156
Ginkgo *(Ginkgo biloba)*, 99; PSI 155
Ginkgo biloba (Ginkgo), PSI 155
Ginseng (Panax quinquefolius), PSI 167
Gleditsia triacanthos (Honey locust), PSI 155
Globe artichoke *(Cynara scolymus)*, PSI 159
Golden bamboo *(Phyllostachyus aurea)*, PSI 162
Golden chinkapin *(Castanopsis chrysophylla)*, native habitat, 100; PSI 154
Golden currant *(Ribes aureum)*, PSI 160
Goldenchain tree *(Laburnum anagyroides)*, in orchard, 105
Goldenrod, as bee forage, 130
Gooseberries *(Ribes spp.)* native species, 90; natural habitat, 111
Gooseberry *(Ribes divaricatum)*, PSI 160
Gossypium hirsutum (Upland cotton), PSI 166
Grain cleaners, *see* By Hand & Foot, Ltd., 68; Sears, 45
Grains, perennial, 50-51; *see also* Kusa Research Foundation, 44; Logsdon, 50
Grand fir *(Abies grandis)*, in Ohme Garden, 23; native habitat, 110, 112; PSI 154
Grapes, bibliography, 94; in mixed orchards, 105, 106; for wildlife, 127; *see also* North Willamette Agricultural Experiment Station, 43
Grassland farming, 60-61
Grasshopper control, *see* Rangeland Insect Laboratory, 43
Grazing, in forests, 61, 123; in pastures, 60-61; in rangeland, 62
The Green Front, 147
Green manure, for nitrogen-fixation, 37, 38; in dryland farming, 30; trees and shrubs as, 37, 116
Greenhouses, bibliography, 46, 47, Seattle Tilth, 149
Ground beetles *(Lebia spp.)* attracting with weeds, 33
Ground cherry *(Physalis pruinosa)*, PSI 167
Guard dogs, 134
Guinea pigs, as food, 127

Habitat, definition, 182; improvement 123, 126-128, 135
Hairy vetch *(Vicia villosa)*, PSI 169
Hardwoods, alder, 124; chestnuts, 101; manual release, 122; maple, 124; walnuts, 102
Hawthorn, *(Crataegus spp.)*, in pastures, 61
Hazelnuts (corylus spp.), as coppice, 122; native habitat, 110; species descriptions, 100; for wildlife, 127
Heartsease, as bee forage, 130
Heather, as bee forage, 130; as hedgerow plant, 59
Hedgerows, 58-59; benefits of, 172; as livestock food, 61; as wildlife shelter, 61; for attracting beneficial insects, 32
Helianthus annuus (Sunflower), PSI 166; *H. tuberosus* (Jerusalem artichoke), PSI 159
Hemerocallis fulva (Tawny daylily), as edible flower, 76; PSI 160
Hemoglobin, in legume nodules, 36
Hemp *(Cannabis sativa)*, PSI 165
Henon bamboo *(P. henonis, P. nigra* cv *henon)*, PSI 163
Herbal ley, definition, 182; for livestock, 60-61
Herbs, bibliography, 79; forests as source of, 120, 121
Hickories *(Carya* spp.), species descriptions, 100
Highbush blueberry *(Vaccinium corymbosum)*, PSI 161
Highbush cranberry *(Virburnum opulus)*, 92
Himalayan blackberry *(Rubus procerus)*, in Maritime lowlands, 92; PSI 161
Hippophae rhamnoides (Buckthorn), PSI 160
Holistic forestry, 116-119
Holly, as hedgerow plant, 59
Holly oak *(Quercus ilex)*, PSI 155
Holodiscus discolor (Ocean spray), PSI 160
Homesteading, in forests, 119; history of Franklin County, 49-52; *see also* Belanger, 131; Kern, 50; Seymour, 51; Thomas, 51
Honey, as market crop, 29; from forests, 121; *see also* Beekeeping
Honey locust *(Gleditsia triacanthos)*, in pastures, 61; PSI 155
Honeysuckle, as hedgerow plant, 59
Hops *(Humulus lupulus)*, PSI 162
Hordeum vulgare (Barley), PSI 166
Hornets, insectary plants for, 34-35
Horseradish *(Armoracia rusticana)*, PSI 164
Horsetail *(Equisetum arvense)*, PSI 165
Horses, as draft animals, 131; logging with, 118-119, 133
Horticulture therapy, 74
Huckleberries *(Vaccinium* spp.), as bee forage, 130; management with sheep, 123; native habitat, 110; descriptions, 91
Humulus lupulus (Hops), PSI 162

Ilex aquifolium (English holly), PSI 157
Imported plants, 171
Incense cedar *(Calocedrus decurrens)*, description, 113; habitat, 110; PSI 154
Indian hemp *(Apocynum cannabinum)*, description of uses, 46
Indian mallow *(Abutilon theophrasti)*, PSI 163
Indian plum *(Oemleria cerasiformis)*, PSI 160
Indians, see Native Americans
Insectary plants, 31-35; definition, 182
Insects, as chicken feed, 41; beneficial, 32-35; in weed control, *see* Altieri and Story, 56; *see also* Pests
Integrated Pest Management (IMP), 31-35, 97; definition, 182
Intercropping, definition, 182; agrisilviculture, 120-121; in gardens, 72; see also Polyculture
Interior Northwest, definition, 182; physiographic description, 16
Irish Dexter cattle, 133
Irrigation, drip, 52; large-scale developments, 64; *see also* Salter, 77
Italian broccoli *(Brassica oleracea* var *italica)*, PSI 164
Italian stone-pine *(Pinus pinea)*, as source of pine-nuts, 101
Ivy *(Hedera* spp.), as insectary plants, 35; as winter forage, 59

Japanese chestnut *(Castanea crenata)*, blight resistance, 101; for shiitake growing, 120
Japanese cryptomeria *(Cryptomeria japonica)*, PSI 155
Japanese persimmon *(Diospyros kaki)*, PSI 157
Japanese plum *(Prunus salicina)*, PSI 160
Japanese quince *(Chaenomeles lagenaria)*, 96
Japanese sand pear *(Pyrus pyrifolia* var *culta)*, PSI 158
Japanese walnut *(Juglans sieboldiana)*, 102
Japanese wisteria *(Wisteria floribunda)*, PSI 162
Jeffrey pine, native habitat, 110
Jerusalem artichoke *(Helianthus tuberosus)*, as fodder crop, 61; as fuel crop, 65; in shelter belts, 127; PSI 159
Juglans nigra *(Black walnut)*, PSI 155; J. regia *(English walnut)*, PSI 155
Juniper forests, description, 113-114; grazing, 61; preservation efforts, 114

Kale *(Brassica oleracea* var *acephala)*, as fodder crop, 61; from Europe, 178; in winter garden, 80; PSI 164
Kama, tool for natural farming, 13, 82
Kellogg oak, native habitat, 111
Keyline, definition, 182; see Yeomans, 64
Kiwi *(Actinidia chinensis)*, species description, 94; in Puget Sound area, 94, 104-105; PSI 162
Klamath Mountains, vegetation, 111
Klamath plum *(Prunus subcordata)*, PSI 158
Korean Stone Pine *(Pinus koreanis)*, as source of pine-nuts, 101

Laburnum alpinum *(Scotch laburnum)*, PSI 157
Ladybugs, insectary plants for, 34-35
Lacewings, insectary plants for, 34-35
Lambs' quarters *(Chenopodium album)* as calcium source, 83; in wild garden, 82; PSI 165
Lance-leaved springbeauty *(Claytonia lanceolata)*, PSI 165
Land preservation, 142-144
Land trust, definition, 182; bibliography, 143; organizations, 142, 144
Landscape design, 40-41; *see also* Edible landscaping
Lavandula angustifolia (English lavendar) PSI 160
Leek *(Allium ampeloprasum* var *porrum)*, PSI 163
Legumes, definition, 182; and hemoglobin, 36; as green manure, 30; as living mulch, 55; in agrisilviculture, 121; in grassland agriculture, 60
Lemna minor (Duckweed), PSI 166
Lemon balm *(Melissa officinalis)*, PSI 166
Leopold, Aldo, 56
Leucana, 127; in urban forestry, 109
Ley, *see* Herbal ley
Lilac *(Syringa vulgaris)*, PSI 161
Lilies, conservation of, 21; camas, 83
Limber pine *(Pinus flexilis)*, as source of pine-nuts, 101
Linden, in pasture, 61
Linum usitatissimum (Flax), PSI 166
Livestock, bibliography, 131-132; in cities, 72-73; guard dogs, 134; management, 131; winter food for, 59, 61
Lodgepole pine *(Pinus contorta* var. *latifolia)*, native habitat, 110-112; PSI 155
Lotus corniculatus (Bird's foot trefoil), PSI 166
Lowbush blueberry *(Vaccinium angustifolium)*, PSI 168
Lupinus spp., as green manure, 30; *L. albifrons* (Silver lupine), PSI 160; *L. albus* (White lupine), PSI 166; *L. littoralis* (Seashore lupine), PSI 166; *L. luteus* var. *sativas* (Yellow lupine), PSI 166; *L. polyphyllus* (Big-leaved lupine), PSI 160

Madrone *(Arbutus menziesii)*, native habitat, 110; PSI 154
Mahala mat *(Ceanothus prostratus)*, PSI 165
Mail order, as marketing technique, 67
Manchurian walnut *(Juglans manchurica)*, 102
Mangle beet, as fodder crop, 61
Manual release, *vs.* herbicides, 122
Manure, as pond fertilizer, 136; livestock, 131; value in city gardens, 73
Map making, 41
Maple *(Acer* spp.), as bee forage, 129;

as hardwood, 124; as wind shelter, 59; for pastures, 61
Mariculture, 136, 137-138
Mariposa Foundation, 21, 83
Maritime, definition, 182; physiographic description, 16
Maritime forests, 110
Marketing, 66-67; and horticulture therapy, 74; new crops, 65
Marketing cooperatives, 67; Tilth Producers' Coop, 150
Matilija poppy *(Romneya coulteri)*, PSI 161
Maypop *(Passiflora incarnata)*, PSI 162
Meadow fescue, for pastures, 61
Meadow foam, as oil crop, 65; see also Oregon Agriculture Station, 43
Madake, Giant timber bamboo *(P. bambusoides)*, PSI 162
Madake, Simon bamboo, *(Arundinaria simonii)*, PSI 162
Medicago sativa (Alfalfa), PSI 166
Medicinal gardens, 79; plants, 79, 120-121
Medlar *(Mespilus germanica)*, as hedge plant, 59; description, 96
Melissa officinalis (Lemon balm), PSI 166;
Mentha piperita (Peppermint), PSI 166; *M. spicata* (Spearmint), PSI 166
Metasequoia glyptostroboides (Dawn redwood), 105; PSI 155
Methane, *see* Merrill, 45; Pain, 46; FAO, 49
Microclimate, definition, 182; *see also* Design, 41
Milkweed *(Asclepias cornuti)*, PSI 164
Millet, *(Panicum miliaceum)*, wildlife, 127; PSI 167
Mimulus guttatus (Monkey flower), PSI 166
Miner's lettuce *(Montia perfoliata)*, PSI 166
Mint, as bee forage, 130
Minute pirate bugs, insectary plants for, 34-35
Mollison, Bill, permaculture concepts, 11-13; permaculture design, 40-41
Monkey flower *(Mimulus guttatus)*, PSI 166
Monterey pine *(Pinus radiata)*, PSI 155
Montia perfoliata (Miner's lettuce), PSI 166
Morus alba (White mulberry), PSI 155; *M. nigra* (Black mulberry), PSI 157; *M. rubra* (Red mulberry), PSI 155
Moso bamboo *(Phyllostachys pubescens)*, edible shoots, 87; PSI 163
Mountain alder *(Alnus tenuifolia)*, PSI 156
Mountain ash, as wind shelter, 59
Mountain avens *(Dryas octopetala)*, PSI 165
Mountain hemlock, native habitat, 110
Mountain mahogany *or* Plume tree *(Cercocarpus montanus)*, PSI 159; see also Vallentine, 62
Mulberries *(Morus* spp.*)*, description, 97; as chicken feed, 41; for pastures, 61; for shelterbelts, 59; for wildlife, 127
Mulch, definition, 183; for dryland farming, 52; for tender plants, 105; gardening, 78; living mulch, 54-55
Mullein *(Verbascum thapsus)*, PSI 169
Multiple product forestry, 121, 122-123
Mushrooms, bibliography, 88; supplies, 120; in integrated forestry, 120; sources, 88, 120; truffles, 115
Mycorrhizae, definition, 183; in forests, 115; *see also* Forest Sciences Laboratory, 43
Myrica californica (California wax myrtle), PSI 157; *M. pennsylvanica* (Northern bayberry), PSI 157
Myrrhis odorata (Sweet cicely), PSI 166

Nanking cherry *(Prunus tomentosa)*, PSI 160
Narihara bamboo *(Semiarundiunaria fastuosa)*, PSI 163
Nasturtium *(Tropaeolum majus)*, 76; in wild garden, 82; PSI 168
Nasturtium officinale (Watercress), PSI 166
Native Americans, 90-93; food plants of, 20; *see also* Turner, Nancy
Native plants, bibliography and organizations, 21; sources of, 171
Natural farming, 11-13; grains, 54-55; vegetables, 82
Neighborhood gardens, 72-75
Nematodes, controlling, 31
Netleaf hackberry *(Celtis reticulata)*, PSI 157
Nettles, *see* Stinging nettle
New crops, bibliography, 65
New Jersey tea *(Ceanothus americanus)*, PSI 159
Nitrogen, from urine, 73; in agriculture, 36-38; *see also* Nitrogen-fixing plants
Nitrogen-fixing plants, 36-38; bibliography 37-38; as green manure, 30; in agrisilviculture, 121; in pastures, 61; in shelterbelts, 59
Nitrogen-fixing plants in Species Index: *Albizia*, 156; *Alnus*, 154, 156; *Baptisia*, 159; *Caragana*, 159; *Ceanothus*, 159, 165; *Cytisus*, 159; *Dryas* 165; *Elaeagnus*, 157, 159; *Genista*, 166; *Hippophae*, 160; *Laburnum*, 157; *Lotus*, 166; *Lupinus*, 160, 166; *Medicago*, 166; *Myrica*, 157, 160; *Onobrychis*, 167; *Pisum*, 167; *Robinia*, 156, 160; *Sheperdia*, 161; *Trifolium*, 168; *Vicia*, 169
Noble fir *(Abies procera)*, native habitat, 110; PSI 154
Nootka rose, *see* Wild Rose
Nori, 138
Northern bayberry *(Myrica pennsylvanica)*, PSI 160
Northern wild rice *(Zizania aquatica)*, PSI 169
No-till, 54-55; bibliography, 54-55; farming techniques, 13; gardening, 78; in the city, 75; *see also* Rodale Research Center, 44
Nuphar polysepalum (Spatterdock), PSI 167
Nursery management, bibliography, 76; inoculation of seedlings, 115
Nut trees, 99-103, 106; bibliography, 102-103; organizations, 103
Nutrient cycles, 53, 60, 73, 131

Oaks *(Quercus* spp.*)*, and truffles, 115; as wind shelter, 59; description of species, 103; for shiitake, 120; in pastures, 61; native habitats, 111
Oats *(Avena sativa)*, as green manure, 30; PSI 164
Ocean spray *(Holodiscus discolor)*, native habitat, 110; PSI 160
Oemleria cerasiformis (Indian plum), PSI 160
Ohme Garden, 23
Oil, as market crop, 29, 65
The One-Straw Revolution, 11, 55, 81
Onobrychis vicifolia (Sainfoin), PSI 167
Opium poppy *(Papaver somniferum)*, PSI 167
Opuntia polyacantha (Plains prickly pear), PSI 167
Orchard grass, for pastures, 61
Oregon ash, as hardwood, 124; native habitat, 110
Oregon crabapple *(Pyrus fusca)*, PSI 158
Oregon grape *(Berberis nervosa)*, description, 93; native habitat, 110; 111; PSI 164
Oregon white oak *(Quercus garryana)*, as hardwood, 124; as swine forage, 112; native habitat, 110, 111; PSI 155
Oregon certification programs, 67; Tilth Producers' Coop, 150
Organic farming, bibliography, 49-52; research 30, 42-44
Oryza sativa (Rice), PSI 167
Owls, for rodent control, 106
Oxen, as draft animals, 131

Pacific blackberry *or* Dewberry *(Rubus ursinus)*, 92; in clearcuts, 123; PSI 167
Pacific madrone *(Arbutus menziesii)*, native habitat, 110; PSI 154
Pacific Northwest, arboreta and botanic gardens, 22; boundaries, 16; natural history bibliography, 19; physiographic regions map, 14
Pacific silver fir *(Abies amabilis)*, native habitat, 110; PSI 154
Pacific yew, native habitat 112
Papendick, Robert, 49, 50, organic farming research, 30
Panax quinquefolius (Ginseng), PSI 167
Panicum miliaceum (Millet), PSI 167
Papaver somniferum (Opium poppy), PSI 167
Passiflora incarnata (Maypop), PSI 162
Pasture management, 60-61
Pawpaw *(Asimina triloba)*, PSI 156
Peach *(Prunus persica)*, in mixed orchards, 105, 106; preferred habitats, 96; PSI 158
Peacocks, 106
Peanut *(Arachis hypogaea)*, 65; PSI 164
Pear *(Pyrus communis)*, as wind shelter 59; in mixed orchard, 106; in pastures, 61; preferred habitats, 95; wild, 102; PSI 15
Peas, as legume crop, 30, 60; nitrogen-fixing, 36-38; Peas, Please, 36; virus-resistant varieties, 43
Pecans *(Corya* spp.*)*, 100
Peppermint *(Mentha piperita)*, PSI 166
Perennial grains, development of, 50-51, 54; for wildlife, 127
Perigord truffles *(Tuber melanosporum)*, 115
Permaculture, definition, 12, 183; design, 40; organizations, 12; permaculture controversy, 12
Persian walnut *(Juglans regia)*, 102, PSI 155
Persimmons *(Diospyros* spp.*)*, description, 96; in pastures, 61; in Puget Sound area, 105; wild, 102; PSI 157
Pest management, 31-35; bibliography, 31-33, 35, 97; *see also* Brunner, 39; Research Centers, 42-44
Pesticides, bibliography and organizations, 146; *vs.* manual release, 122
Pests, definition, 183; control with chickens, 41; *see also* Pest Management
Phaseolus coccineus (Scarlet runner bean), PSI 162
Pheremone traps, *see* Bio Lure, 97
Photosynthesis, definition, 183 and magnesium, 36
Phyllostachys aurea (Golden bamboo), PSI 162; *P. aureosulcata* (Yellow-groove bamboo), PSI 162; *P. bambusoides* (Madake, Giant timber bamboo), PSI 162; *P. dulsis* (Sweetshoot bamboo), PSI 162; *P. henonis* (Henon bamboo), PSI 163; *P. meyeri* (Meyer bamboo), PSI 163; *P. nigra* (Black bamboo), PSI 163; *P. nigra* cv *henon* (Henon bamboo), PSI 163; *P. nuda*, PSI 163; *P. pubescens*, (Moso), PSI 163; *P. vivax* (Smoothsheath), PSI 163; *P. viridi glaucescens*, PSI 163; *P. viridis*, PSI 163
Physalis pruinosa (Ground cherry), PSI 167
Phytolacca americana (Pokeweed), PSI 160
Picea brewerana (Weeping spruce), PSI 155; *P. engelmanii* (Engelmann spruce), PSI 155; *P. sitchensis* (Sitka spruce), PSI 155
Pigweed *(Amaranthus retroflexus)*, PSI 164
Pine-nuts, 101
Pinus contorta var. *latifolia* (Lodgepole pine), PSI 155; *P. edulis* (Pinyon pine), PSI 157; *P. lambertiana* (Sugar pine), PSI 155; *P. monticola* (Western white pine), PSI 155; *P. ponderosa* (Ponderosa pine), PSI 155; *P. radiata* (Monterey pine), PSI 155;
Pinyon pine *(Pinus edulis)*, PSI 157
Pisum arvense (Field pea), PSI 167
Plains prickley pear *(Opuntia polyacantha)*, PSI 167
Planet Drum, 17
Plant preservation, *see* Genetic preservation

Plantago major (Common plantain), PSI 167
Plantain *(Plantago major)*, for pastures, 61; PSI 167
Planter flats (Todd Speedling trays), source, 68; use, 78
Plants, as bee forage, 129-130; domestication of, 70; importing, 171; patenting, 180; *see also* Genetic preservation
Plum *(Prunus* spp.*)*, description, 96; in mixed orchards, 105, 61; PSI 157
Plume tree *(Cercocarpus montana)*, PSI 159
Pokeweed *(Phytolacca americana)*, PSI 160
Politics, bibliography 145
Polyculture, definition, 183; aquatic, 135-138; in orchard and garden, 105; with insectary plants, 131-133; with nut trees, 106; with perennial grains, 51
Polyconum aubertii (Silver lace vine), PSI 162
Ponderosa pine *(Pinus ponderosa)*, for grazing livestock, 61; native habitat, 110, 111; in Ohme Garden, 23; PSI 155
Ponds, aquaculture, 136-137
Poplar *(Populus* spp.*)* as fuel crop, 65, 176; in China, 120; in coppiced woodlots, 122; in pastures, 61; PSI 155 *(P. nigra* var. *italica)*
Populus nigra var. *italica* (Lombardy poplar), PSI 155; *P. tremuloides* (Quaking aspen) PSI 157; *P. trichocarpa* (Black cottonwood), PSI 155
Port Orchard cedar *(Chamaecyparis lawsoniana)*, description, 113; native habitat, 111; PSI 154
Portulaca oleracea (Purslane), PSI 167
Potentilla anserina (Silverweed), PSI 167
Predators, control with guard dogs, 134; fencing, 61
Prickly currant *(Ribes lacustre)*, PSI 160
Prince's feather *(Amaranthus hypochondriacus)*, PSI 164
Protein, nuts as a source of, 99
Prunus spp., as wind shelter, 59; *P. americana* (American plum), PSI 157; *P. amygdalus* (Almond), PSI 157; *P. armeniaca* (Apricot), PSI 157; *P. avium* (Sweet cherry), PSI 155; *P. besseyi* (Western sand cherry), PSI 160; *P. cerasus* (Sour cherry), PSI 157; *P. domestica* (Common plum), PSI 157; *P. emarginata* (Bitter cherry), PSI 157; *P. persica* (Peach), PSI 158; *P. serotina* (Black cherry), PSI 155; *P. subcordata* (Klamath plum), PSI 158; *P. tomentosa* (Nanking cherry), PSI 160; *P. virginiana* (Chokecherry), PSI 158
Pseudotsuga menziesii (Douglas-fir), PSI 155
Public Gardens, 22-23
Purple amaranth *(Amaranthus cruentus)*, PSI 164
Purslane *(Portulaca oleracea)*, PSI 167
Pyrus baccata (Siberian crabapple), PSI 158; *P. communis* (Pear), PSI 158; *P. fusca* (Oregon crabapple), PSI 158; *P. pumila* (Apple), PSI 158; *P. pyrifolia* var. *culta* (Japanese sand pear), PSI 158

Quaking aspen *(Populus tremuloides)*, PSI 157
Quercus chrysolepsis (Canyon live oak), PSI 158; *Q. garryana* (Oregon white oak), PSI 155; *Q. ilex* (Holly oak), PSI 155; *Q. lobata* (Valley oak), PSI 155
Quince *(Cydonia oblonga)*, description, 96; as hedgerow plant, 59, PSI 159

Rabbits, 43, 72, 73, 132
Radish *(Raphanus sativus)*, 75, PSI 167
Rain Community Resource Center, 18
Rain magazine 18, 47, 143
Ramanus rose *(Rosa rugosa)*, PSI 161
Range management, 62; see also Research Centers, 42-44
Rape *(Brassica napus)*, PSI 164
Raphanus sativus (Radish), PSI 167
Rare breeds, 133
Red alder *(Alnus rubra)*, for pastures, 61; hardwood, 124; in association with Douglas-fir, 37, 124; in coppiced woodlots, 122; in reforestation, 124; native habitat, 110; nitrogen-fixing plant, 37, 124; PSI 154; *see also* Alder
Red clover *(Trifolium pratense)*, 76; as insectary plant, 33, 34; as living mulch, 55; PSI 168
Red elderberry *(Sambucus racemosa)*, 92
Red huckleberry *(Vaccinium parvifolium)*, description, 91; PSI 161
Red mulberry *(Morus rubra)*, PSI 155; *see also* Mulberries
Red raspberry *(Rubus idaeus* var. *strigosus)*, PSI 161
Red tide, 138
Red-flowering currant *(Rhamnus sanguineum)*, description, 90; PSI 160
Red-osier dogwood, native habitat, 113
Reforestation, 119; community forestry, 120; holistic forestry, 116; importance of micorrhizae, 115; in Oregon, 114; with red alder, 124
Reinhabitation, 15, 17
Research centers, 42-44
Rhamnus californica (California coffeeberry), PSI 160; *R. cathartica* (European buckthorn), PSI 158; *R. purshiana* (Cascara sagrada), PSI 158
Rheum rhabarbarum (Rhubarb), PSI 167
Rhizobia, definition, 183; in nitrogen-fixation, 36-37
Rhododendron, native habitat, 110
Rhubarb *(Rheum rhabarbarum)*, PSI 167
Rhus glabra (Smooth sumac), PSI 158
Ribes spp., descriptions, 90; *R. aureum* (Golden currant), PSI 160; *R. cereum* (Wax currant), PSI 160; *R. divaricatum* (Common gooseberry), PSI 160; *R. lacustre* (Prickly currant), PSI 160; *R. nigrum* (Black currant), PSI 160; *R. sanguineum* (Red-flowering currant), PSI 160
Rice *(Oryza sativa)*, and alder, 37; no-till, 54-55; PSI 167; *see also* Wild rice
Riparian, definition 183; stream rehabilitation, 135
Robinia hispida (Rose acacia), PSI 160; *R. pseudoacacia* (Black locust), PSI 156
Rock onion *(Allium acuminatum)*, PSI 163
Rocky Mountains, vegetation, 112
Rodale Research Center, 44, 54-55
Rogue River Valley, climate, 16; vegetation, 111
Romneya coulteri (Matilija poppy), PSI 161
Rosa spp., descriptions, 92; edible flowers, 76; *R. canina* (Dog rose), PSI 160; *R. nutkana* (Wild rose), PSI 161; *R. rugosa* (Ramanus rose), as hedgerow plant, 59; PSI 161
Rose acacia *(Robinia hispida)*, PSI 160
Rubus spp., descriptions, 92; *R. procerus* (Himalyan blackberry), PSI 161; *R. spectabilis* (Salmonberry), PSI 161; *R. idaeus* var. *strigosus* (Red raspberry), PSI 161; *R. laciniatus* (Evergreen blackberry), PSI 161; *R. occidentalis* (Black raspberry), PSI 161; *R. ursinus* (Pacific blackberry, Trailing blackberry, Dewberry), 92, 123; PSI 167
Rue *(Ruta graveolens)*, as insectary plant, 34
Rumex crispus (Yellow dock), PSI 167
Russian olive *(Elaeagnus angustifolia)*, as nitrogen-fixing plant, 37; as wind shelter, 59; for wildlife, 127; PSI 157
Rutabaga, as fodder crop, 61
Rye *(Secale cereale)*, PSI 168

Safflower *(Carthamus tinctorius)*, PSI 165
Sagebrush, as bee forage, 130; native habitat, 111; see also Vallentine, 62
Sagebrush rebellion, opposition to, *see* Hart, 145; Save Our Public Lands, 146
Sagittaria latifolia (Wapato), PSI 167
Sainfoin *(Onobrychis vicifolia)*, in grassland agriculture, 60; PSI 167
Salal *(Gaultheria shallon)*, description, 93, native habitat, 110, 111; PSI 159
Salicornica virginica (Virginia glasswort), PSI 167
Salix alba (White willow), PSI 156; *S. babylonica* (Weeping willow), PSI 156; *S. laseolepis* (Arroyo willow), PSI 158
Salmonberry *(Rubus spectabilis)*, description, 92; PSI 161
Salmonids, 135
Salsify *(Tragopogon porrifolius)*, PSI 168
Saltbush *(Atriplex* spp.*)*, as insectary plant, 35, see also Vallentine, 62
Salvia sclarea (Clary), PSI 167
Sambucus spp., as hedgerow plants, 59; species descriptions, 92; *S. caerulea* (Blue elderberry), PSI 161; *S. canadensis* (American elder), PSI 161
Sand pear *(Pyrus pyrifolia)*, 97
Santolina *(Santolina virens)*, PSI 167
Sasa palmata, PSI 163
Saskatoon serviceberry *(Amelanchier alnifolia)*, description, 91; PSI 156
Satureja douglasii (Yerba buena), PSI 167; *S. montana* (Winter savory), PSI 167
Savanna, 114; definition, 183
Scallions *(Allium fistulosum)*, PSI 163
Scarlet runner bean *(Phaseolus coccineus)*, PSI 162
Scirpus acutus *(Viscid bulrush)*, PSI 161
Scouring rush *(Equisetum hyemale)*, PSI 165
Scorzonera hispanica (Black salsify), PSI 168
Scotch laburnum *(Laburnum alpinum)*, PSI 157
Scythes, see By Hand & Foot, Ltd., 68
Sea buckthorn *(Hippophae rhamnoides)*, PSI 160
Seashore lupine *(Lupinus littoralis)*, PSI 166
Seaweed, as food and fertilizer, 138; sources, 68
Secale cereale (Rye), PSI 168
Seed saving, 84-85, 173; see also Genetic preservation
Seeds, and the crisis in agriculture, 178-180; animal sowing, 115, 176; bibliography, 85; fluid sowing, 76; organizations, 85
Sego lilies, conservation of, 21
Semiarundinaria fastuosa (Narihira bamboo), PSI 163
Sequoia sempervirens (Coast redwood), PSI 156
Sequoiadendron giganteum (Giant sequoia) PSI 156
Service tree *(Sorbus domestica)*, as hedgerow plant, 59
Serviceberries *(Amelanchier* spp.*)*, descriptions, 91; native habitat, 111
Shadblow serviceberry *(Amelanchier canadensis)*, PSI 156
Shagbark hickory *(Carya ovata)*, PSI 154
Sheep, fencing, 61; for brush control, 123; protection from predators 134; see also Livestock bibliography, 131-132; United States Sheep Experiment Station, 42
Sheep's parsley, for pasture, 61
Shellbark hickory *(Carya laciniosa)*, PSI 154
Shellfish, 138
Shelterbelts, definition 183; description and use, 58-59, 172; the Great Plains Shelterbelt, 147; plants used for, 59
Sheperdia canadensis (Canadian buffaloberry), PSI 161
Shiitake mushroom, *(Lentinus edodes)*, 88, 120
Siberian crabapple *(Pyrus baccata)*, PSI 158
Siberian hazel *(Corylus heterophylla)*, PSI 159
Siberian montia *(Claytonia siberica)*, PSI 165
Siberian peashrub *(Caragana arborescens)*, for pasture, 61; in shelterbelts, 59; PSI 159
Siberian stone pine *(Pinus siberica)*, 101

Silk tree *(Albizia julibrissin)*, in orchard, 105; PSI 156
Silver lace vine *(Polygonum aubertii)*, PSI 162
Silver lupine *(Lupinus albifrons)*, PSI 160
Silverberry *(Elaeagnus commutata)*, as fiber plant, 46; PSI 159
Silverweed *(Potentilla anserina)*, PSI 167
Siskiyou Mountains, vegetation 111
Site-specific forestry, 116-119, 122-123; huckleberry management, 91
Sitka alder *(Alnus sinuata)*, native habitat, 110; PSI 156
Sitka spruce *(Picea sitchensis)*, native habitat, 110; PSI 155
Slugs, biology and management, see McLeod, 33; in cold frames, 81
Smooth sumac *(Rhus glabra)*, PSI 158
Smoothsheath *(P. vivax)*, PSI 163
Snowberry *(Symphoricarpos* spp.*)*, as insectary plant, 35; native habitat, 111; *(S. albus)*, PSI 161
Snowbush *(Ceanothus cordulatus)*, PSI 159
Soaproot *(Chlorogalum pomeridianum)*, PSI 165
Soil, bacteria and fungi, 115; bibliography, 53; fertilization, 36-38; green manuring, 30
Soil conservation, with bamboo, 86; with grass and legumes, 60, 131; no-till, 55; with Persian walnuts, 102; with trees, 121, 124
Soil testing laboratories, 53
Solanum spp., as insectary plants, 33
Sorbus aucuparia (European mountain ash), PSI 158
Sour cherry *(Prunus cerasus)*, preferred habitat, 96; PSI 157
Spanish chestnut *(Castanea sativa)*, PSI 154
Spatterdock *(Nuphar polysepalum)*, PSI 167
Spearmint *(Mentha spicata)*, PSI 166
Species introduction, 22; cautionary note, 171; permaculture controversy, 12; introduction and spread of weeds, 56
Spirulina *(Spirulina plantensis)*, 137
Squashberry *(Viburnum edule)*, 92
Stealhead, 135
Stellaria media (Common chickweed), PSI 168
Stinging nettle *(Urtica dioica)*, as food and fiber, 82; compared to Indian hemp, 46; PSI 161
Strawberry clover *(Trifolium fragiferum)*, PSI 168
Strawberry tree *(Arbutus unedo)*, description, 97; PSI 158
Stream rehabilitation, 135
Stubble drills, 55
Subterranean clover, *(Trifolium subterraneum)*, in grassland agriculture, 60; PSI 168
Sugar beet, as fodder crop, 61
Sugar maple *(Acer saccharum)*, PSI 154
Sugar pine *(Pinus lambertiana)*, as source of pine-nuts, 101; habitat, 110; PSI 155
Sunflower *(Helianthus annuus)* as bee forage, 130; in wild garden, 82; PSI 166
Sustainable agriculture, definition, 183; components, 25-26; bibliography, 26-27
Swamp cranberry *(Vaccinium oxycoccos)*, PSI 168
Swamp milkweed *(Asclepias incarnata)*, PSI 164
Sweet cherry *(Prunus avium)*, preferred habitats, 96; PSI 155
Sweet cicely *(Myrrhis odorata)*, PSI 166
Sweet violet *(Viola odorata)*, edible flower, 76; PSI 169
Sweetclover *(Melilotus* spp.*)*, as green manure, 30; for wildlife, 127; in grassland agriculture, 60
Sweet-pit apricots, 100
Sweetshoot bamboo *(Phyllostachys dulcis)*, PSI 162
Swine, foraging, 112
Swiss chard *(Beta vulgaris* var. *cicla)*, as fodder crop, 61; PSI 164
Swiss stone pine *(Pinus cembra)*, 101
Symphoricarpos albus (Common snowberry), PSI 161
Symphytum officinale (Comfrey), PSI 168
Syringa vulgaris (Common lilac), PSI 161

Tachina flies, insectary plants for, 34-35
Tagetes erecta (African marigold), PSI 168
Tall Oregon grape *(Berberis aquifolium)*, PSI 159
Taraxacum officinale (Common dandelion), PSI 168
Tawny daylily *(Hemerocallis fulva)*, edible flowers, 76; PSI 160
Taxus brevifolia (Western yew), PSI 158
Thimbleberry *(Rubus parviflorus)*, description, 92; native habitat, 111
Thin leaf huckleberry *(Vaccinium membranaceum)*, 91, PSI 161
Thuja plicata (Western red cedar), description, 113; PSI 156
Thyme *(Thymus* spp.*)*, in Ohme Garden, 23; *T. serpyllum* (Creeping thyme), PSI 168; *T. vulgaris* (Common thyme), PSI 168
Tilia americana (American linden), PSI 156
Tilth, of soil, definition, 183; improved with green manure, 30
Tilth, organization, history and current activities, 149-150, 192, inside back cover
Tilth Producers' Cooperative, 67, 150
Timothy *(Phleum pratense)*, for pastures, 61
Tomatoes, in blue-green algae research, 38; fluid sowing, see Salter, 76; no-till, 55; varietal testing, 84
Tools, sources of, 45, 68
Tonkin cane *(Arundinaria amabilis)*, PSI 162
Tragopogon porrifolius (Salsify), PSI 168
Trailing blackberry *(Rubus ursinus)*, description, 92; use, 123; PSI 167
Tree fruits, 95-98; bibliography 98; organizations, 97; pest management, 97
Tree of heaven *(Ailanthus altissima)*, PSI 154
Tree planting festival, 114
Trees, as green manure, 116; forests, 107-124; fruit trees, 95-98; nut trees, 99-103; mixed orchards, 104-106; for pastures, 61; native Northwest, 110-114; value of, 177
Trefoil, see Bird's-foot trefoil
Trichgramma parasitic wasps, insectary plants for, 34-35
Trifolium frugiferum (Strawberry clover), PSI 168; *T. hybridum* (Alsike clover), PSI 168; *T. incarnatum* (Crimson clover), PSI 168; *T. pratense* (Red clover), PSI 168; *T. repens* (White clover), PSI 168; *T. subterraneum* (Subterraneun clover), PSI 168
Trigonella foenum-graecum (Fenugreek), PSI 168
Triticum aestivum (Winter wheat), PSI 168
Tropaeolum majus (Garden nasturtium), PSI 162, 168
Trout, 135, 136
Truffles, native, 115
Tsuga heterophylla (Western hemlock), PSI 156
Turkish hazel *(Corylus colurna)*, description, 100; in mixed orchard, 106; PSI 154
Turner, Nancy, 20, 46, 83, 91, 95, 100
Turnip *(Brassica rapa* var. *septiceps)*, as fodder crop, 61; PSI 164
Typha latifolia (Broad-leaved cattail) PSI 161
U-pick, marketing, 67
Upland cotton *(Gossypium hirsutum)*, PSI 166; see also fibers, 47
Urban agriculture, 69, 72-75, 109
Urban Agriculture Center, Seattle, 81, 149, 150
Urban forestry, 108-109
Urban rural cooperation, 75
Urine, as fertilizer, 73
Urtica dioica (Stinging nettle), as fiber, 46, 47; as food and fiber, 82; PSI 161
Vaccinium spp., description, 91, (*see also* Huckleberries); *V. angustifolium* (Low-bush blueberry), PSI 168; *V. corymbosum* (Highbush blueberry), PSI 161; *V. macrocarpon* (Cranberry), PSI 168; *V. membranaceum* (Thin leaf huckleberry), PSI 161; *V. ovatum* (Evergreen huckleberry), PSI 161; *V. oxycoccos* (Swamp cranberry), PSI 168; *V. parvifolium* (Red huckleberry), PSI 161; *V. vitis-idaea* (Cranberry), PSI 169
Valley oak *(Quercus lobata)*, PSI 155
Varietal testing, 84, 150
Vegetables, gardening, 69-83; winter varieties, 81
Verbascum thapsus (Common mullein); PSI 169
Veronica anagallis (Water speedwell), PSI 169
Vetch *(Vicia sativa)*, PSI 169
Vetch *(Vicia* spp.*)*, as green manure, 30; in grassland agriculture, 60; for wildlife, 127; in wild garden, 82
Viburnum trilobum (Cranberry bush), description, 92; PSI 161
Vicia faba (Broad bean,) PSI 169; *V. sativa* (Common vetch), PSI 169; *V. villosa* (Hairy vetch), PSI 169
Viscid bulrush *(Scirpus acutus)*, PSI 161
Viet Nam, use of azolla, 38
Vine maple *(Acer circinatum)*, native habitat, 110; PSI 156
Viola spp., edible flowers, 76; *V. cornuta* (Viola), PSI 169; *V. odorata* (Sweet violet), PSI 169
Virginia glasswort *(Salicornica virginica)*, PSI 167
Virginia strawberry *(Fragaria virginiana)* PSI 165
Vitis labrusca (Fox grape), PSI 162; *V. vinifera* (European grape), PSI 162
Walnut *(Juglans* spp.*)*, breeding, 99, 106; description, 102; in association with autumn olive, 37; in mixed plantings, 105, 106, 120; in pastures, 61; PSI 155
Wapato "Indian potato" *(Sagittaria latifolia)*, as wildlife plant, 127; native harvest, 20; PSI 167
Warminster broom *(Cytisus praecox)*, PSI 159
Wasps, insectary plants for, 34-35
Water birch, native habitat, 113
Water reservoirs, in landscape design, 41; ponds, 136-137
Water speedwell *(Veronica anagallis)*, PSI 169
Watercress *(Nasturtium officinale)*, as calcium source, 83; PSI 166
Waterlilies, as wildlife plant, 127
Wax currant *(Ribes cereum)*, PSI 160
Wax myrtle *(Myrica californica)*, native habitat, 111
Weed control, bibliography, 35, 56; in perennial grains, 50-51; with azolla, 38; with geese, 57; with green manure, 30; with kama, 13; with mulch, 78; with sheep, 123; with white clover, 54-55
Weeder geese, 57
Weeds, as food source, 83; bibliography, 56; importance for beneficial insects, 33
Weeping spruce *(Picea brewerana)*, PSI 155
Weeping willow *(Salix babylonica)*, PSI 156
Western chokecherry *(Prunus demissa)*, 97
Western hazel *(Corylus cornuta* var. *californica*, PSI 159
Western hemlock *(Tsuga heterphylla)*, native native habitat, 110, 112; PSI 156
Western red cedar *(Thuja plicata)*, description, 113; in Ohme Garden, 23; native habitat, 110; PSI 156
Western sand cherry *(Prunus besseyi)*, PSI 160
Western white pine *(Pinus monticola)*, native habitat, 110, 112; PSI 155
Western yew *(Taxus brevifolia)* PSI 158
Wheat *(Triticum aestivum)*, farming, 30, 49-52; in development of perennial grains, 50-51; in nitrogen-fixing bacteria research, 38; no-till, 55; preservation of pioneer varieties, see Kusa Research Foundation, 44
Wheatgrass *(Agropyron* spp.*)*, in development of perennial grains, 50-51; native habitat, 56

White alder *(Alnus rhombifolia)*, native habitat, 111, 113; PSI 154
White clover *(Trifolium repens)*, as bee forage, 130; as insectary plant, 34; for pest control, 33; in no-till farming, 11, 12, 54-55; in wild garden, 82; PSI 168
White fir *(Abies concolor)*, PSI 154
White lupine *(Lupinus albus)*, PSI 166
White mulberry *(Morus alba)*, PSI 155; see also Mulberries
White mustard *(Brassica hirta)*, as insectary plant, 35
White pine blister rust, host plants, 90; spread of, 56
White sweet clover *(Melilotus alba)*, as insectary plant, 34
White willow *(Salix alba)*, PSI 156; see see also Willow
Whitebark pine *(Pinus albicaulis)*, 101
Whitebeam *(Sorbus moravica)*, as hedgerow plant, 59
Wild buckwheats *(Eriogonum* spp.*)*, as insectary plants, 35; PSI 165
Wild carrot *(Daucus carota)*, as insectary plant, 35
Wild cherry *(Prunus* spp.*)*, native habitat, 110
Wild crabapple , native use, 95
Wild foods, 82-83
Wild ginger *(Asarum caudatum)*, PSI 164
Wild gooseberry *(Ribes divaricatum)*, 90
Wild rice *(Zizania aquatica)*, as wildlife plant, 127; Amity fact sheet, 136
Wild rose or Nootka rose *(Rosa nutkana)*, description, 92; PSI 161
Wildflowers, 21
Wildlife, bibliography, 123, 126-128; in forests, 122-123; see also Research Centers, 42-44
Willamette Valley, vegetation, 111, 112
Willow *(Salix* spp.*)*, as bee forage, 129; for pastures, 61; for weaving, 46; in China, 120; in hedgerow, 59; native habitat, 111, 112, 113, PSI 156, 158
Windbreaks, 58-59, 172; definition, 183; in design, 41
Winter gardening, 80-81
Winter savory *(Satureja montana)*, PSI 167
Winter wheat *(Triticum aestivum)*, PSI 168
Winterfat *(Eurotia lanata)*, PSI 165; see also Vallentine, 62
Wintergreen *(Gaultheria procumbens)*, PSI 166
Wisteria floribunda *(Japanese wisteria)*, PSI 162
Women, see Oregon Women's Land Trust, 144; Thomas, 51; Women in Farm Economics, 30
Woodlot, definition 183; design considerations, 40
Woods rose *(Rosa woodsii)*, 92
Worms, see Earthworms

Yampah *(Perideridia gairdneri)*, PSI 167
Yarrow *(Achillea* spp.*)*, as insectary plant, 34; for pastures, 61; PSI 163
Yellow avalanche lily *(Erythronium grandiflorum)*, native harvest, 20
Yellow cedar, see Alaska cedar
Yellow dock *(Rumex crispus)*, PSI 167
Yellow fritillary *(Fritillaria pudica)*, PSI 165
Yellow jackets, insectary plants for, 34-35
Yellow lupine *(Lupinus luteus* var. *sativus)*, PSI 166
Yellow-groove bamboo *(Phyllostachys aureosulcata)*, PSI 162
Yerba buena *(Satureja douglasii)*, PSI 167
Yucca *(Yucca* spp.*)*, edible flowers, 76; in garden, 105

Zizania aquatica (Northern wild rice), PSI 169
Zonation, in landscape design, 40-41

Tilth - Back Issues Available

Tilth, Biological Agriculture in the Northwest is the quarterly journal of the Tilth Association.

Summer 1978; Volume 4, Number 2 $1.50
Urban Agriculture; Reclamation and Irrigation; Bees; The Farm Strike, An Analysis; The New Agribusiness/Industry/Energy Complex; Weeds as Indicators of Soil Conditions.

Fall 1978; Volume 4, Number 3 $1.50
Marketing; The Jetmaster Transplanting System; Goat Husbandry; Asian Farmers in the Northwest; Winter Gardening; Solar Greenhouse Guide.

Spring 1979; Volume 5, Number 1 $2.00
Potato Production; Tilth People; School Gardens; The Clover Project; Factors Affecting the Profitability of Farmers' Markets; Farm Management; Country Women; Goat Husbandry; Small Farm Resources.

Summer 1979; Volume 5, Number 2 $2.00
The Clover Project; Montana Land Reliance; Women and Technology; History of the Grange, Part 1; Strawberry Production; Marketing; Agricultural Economics; Cultivation Tools; Small Farm Assistance; Seeds; Horse Farming.

Fall/Winter 1979; Volume 5, Numbers 3 & 4 $2.00
Organic Certification; The Master of Natural Farming Visits North America; Pelletizing Seed for No-Till Rotations; Intercropping with Clover; History of the Grange, Part 2; Land Preservation Programs; Winter Gardening; Horticulture as Therapy; Control Weeds Without Chemicals; Biological Management of Tansy Ragwort; Money for Small Farmers; Marketing. Seed Supplement: The Plant Patenting Controversy; Growing Your Own; Breeding for Profit.

Spring 1980; Volume 6, Number 1 $2.00
The Clover Project; People Stopping Poisons; Dryland Orcharding; Socially Equitable Agriculture; Women and Agriculture; Talkin' Tractors; Marketing; Reviews; Tilth Information Service.

Summer 1980; Volume 6, Number 2 $2.00
Varietal Testing; Friends of the Trees; Women in Agriculture; The Urban Homestead; Weed Management in Organic Orchards; Alcohol; Reviews.

Fall 1980; Volume 6, Number 3 $2.00
Bulk Seed Buying; Organic Farming Report; Raising Urban Chickens; Solar Grain Dryer; Natural Farming Tools and Techniques; Alan Chadwick.

Winter 1980; Volume 6, Number 4 $2.00
Special Demeter Conference Report: Open Letter on Racism; Using Cold Frames; Solar Greenhouse Design; Integrating Crops and Livestock; Successful Produce Marketing; Is AID Helping Yemen?; Women and Hard Physical Labor.

Spring 1981; Volume 7, Number 1 $2.50
People and Land; Tilth Producers' Coop; Friends of the Trees; The Evergreen Organic Farm; Cornucopia Project; Blueberry Production; The Tunnel Greenhouse; Codling Moth Control.

Summer 1981; Volume 7, Number 2 $2.50
Permaculture Conference Report; Agricultural Ecology; The Future is Abundant; Livestock Poisoning Prevention; Weed Management in Row Crops; Weeder Geese; Kiwi Fruit.

Fall 1981; Volume 7, Number 3 $2.50
Tree Sales; Cold Frames and Year Round Gardening; Flax; Principles of Permaculture; Aquaculture; Seaweed Culture; Medfly Madness; Alcohol Fuels; Horticulture Therapy; Quality Hay.

Winter 1981; Volume 7, Number 4 $2.50
Organic Certification; Oriental Pears; Gypsy Moth; Bamboo; Farm Apprenticeships; Work Horse Handbook; Wendell Berry; Year-Round Gardening; Variety Trials; Tomatoes in the Coast Range; No-Till Gardening.

Special Publications

Nitrogen-Fixing Trees and Shrubs for the Pacific Northwest, by Mark Musick. Review of current research with bibliography, seed sources, and list of 21 species of nitrogen-fixing trees and shrubs.
Resources for Permaculture, by Michael Pilarski. Overview of permaculture concepts with brief articles and references to organizations, books, seeds and plants.

ORDER FORM

PERSONAL ORDER

Name ______________________

Address ______________________

______________________ ZIP ________

GIFT ORDER
(Gift note will be sent)

Name ______________________

Address ______________________

______________________ ZIP ________

Make check or money order payable to Tilth.

Mail to—

Tilth, 13217 Mattson Rd, Arlington, WA 98223.

____ Copies of *The Future is Abundant* @ $11.95 ________
☐ Sales Tax (WA residents only) 70¢/book ________
☐ Postage (1-2 Bks $1, 3-4 Bks $2, etc) ________
☐ Tilth Membership $10/yr U.S. ________
Renewal? Yes ☐ No ☐ $15/Foreign ________
☐ Contributing Member $15-30/year ________
☐ Sustaining Member $35 and up/year ________
☐ Tilth Back Issues (list separately) ________
☐ Tilth, Set of Back Issues $25.00 ________
☐ *Nitrogen-Fixing Trees and Shrubs* $ 2.00 ________
☐ *Resources for Permaculture* $ 1.50 ________
TOTAL ________